Computational and Applied Mathematics

Volume I

Computational and Applied Mathematics Volume I

Edited by **Lucas Lincoln**

Clanrye International

New Jersey

Published by Clanrye International,
55 Van Reypen Street,
Jersey City, NJ 07306, USA
www.clanryeinternational.com

Computational and Applied Mathematics: Volume I
Edited by Lucas Lincoln

International Standard Book Number: 978-1-63240-341-4 (Hardback)

Printed in the United States of America.

Contents

Preface

Computational mathematics is a field that focuses on doing mathematical research in areas of science where computing plays a critical and crucial role and emphasizes on algorithms as well as numerical and symbolic methods. Computation in the field of research is prominent. Computational Mathematics is a pioneering and innovative multidisciplinary program where the focus lies in the intersection of computer science and mathematics. Researchers in this field are able to effectively deploy a wide range of computational and mathematical techniques and methodologies to solve large scale problems in industry, science and commerce. Such tools are also helpful in developing, enhancing and maintaining the software tools as well as to communicate results of complex modelling and simulation to end-users. Because of the rapidly increasing complexity of mathematical models and the exponentially rising power of computers there is now a considerable need for specialists who can intricately understand theoretical problems in computational mathematics. The practice of computational mathematics has spread far beyond its humble roots in science to straddle issues arising in all areas of society. The results of such simulations are plots, charts, numerical answers, data sets, formulae and images that can assist us in understanding the nature of the world surrounding us and allow us to influence and predict the future.

This book is an attempt to compile and collate all available research on computational mathematics under one umbrella. I am grateful to those who put their hard work, effort and expertise into these research projects as well as those who were supportive in this endeavour.

Editor

1

Jovian Problem: Performance of Some High-Order Numerical Integrators

Shafiq Ur Rehman[1,2]
[1]Department of Mathematics, The University of Auckland, Auckland, New Zealand,
[2]Department of Mathematics, University of Engineering and Technology, Lahore, Pakistan

ABSTRACT

N-body simulations of the Sun, the planets, and small celestial bodies are frequently used to model the evolution of the Solar System. Large numbers of numerical integrators for performing such simulations have been developed and used; see, for example, [1,2]. The primary objective of this paper is to analyse and compare the efficiency and the error growth for different numerical integrators. Throughout the paper, the error growth is examined in terms of the global errors in the positions and velocities, and the relative errors in the energy and angular momentum of the system. We performed numerical experiments for the different integrators applied to the Jovian problem over a long interval of duration, as long as one million years, with the local error tolerance ranging from 10^{-16} to 10^{-8}.

Keywords: *N*-Body Simulations; Jovian Problem; Numerical Integrators; CPU-Time

1. Introduction

Computational astronomers make extensive use of accurate *N*-body simulations when studying the dynamics of the planets, asteroids and other small celestial bodies in the Solar System. These simulations are performed by first deriving a set of differential equations for the acceleration of the *N* bodies in the simulation, and specifying the initial positions and velocities of the bodies at time $t = t_0$. Generally, the initial value problems (IVPs) that occur for *N*-body simulations are a mixture of first- and second-order differential equations, but the sort of problems we are considering are of the form,

$$y''(t) = f(t, y(t)),\ y(t_0) = y_0,\ y'(t_0) = y'_0, \qquad (1.1)$$

where $y_0 \in \mathbb{R}^k$ and $y'_0 \in \mathbb{R}^k$ denote the initial positions and velocities, the operator denotes differentiation with respect to time *t*, and $f : \mathbb{R} \times \mathbb{R}^k \to \mathbb{R}^k$ is a sufficiently smooth function. Here, k is the dimension of the IVP, which in some cases may change over time, as bodies are added or removed in the simulations. In some cases, these equations can be solved analytically, but mostly the differential equations are too complicated to find analytical solutions, necessitating the use of approximation techniques to find the numerical approximate solution. A wide range of integrators, for example, Runge-Kutta [3,4], Linear multistep [5], Runge-Kutta-Nyström [6], and Störmer [7] are used to find a numerical solution to the differential equations at $t = t_0 + ih$, with $i = 1, 2, \cdots$ and time-step *h*, which can depend on *i*.

2. Jovian Problem

The Jovian problem (see, for example, [1]) models the orbital motion of the Sun and the four Gas giants, Jupiter, Saturn, Uranus and Neptune, interacting through Newtonian gravitational forces. The Jovian problem is often used in numerical experiments, because the Gas giants collectively drive much of the dynamics of the Solar System. Let $r_i = [x_i, y_i, z_i]^T, i = 1, \cdots, 5$, be the position vector of the ith body of the Jovian problem, where the bodies are ordered from Sun to Neptune and the coordinate system is the three-dimensional Cartesian system with the origin at the barycentre (centre of mass) of the bodies. Then the equations of motion for the ith body can be written as

$$r_i''(t) = \sum_{j=1, j\neq i}^{5} \frac{\mu_j \left(r_j(t) - r_i(t)\right)}{\left\| r_j(t) - r_i(t) \right\|_2^3}, \quad i = 1, \cdots, 5, \qquad (1.2)$$

where $\|\cdot\|_2$ denotes the L_2-norm and μ_j is the gravitational constant G times the mass m_j of the jth body, *i.e*, $\mu_j = Gm_j$. For each body we have a second-order differential equation for the *x*-, *y*-, and *z*-component, giving us fifteen second-order differential equations in total. We express units of distance in astronomical units, the independent variable t in Earth days and the mass m_j in Solar mass.

We use the symmetry of interactions to reduce the number of calculations in the subroutine for evaluating the acceleration for the Jovian problem. Consider the individual terms in the summation and observe

$$\frac{\left(r_j(t)-r_i(t)\right)}{\left\|r_j(t)-r_i(t)\right\|_2^3}=\frac{-\left(r_i(t)-r_j(t)\right)}{\left\|r_j(t)-r_i(t)\right\|_2^3}.$$

Hence, once this term for r_j is calculated, we can update the acceleration for the second body by symmetry. Using this symmetry, we found that the subroutine for the evaluation of the force term for the Jovian problem reduces to approximately half of the CPU-time.

Unlike the Kepler problem, an analytical solution for the Jovian problem is unavailable. Therefore, numerical experiments using the Jovian problem require a reference solution in order to obtain an estimate of the error in the position and velocity. The reference solution has to be more accurate than the numerical solution. Since we plan to test the numerical integrators near the limit of double-precision arithmetic (2.2×10^{-16}), it is essential to use quadruple-precision arithmetic for the reference solution. Therefore, for long-term simulations, obtaining a reference solution can require considerable CPU-time.

Different types of errors are discussed throughout this paper. The global error is of major importance in the measurement of the quality of the numerical solution. We measure this global error in position and velocity, and also measure the relative error in energy and angular momentum. For the total error in the system the main source of error is the integration error, which consists of a truncation and round-off error. While performing accurate simulations, the round-off error contributes significantly to the global error because computers store numbers to only a certain precision. So, there will always be a loss of accuracy when performing long-term simulations. For fixed-step-size schemes, Brouwer [8] showed that, if the step-size is smaller than a prescribed value, the round-off error for conserved quantities, such as total energy and angular momentum, grows as $t^{\frac{1}{2}}$ and for other dynamical variables, such as coordinates of particles, as $t^{\frac{3}{2}}$. This error growth is known as Brouwer's law in the literature; see, for example, [9,10]. In contrast, when the round-off error is systematic, the power laws become t and t^2, respectively. In addition to these aspects, we investigate other effects of the round-off error here.

First we define the types of errors used in this paper. Let y_n and y_t be the vectors of the solution calculated numerically and the reference solution, respectively, and y_n' and y_t' are the vectors of the derivative to the numerical and reference solutions, respectively. Then the norm of the global errors in the position and the velocity are given by

$$E_r(t)=\left\|y_n-y_t\right\|_2,\ E_v(t)=\left\|y_n'-y_t'\right\|_2,$$

where, $\|\cdot\|_2$ is the unweighted L_2-norm.

Physical systems often have conserved quantities, for example, the total energy H or the total angular momentum L as for Kepler's two-body problem and the Jovian problem. Usually, these quantities will not be conserved exactly by the numerical solution and this derivation provides assessment about the accuracy of the solution. The total energy is defined as

$$H=\frac{1}{2}\sum_{i=1}^{N}m_i v_i^2-\sum_{j=1}^{N-1}\sum_{i=j+1}^{N}G\frac{m_i m_j}{d_{ij}},$$

where G is the gravitational constant, m_i the mass of the ith body, v_i its velocity, and $d_{ij}=\| r_i-r_j \|_2$ the distance between the ith and j^{th} bodies. The relative error in the energy can be calculated as

$$H_{rel}=\left|\frac{H_0-H}{H_0}\right|,$$

where H_0 is the total energy at the initial time $t=0$. However, we use

$$H_{rel}=\left|\frac{GH_0-GH}{GH_0}\right|,$$

to calculate H_{rel}, because the value of $\mu=Gm$ is known more accurately than m.

The total angular momentum L and the relative error of the angular momentum L_{rel} are defined as

$$L=\sum_{i=1}^{N}m_i r_i\times v_i,\ L_{rel}=\frac{\left\|L_0-L\right\|_2}{\left\|L_0\right\|_2},$$

where L_0 is the angular momentum at the initial time $t=0$. Note that a reference solution is not required to calculate H_{rel} and L_{rel}, unlike for the errors in the position and velocity. Hence, less computing resources are needed to measure the performance of the integrators here. However, H_{rel} and L_{rel}, being scalar quantities, impose only one constraint; if we look at the error in position or velocity then every component of E_r or E_v has to be small.

The phase error is the difference between the phase angle of the numerical solution and the reference solution. The phase error is defined by

$$\cos(\theta)=\frac{y_n\cdot y_t}{\left\|y_n\right\|_2\left\|y_t\right\|_2}.$$

where θ is the angle between the numerical solution and the reference solution.

3. Numerical Methods and Integrators

Explicit Runge-Kutta-Nyström methods (ERKN) were introduced by E. J. Nyström in 1925 [6]. The efficiency of an ERKN method depends upon the approach for controlling the error in the numerical approximations. One way of controlling the error is to use an adaptive step-size technique. In order to control the local error of a single step, a pair of formulae of different orders is used in such a way that the function evaluations of the two methods are identical. If the numerical solution y_n is obtained by using the lower-order formula, then the pair is said to be implemented in lower-order mode. However, it is recommended for efficiency reasons that the solution y_n be obtained using the higher-order formula for the next step [11], and the pair operated in this fashion is said to be implemented in higher-order mode or local extrapolation. In this paper we are using two variable-step-size ERKN integrators: Integrator *ERKN*689 is a nine stage, 6-8 FSAL pair [12] and integrator *ERKN*101217 is a seventeen stage, 10-12 non-FSAL pair [12].

3.1. Round-Off Error Control for ERKN Integrators

In this paper, we perform experiments with tolerance close to the machine precision (2.2×10^{-16}). Therefore, we investigate the possibility of reducing the growth of round-off error in the explicit Runge-Kutta-Nyström integrators using the technique known as compensated summation [13]. The idea of compensated summation is based on estimating the dominant contribution term of the round-off error. To explain the round-off error control technique, we consider the following solution formula

$$y_n = y_{n-1} + hy'_{n-1} + h^2\sum_{j=1}^{s} b_j K_j. \qquad (1.3)$$

Equation (1.3) contains three types of errors; the integration error already in y_n from the previous time-step, the round-off error in the formation of hy'_{n-1} and $h^2\sum_{j=1}^{s} b_j K_j$, and the round-off error caused by adding terms on the right-hand side of (1.3). If the integration time-step is small then $hy'_{n-1} + h^2\sum_{j=1}^{s} b_j K_j$ is small compared to y_{n-1}. Hence, the round-off error will be dominated by adding $hy'_{n-1} + h^2\sum_{j=1}^{s} b_j K_j$ to y_{n-1}. In each time-step we estimate the round-off error caused by adding $hy'_{n-1} + h^2\sum_{j=1}^{s} b_j K_j$ to y_{n-1} and then update the solution as follows; First, calculate

$$\tau = hy'_{n-1} + h^2\sum_{j=1}^{s} b_j K_j - \delta,$$

where δ is the estimated round-off error on the previous time-step (at the start of the integration $\delta = 0$). Since $hy'_{n-1} + h^2\sum_{j=1}^{s} b_j K_j$ and δ are small compared to y_{n-1}, the error caused in the formation of τ is negligible. The solution is then updated to

$$Y_n = y_{n-1} + \tau,$$

and the estimated round-off error for the time-step is calculated as

$$\delta = Y_n - y_n - \tau \qquad (1.4)$$

The solution is then updated as $y_n = Y_n$. The velocity formula also uses the same concept to control the round-off error.

We used the round-off error control technique to investigate the maximum error in position (E_r) and velocity (E_v) for the Jovian problem described in Section 2. The integration was performed over 10^6 years using $TOL = 10^{-2i}$, for $i = 4,5,\cdots,8$. **Table 1** shows the maximum values of E_r and E_v for the explicit Runge-Kutta-Nyström integrators *ERKN*689 and *ERKN*101217. The column labelled *With* contains E_r and E_v calculated when the integration is performed with round-off error control, whereas the column labeled *Without* contains the percentage variation corresponding to the values in column *With* when calculated E_r and E_v by performing integration with-out round-off error control.

For *ERKN*689, the maximum values for E_r and E_v with round-off error control are always less than E_r and E_v without round-off error control. The only exception is for $TOL = 10^{-10}$, where the values of E_r and E_v in the

Table 1. The maximum values of E_r and E_v for *ERKN*689 and *ERKN*101217 obtained with and with-out round-off error control applied to the Jovian problem over one million years for the local error tolerances 10^{-8}, 10^{-10}, 10^{-128}, 10^{-14}, 10^{-16}.

	*ERKN*689				*ERKN*101217			
	E_r		E_v		E_r		E_v	
TOL	*With*	*Without*	*With*	*Without*	*With*	*Without*	*With*	*Without*
10^{-8}	4.37×10^{-2}	+0.02%	6.31×10^{-5}	+0.02%	4.70×10^{-1}	+0.21%	6.78×10^{-4}	+0.29%
10^{-10}	1.11×10^{-4}	−0.91%	1.60×10^{-7}	−0.63%	1.63×10^{-3}	−0.62%	2.35×10^{-6}	−0.86%
10^{-12}	1.70×10^{-6}	+71.8%	1.94×10^{-9}	+68.9%	4.82×10^{-5}	−0.21%	6.63×10^{-8}	−0.15%
10^{-14}	9.74×10^{-7}	+86.0%	9.35×10^{-10}	+84.4%	2.42×10^{-5}	−6.61%	3.33×10^{-8}	−7.77%
10^{-16}	2.28×10^{-6}	+71.0%	1.58×10^{-9}	+78.5%	8.71×10^{-6}	−7.40%	1.19×10^{-8}	−6.25%

column *Without* are close to zero and insignificant compared with values for smaller tolerances. The maximum difference was observed for $TOL = 10^{-14}$. Here, the maximum values for E_r and E_v obtained with round-off error control were approximately 86% and 84% less than those obtained without round-off error control. For *ERKN*101217, except for $TOL = 10^{-8}$, we found that the round-off error control technique is not very effective, because the errors in the position and velocity obtained with round-off error control are not always less than E_r and E_v without round-off error control.

This could be because the average time-step for *ERKN*101217 over 10^6 years is quite large. For example, with $TOL = 10^{-14}$, *ERKN*101217 takes a time-step of approximately 144 days on average over 10^6 years, and hence, the assumption that $hy'_{n-1} + h^2 \sum_{j=1}^{s} b_j K_j$ is small relative to y_{n-1} is invalid. Therefore, for *ERKN*101217, using $TOL = 10^{-2i}$, with $i = 5,6,7,8$, it is not recommended to use the round-off error control technique.

3.2. *ODEX*2 Integrator

For the direct numerical solution of systems of second-order ordinary differential equations, Hairer, Nørsett and Wanner [14] developed an extrapolation code *ODEX*2 based upon the explicit midpoint rule with order selection and step-size control. The *ODEX*2 integrator is good for all tolerances, especially for high precision, like 10^{-20} or 10^{-30}. To observe the change in results for E_r, we performed experiments with a variety of default settings of *ODEX*2, for example, by setting the parameter *ITOL* used for controlling the local error to 0 or 1. We observed that there is hardly any significant difference in results when applied to the Jovian problem over one million years for $TOL = 10^{-16}$ to 10^{-8}.

3.3. Step-Size Variation

Here, we investigate the step-size variation for the variable-step-size integrators *ERKN*689, *ERKN*101217, and *ODEX*2 applied to the Jovian problem. The eccentricities of the orbits of the Jovian planets are no more than 0.1 and there are no close-encounters between the planets. Therefore, the variable-step-size integrators should require small step-size variation. **Table 2** shows the step-size variation for the above integrators applied to the Jovian problem over one million years for the local error tolerances in the range 10^{-16} to 10^{-8}. The columns h_{mn} and h_{mx} list the percentage variation in the minimum and maximum step-sizes relative to the mean step-size. For example, h_{mn} is calculated as

$$h_{mn} = 100\left(\frac{h_{min} - \bar{h}}{\bar{h}}\right),$$

where, h_{min} is the smallest step-size used and $\bar{h}$ the mean step-size. For these results, we considered the on-scale step-sizes by ignoring the first few step-sizes in a transient region near $t = 0$ as well as the final step-size.

The step-size variation depends both upon the integrator and the tolerance chosen and ranges from approximately −34% to 152%. The largest variation between the maximum and minimum step-sizes occurs for *ERKN*689 with $TOL = 10^{-16}$, where it is a factor of three, with h ranging from $0.89\bar{h}$ to $2.52\bar{h}$. For the purpose of our work, we regard this variation as small. This small step-size variation enables us to add a fixed-step-size integrator $\bar{S}$-13 (Störmer of order 13) in this paper. Therefore, we conclude that *TOL* has little effect on the step-size variation for *ERKN*101217 and *ODEX*2.

To see the effect of round-off error, we also performed integrations with $TOL = 10^{-14}$ in quadruple-precision arithmetic. The percentage variations of h_{mn} and h_{mx} were approximately −18% and 133% for *ERKN*689, −20% and 21% for *ERKN*101217, and −30% and 21% for *ODEX*2. Except for h_{mx} in *ERKN*101217, the step-size variations obtained in quadruple-precision arithmetic have reasonably good agreement with **Table 2**. Hence the round-off error is not significant with $TOL = 10^{-14}$.

3.4. Störmer Methods

Störmer methods are an important class of methods for solving systems of second-order differential equations. Introduced by Störmer [7], the methods have long been utilised for accurate long-term simulations of the Solar System [2]. Grazier [15] recommended an order-13, fixed-step-size Störmer method that uses backward differences

Table 2. Step-size variation for the variable-step-size integrators *ERKN*689, *ERKN*101217, and *ODEX*2 applied to the Jovian problem over one million years, with the local error tolerance *TOL* as specified in the first column.

	*ERKN*689		*ERKN*101217		*ODEX*2	
TOL	h_{mn}	h_{mx}	h_{mn}	h_{mx}	h_{mn}	h_{mx}
10^{-8}	−17%	84%	−20%	23%	−30%	14%
10^{-10}	−17%	99%	−20%	22%	−13%	12%
10^{-12}	−18%	115%	−19%	21%	−30%	31%
10^{-14}	−18%	134%	−20%	32%	−29%	21%
10^{-16}	−18%	152%	−34%	71%	−21%	26%

in summed form, summing from the highest to lowest differences. The test results in [9] for simulations of the Sun, Jupiter, Saturn, Uranus and Neptune in double precision showed that the error in the energy and phase error grows as $t^{1/2}$ and $t^{3/2}$, respectively, to within numerical uncertainty when the step-size is $(\frac{1}{1024})$-th (4.1 days) of Jupiter's orbital period. This choice of step-size ensures that the local truncation error of the Störmer method is well below machine precision. In this paper we consider the fixed-step-size Störmer method of order 13 and refer to it as the $\bar{S}$-13 integrator.

4. Numerical Experiments for Long-Term Simulation

First we consider the error growth in the position and velocity using the variable-step-size integrators *ODEX*2, *ERKN*689, and *ERKN*101217. We obtained the reference solution in quadruple-precision using *ERKN*101217 with $TOL = 10^{-18}$. To justify this particular choice for the reference solution, we integrated the Jovian problem using the *ERKN*101217 integrator with $TOL = 10^{-20}$. The maximum difference between the positions and velocities of these two solutions is no more than 4.61×10^{-13}. We also integrated the Jovian problem in quadruple-precision with the tolerance $TOL = 10^{-18}$, but using the *ERKN*689 integrator and found that the maximum difference with the solution for *ERKN*101217 with $TOL = 10^{-18}$ is no more than 5.11×10^{-13}. This suggests that the *ERKN*101217 integrator with $TOL = 10^{-18}$ is sufficiently accurate to obtain the reference solution.

Figure 1 illustrates the unweighted L_2-norm of the estimation of the maximum global error in the position as a function of tolerance with three variable-step-size integrators *ODEX*2, *ERKN*689, and *ERKN*101217 for the Jovian problem over one million years. In most cases, the maximum global error occurs at the end point of the integration. We chose the range 10^{-16} to 10^{-8} for the local error tolerance, because 10^{-16} is close to machine precision in double-precision arithmetic and tolerances greater than 10^{-8} lead to errors that are too large to be meaningful. The result for *ODEX*2 is an accuracy (maximum global error) that ranges from 7.4×10^{-5} to 1.1×10^{2}. We observe that the maximum accuracy (minimum of the maximum global error) is obtained with $TOL = 10^{-16}$ and the minimum accuracy with $TOL = 10^{-8}$. The graph for *ODEX*2 exhibits three phases: In the middle phase, with $TOL \in [10^{-15}, 10^{-12}]$ the round-off error does not yet affect the global error. Round-off has an effect for $TOL \le 10^{-16}$, which we further investigated by using smaller tolerances, *i.e*, $TOL = 10^{-16+k}$, for k = 0.2, 0.4, 0.6, 0.8, and 1. As TOL decreases further from 10^{-16}, the global error starts to increase again, which indicates the influence of the round-off error. The phase for $TOL > 10^{-11}$, shows a global error of approximately 10^{1} AU, which is the diameter of Jupiter's orbit. Here, the integrator still finds the orbit but at an arbitrary position angle that could deviate as much as 180°. We evaluated the phase error using the formula described in Section 2 and found that it is approximately 172°. This means that the amplitude of the orbit is not changing, but the error in its phase angle may be as large as π.

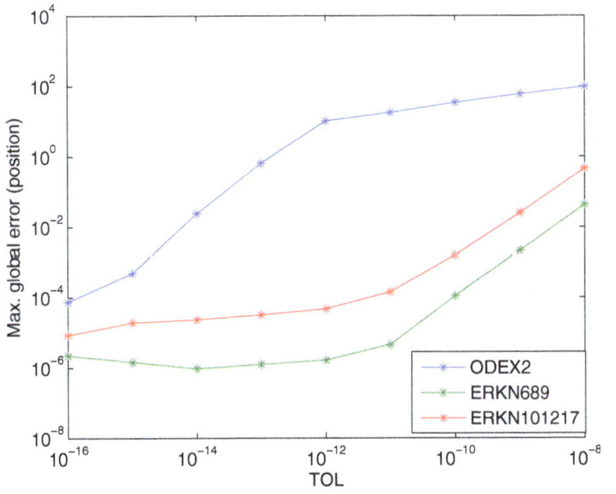

Figure 1. The maximum global error in position for the variable-step-size integrators *ODEX*2, *ERKN*689, and *ERKN*101217 applied to the Jovian problem over one million years for local error tolerances ranging from 10^{-16} to 10^{-8}.

Let us now consider the integrator *ERKN*101217 in **Figure 1**. Here, the accuracy ranges from 8.7×10^{-6} to 4.6×10^{-1}. The maximum accuracy is again obtained with $TOL = 10^{-16}$ and the minimum with $TOL = 10^{-8}$. We observe that from $TOL = 10^{-11}$ to 10^{-16} there is hardly any gain in accuracy. Therefore, if the best accuracy is required then $TOL = 10^{-16}$ should be used, but otherwise, a small sacrifice in accuracy will save a considerable amount of CPU-time.

The integrator *ERKN*689 has an accuracy ranging from 9.7×10^{-7} to 4.4×10^{-2}, with the maximum accuracy obtained at $TOL = 10^{-14}$ and the minimum at $TOL = 10^{-8}$. Therefore, nothing is gained by decreasing the tolerance from 10^{-14} to 10^{-16}. The maximum at $TOL = 10^{-14}$ is an indicator that the round-off error affects the global error when using tolerances between 10^{-14} and 10^{-16}. To measure the possible effect of round-off error, we performed experiments in quadruple-precision. We obtained the maximum global error in the position as a function of tolerance for the local error tolerances 10^{-16} and 10^{-10} using *ERKN*689 and *ERKN*101217. We observed that both curves are straight and maintain a difference of about 1.5 orders of magnitude. In particular, the graph is not bending up for *ERKN*689 using the small tolerance of 10^{-16}. This confirms the effect of round-off error in the double-precision arithmetic. We conclude from **Figure 1** that, for local error tolerances ranging from $TOL = 10^{-16}$ to 10^{-8}, the integrator *ERKN*689 is the most accurate and *ODEX*2 is the least accurate integrator.

Let us now compare this performance with the $\bar{S}$-13 integrator. **Figure 2** shows the error growth in the position for the Jovian problem using the integrators $\bar{S}$-13, *ODEX*2, *ERKN*689, and *ERKN*101217. The integration was performed over 10^6 years and the error was sampled at every 100 years. The integration with the $\bar{S}$-13 integrator was performed in double-precision using a step-size of four days.

We performed two sets of experiments. For the first set of experiments, we maintained a given accuracy of approximately 10^{-4} for the maximum global error in the position over 10^6 years. We set $TOL = 10^{-16}$, 10^{-10}, and 10^{-11} for *ODEX*2, *ERKN*689, and *ERKN*101217, respectively; note that this variation in tolerance is necessary to achieve the prescribed accuracy, as illustrated in **Figure 1**. For small t, *ERKN*689 and *ERKN*101217 are more accurate than the other two integrators, but there is a crossover approximately at 5×10^4 years. We see in **Figure 2** that the three variable-step-size integrators achieve almost the same accuracy for the global error in position at the end of 10^6 years of integration and the fixed-step-size integrator $\bar{S}$-13 achieves almost one order of magnitude better accuracy than the variable-step-size integrators.

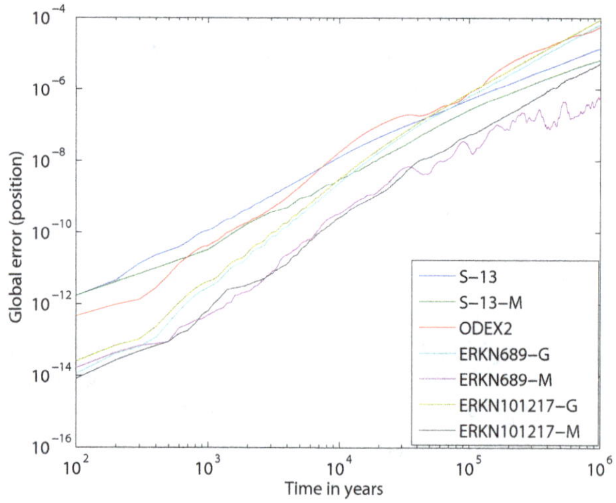

Figure 2. The error growth in position for the integrators $\bar{S}$-13, *ODEX*2, *ERKN*689, and *ERKN*101217 applied to the Jovian problem over one million years. The selection of local error tolerances is subject to a prescribed maximum accuracy.

To gain insight about the error growth depicted in **Figure 2**, we used unweighted linear least squares to fit a power law αt^{β} to E_r. We found that the integration error for the integrators *ERKN*689 and *ERKN*101217 grows approximately as t^2 (quadratic growth), while for *ODEX*2 and $\bar{S}$-13 it is approximately $t^{3/2}$. The error growth for *ODEX*2 is unexpected. Therefore, we repeated the integrations for *ODEX*2 by increasing the tolerance from $TOL = 10^{-16}$ to 10^{-15} and 10^{-14}; then we observe approximately the quadratic error growth.

The second set of experiments for integrators $\bar{S}$-13, *ERKN*689, and *ERKN*101217, labelled *S*-13-M, ERKN689-M and ERKN101217-M in **Figure 2**, respectively, are done such that maximum accuracy is maintained. To attain maximum accuracy, integrators *ERKN*689 and *ERKN*101217 use $TOL = 10^{-14}$ and 10^{-16}, respectively. For $\bar{S}$-13, we performed experiments with step-size variations as shown in **Table 3**. We observe that $\bar{S}$-13 achieves best accuracy with a step-size of approximately 10 days. The performance of the *ODEX*2 integrator at the prescribed accuracy, as shown in **Figure 1**, is also at the maximum accuracy for the local error tolerance of 10^{-16}. When performed at the maximum accuracy, there is no longer a crossover of the $\bar{S}$-13 integrator with the integrators *ERKN*689 and *ERKN*101217. At the end of 10^6 yuracyears of integration, *ERKN*689 achieves the best acc and *ERKN*101217 achieves the next best accuracy.

Some of the plots in these kinds of experiments have high-frequency oscillations. In order to smooth that data, the filter command in Matlab was employed with a window size of 50. The appropriate choice of window size is important. We have experimented (using the experiments illustrated in **Figure 2** with the exclusion of those labelled S-13M, ERKN689-M, and ERKN101217-M) for values of window sizes, 0, 10, 20, and 50 as shown in **Figure 3**. **Figure 3(a)** shows the result without filtering (WS = 0). There are enough oscillations of sufficient amplitude that it is difficult to distinguish the graphs. If the window size is small, as shown in **Figure 3(b)** (WS = 10) then quite a few oscillations are still there and it is not clear which of the integrators is being crossed. A window size of 50 seems to be a sensible value for this set of experiments. As is shown in **Figure 3(d)**, it is quite clear that $\bar{S}$-13 crosses only the integrators *ERKN*689 and *ERKN*101217. We also observed (although not shown)

Table 3. The maximum global error as a function of the step-size for the fixed-step-size integrator $\bar{S}$-13, applied to the Jovian problem over one million years.

Days	4	10	15	20	25	30
Global error in position	1.96×10^{-5}	1.08×10^{-5}	1.89×10^{-5}	3.95×10^{-5}	6.23×10^{-5}	1.05×10^{-4}

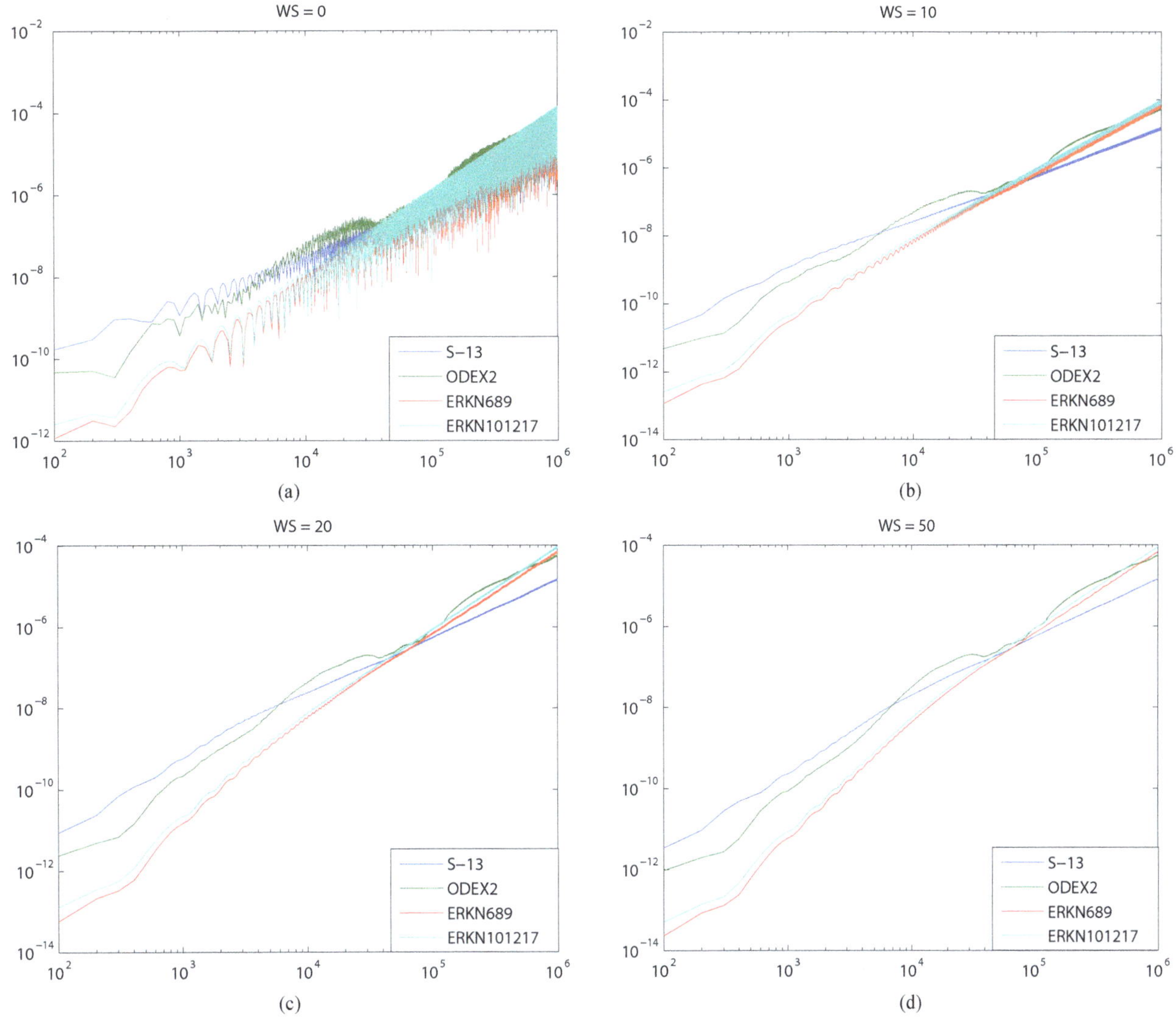

Figure 3. Experiments with a variation of the window size for the Matlab function filter. The window size is set to 0, 10, 20 and 50 in plots (a)-(d), respectively.

that filtering can complicate the interpretation of results for the first WS points, but this effect can be removed by ignoring the first WS points.

Let us now consider the accuracy of the integrators in terms of the relative error in energy and angular momentum. **Figure 4** shows the error growth in the energy for the Jovian problem. The integration has been performed in double-precision over 10^6 years using the same local error tolerances and integrators as for the results shown in **Figure 2**. For this set of experiments, we used the filter command in Matlab with a larger window size of WS = 100, because the oscillations were more pronounced than the set of experiments shown in **Figure 2**. The interval of integration is divided into 10,000 evenly spaced sub-intervals. To see the effect on the performance of the integrator by forcing it to hit every 100 years. We also performed experiments using *ERKN*101217, where we forced the integrator to hit every 50 and 200 years. We found three parallel graphs with a maximum difference in errors at 10^6 years of no more than 3.5×10^{-13}. Using 10,000 sub-intervals, we calculate the L_2-norm of the relative error in energy and angular momentum on the last accepted time step at the end of each sub-interval.

Similar to the set of experiments illustrated in **Figure 2** that attain a given accuracy of 10^{-4}, for the integrators *ERKN*689 and *ERKN*101217 (labeled by ERKN689-G and ERKN101217-G in **Figure 4**, respectively) we observe an error growth proportional to t in energy and angular momentum. For *ODEX*2, the error growth for energy and angular momentum shows some oscillations. The integrations were repeated for *ODEX*2 by increasing the tolerance from $TOL = 10^{-16}$ to 10^{-15} and 10^{-14}, which causes the oscillations to disappear. This indicates that round-off error is the cause of the oscillations. Approximately linear error growth in energy and angular

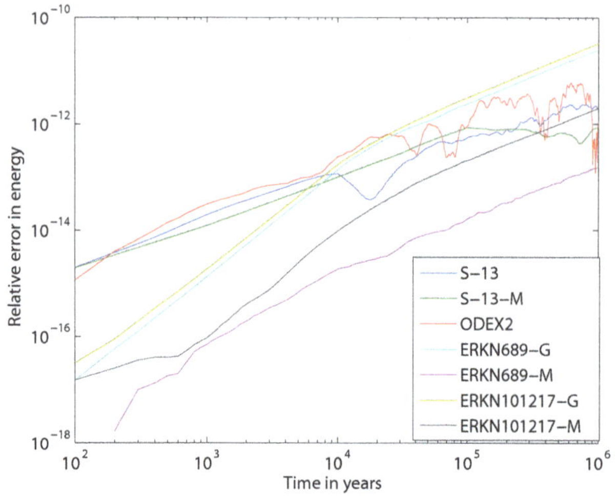

Figure 4. The error growth in the energy for the four integrators $\bar{S}$-13, *ODEX*2, *ERKN*689, and *ERKN*101217 applied to the Jovian problem over one million years. The selection of local error tolerances is subject to attaining the given and maximum accuracy.

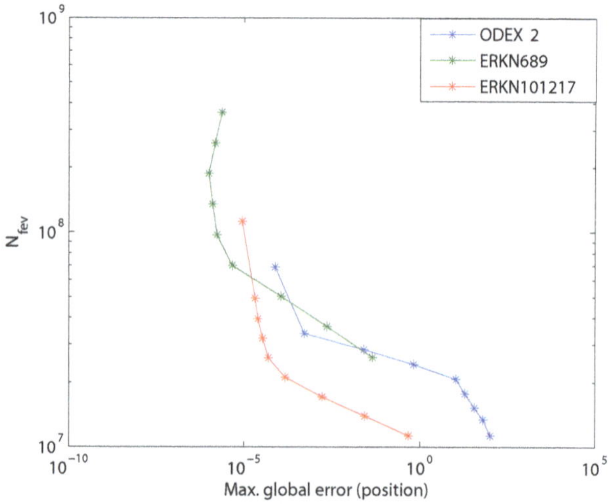

Figure 5. Efficiency plots showing the number N_{fev} of function evaluations against the L_2-norm of the maximum global error in position, obtained for the variable-step-size integrators *ERKN*689, *ERKN*101217, and *ODEX*2, applied to the Jovian problem over one million years with *TOL* ranging from 10^{-16} to 10^{-8}.

momentum was observed particularly for *ODEX*2 with $TOL = 10^{-14}$. As in **Figure 2**, the integrators *ODEX*2 and $\bar{S}$-13 with step-sizes of four days, cross the integrators *ERKN*689 and *ERKN*101217.

However, this crossover for the relative error in energy occurs at a smaller t than for the global error in position. We observe from **Figure 4** that, for the relative error in energy, the integrator ERKN689 using $TOL = 10^{-14}$ (labeled by ERKN689-M) again achieves the best accuracy.

Let us now consider the efficiency of the integrators, which is the amount of work to attain prescribed accuracy. One way of measuring the work of different integrators is to count the number of function evaluations. **Figure 5** shows plots of the number of function evaluations against the maximum global error in position, obtained for the variable-step-size integrators *ERKN*689, *ERKN*101217, and *ODEX*2, and applied to the Jovian problem over one million years with *TOL* ranging from 10^{-16} to 10^{-10}. As described in **Figure 1** the best accuracy for ERKN689 is achieved at $TOL = 10^{-14}$, which needs approximately 1.7 and 2.7 times more function evaluations than *ERKN*101217 and *ODEX*2, respectively. If we consider tolerances such that all three integrators achieve the same accuracy 10^{-4} then *ERKN*101217 is the most efficient, because it uses the least number of function evaluations. The integrator *ERKN*689 is approximately 2.4 and *ODEX*2 approximately 3.3 times more expensive than *ERKN*101217. Our conclusion slightly changes by reducing the accuracy from 10^{-4} to approximately 10^{-3} or 10^{-2}. The integrator *ERKN*101217 again achieves the best accuracy compared to the integrators *ODEX*2 and *ERKN*689. For an accuracy of approximately 10^{-3}, the integrator *ODEX*2 is approximately 1.9 and *ERKN*689 approximately 2.1 times more expensive than *ERKN*101217. In contrast, for an accuracy of approximately 10^{-2}, the integrators *ODEX*2 and *ERKN*689 achieve almost the same accuracy and are approximately 2 times more expensive than *ERKN*101217.

We also investigated the CPU-time taken by the same variable-step-size integrators applied to the Jovian problem over one million years with local error tolerances in the range from 10^{-16} to 10^{-8}. For $TOL = 10^{-10}$, we found that *ODEX*2 and *ERKN*101217 take almost the same CPU-time, but *ERKN*101217 has approximately four orders of magnitude better accuracy than *ODEX*2. For the same tolerance, *ERKN*689 is almost three times more expensive than *ERKN*101217 and *ODEX*2, but has approximately one and five orders of magnitude better accuracy, respectively. For a given accuracy of approximately 10^{-4}, 10^{-3}, and 10^{-2}, *ERKN*101217 takes the least CPU-time. Hence, the integrator *ERKN*101217 is the cheapest option.

For the given range of tolerances from 10^{-16} to 10^{-10}, we found that *ERKN*689 achieves the best accuracy (at $TOL = 10^{-14}$), which is approximately one and two orders of magnitude better than the best accuracies achieved by *ERKN*101217 and *ODEX*2, respectively. At the same point in-time, *ERKN*689 is almost 1.6 and 2.4 times more expensive than *ERKN*101217 and *ODEX*2, respectively. These results clearly illustrate a trade-off between accuracy and efficiency.

5. Conclusions

The main objective of this paper was to analyse and

compare the efficiency and the error growth for different numerical integrators applied to the realistic problem involving the Sun and four Gas-giants. Throughout the paper, we examined the growth of the global error in the positions and velocities of the bodies, and the relative error in the energy and angular momentum of the system. The simulations were performed over as much as 10^6 years.

For long-term simulations, we performed experiments to observe the error growth in the positions and velocities using the variable-step-size integrators *ODEX*2, *ERKN*689, and *ERKN*101217, applied to the Jovian problem over one million years for local error tolerances in the range 10^{-16} to 10^{-8}. We observed that the integrators *ODEX*2, *ERKN*689, and *ERKN*101217 attained maximum accuracy with $TOL = 10^{-16}, 10^{-14}$, and 10^{-16}, respectively. Overall, we observed that for the local error tolerances in the range $TOL = 10^{-16}$ to 10^{-8}, the integrator *ERKN*689 is the most accurate and *ODEX*2 is the least accurate. We also observed that the integration error for the integrators *ERKN*689 and *ERKN*101217 grows approximately as t^2, while it grows as $t^{3/2}$ for *ODEX*2 and $\overline{S}$ -13. The error growth for *ODEX*2 was unexpected. Therefore, integrations were repeated for *ODEX*2 by increasing the tolerance from $TOL = 10^{-16}$ to 10^{-15} and 10^{-14}, for which we did observe the quadratic error growth.

We then investigated the efficiency of the integrators by counting the number of function evaluations against the maximum global error. We observed that the best accuracy achieved by *ERKN*689 uses approximately 1.7 and 2.7 times more function evaluations than *ERKN*101217 and *ODEX*2, respectively. Instead, if we require approximately the same accuracy of 10^{-4} achieved by all three integrators, the *ERKN*101217 is the most efficient, because it uses the least number of function evaluations. The integrator *ERKN*689 is approximately 2.4 and *ODEX*2 approximately 3.3 times more expensive than *ERKN*101217. We then investigated the CPU-time and observed that for a given accuracy of 10^{-4}, the number of function evaluations is proportional to the CPU-time. Hence, also in terms of CPU-time *ERKN*101217 is the cheapest option, which is approximately 2.4 and 3.3 times more efficient than *ERKN*689 and *ODEX*2, respectively. For the given range of tolerances from 10^{-16} to 10^{-8}, the integrator *ERKN*689 achieved best accuracy, which is approximately one and two orders of magnitude better than the best accuracy achieved by *ERKN*101217 and *ODEX*2, respectively. At the same point in time, *ERKN*689 is almost 1.6 and 2.4 times more expensive than *ERKN*101217 and *ODEX*2, respectively. These results clearly illustrate a trade-off between the accuracy and the efficiency.

We also measured the accuracy of the integrators by obtaining the relative error in energy and angular momentum. For the integrators *ERKN*689 and *ERKN*101217, the error growth is proportional to t, and for *ODEX*2 with $TOL = 10^{-14}$, we observe approximately linear error growth in energy and angular momentum.

6. Acknowledgements

The author is grateful to the Higher Education Commission (HEC) of Pakistan for providing the funding to carry out this research. Special thanks go to Dr. P. W. Sharp and Prof. H. M. Osinga for their valuable suggestions, discussions, and guidance throughout this research.

REFERENCES

[1] P. W. Sharp, "*N*-Body Simulations: The Performance of Some Integrators," *ACM Transactions on Mathematical Software*, Vol. 32, No. 3, 2006, pp. 375-395.

[2] K. R. Grazier, W. I. Newman, W. M. Kaula and J. M. Hyman, "Dynamical Evolution of Planetesimals in Outer Solar System," *Icarus*, Vol. 140, No. 2, 1999, pp. 341-352.

[3] K. Heun, "Neue Methode zur Approximativen Integration der Differentialgleichungen einer Unabhängigen Veränderlichen," *Mathematical Physics*, Vol. 45, 1900, pp. 23-38.

[4] M. W. Kutta, "Beitrag zur Näherungsweisen Integration totaler Differentialgleichungen," *Mathematical Physics*, Vol. 46, 1901, pp. 435-453.

[5] F. T. Krogh, "A Variable Step Variable Order Multistep Methods for Ordinary Differential Equations," *Information Processing Letters*, Vol. 68, 1969, pp. 194-199.

[6] E. J. Nyström, "Über die Numerische Integration von Differentialgleichungen," *Acta Societas Scientiarum Fennicae*, Vol. 50, No. 13, 1925, pp. 1-54.

[7] C. Störmer, "Sur les Trajectoires des Corpuscles Électrisés," *Acta Societas Scientiarum Fennicae*, Vol. 24, 1907, pp. 221-247.

[8] D. Brouwer, "On the Accumulation of Errors in Numerical Integration," *Astronomical Journal*, Vol. 46, No. 1072, 1937, pp. 149-153.

[9] K. R. Grazier, W. I. Newman, J. M. Hyman and P. W. Sharp, "Long Simulations of the Outer Solar System: Brouwer's Law and Chaos," In: R. May and A. J. Roberts, Eds., *Proceedings of* 12*th Computational Techniques and Applications Conference* CTAC-2004, *ANZIAM Journal*, Vol. 46, 2005, pp. C1086-C1103.

[10] E. Hairer, R. I. McLachlan and A. Razakarivony, "Achieving Brouwer's Law with Implicit Runge-Kutta Methods," *BIT Numerical Mathematics*, Vol. 48, No. 2, 2008, pp. 231-243.

[11] W. H. Enright, D. J. Higham, B. Owren and P. W. Sharp, "A Survey of the Explicit Runge-Kutta Method," Technical Report, 291/94, Department of Computer Science, University of Toronto, Toronto, 1994.

[12] J. Dormand, M. E. A. El-Mikkawy and P. Prince, "Higher

Order Embedded Runge-Kutta-Nyström Formulae," *IMA Journal of Numerical Analysis*, Vol. 7, No. 4, 1987, pp. 423-430.

[13] L. F. Shampine and M. K. Gordon, "Computer Solution of Ordinary Differential Equations," W. H. Freeman, San Francisco, 1975.

[14] E. Hairer, S. P. Nørsett and G. Wanner, "Solving Ordinary Differential Equations I: Nonstiff Problems," Springer-Verlag, Berlin, 1987.

[15] K. R. Grazier, "The Stability of Planetesimal Niches in the Outer Solar System: A Numerical Investigation," PhD Thesis, University of California, 1997.

2

A Family of 4-Point n-Ary Interpolating Scheme Reproducing Conics

Mehwish Bari, Ghulam Mustafa
Department of Mathematics, The Islamia University of Bahawalpur, Bahawalpur, Pakistan

ABSTRACT

The n-ary subdivision schemes contrast favorably with their binary analogues because they are capable to produce limit functions with the same (or higher) smoothness but smaller support. We present an algorithm to generate the 4-point n-ary non-stationary scheme for trigonometric, hyperbolic and polynomial case with the parameter for describing curves. The performance, analysis and comparison of the 4-point ternary scheme are also presented.

Keywords: Interpolation; Non-Stationary; Univariate Ternary Refinement; Continuity; Conic Section

1. Introduction

Subdivision is a method for making smooth curves/surfaces, which first emerged an addition of splines to arbitrary topological control nets. Effectiveness of subdivision algorithms, their flexibility and ease make them appropriate for many relative computer graphics applications. The schemes generating curves are considered to be the basic tools for the design of schemes generating surfaces.

A general form of univariate n-ary subdivision scheme S which maps a control polygon $f^k = \{f_i^k\}_{i\in\mathbb{Z}}$ to a refined polygon $f^{k+1} = \{f_i^{k+1}\}_{i\in\mathbb{Z}}$ is defined by

$$f_{ni+s}^{k+1} = \sum_{j\in\mathbb{Z}} a_{nj+s} f_{i-j}^k,\ \ S = 0,1,2,\cdots,n-1,$$

where the set $a = \{a_i | i \in \mathbb{Z}\}$ of coefficients is called mask of the subdivision scheme. The set of coefficients $a^k = \{a_i^k : i \in \mathbb{Z}\}$ determines the subdivision rule at level k and is termed as the mask at k-th level. If the mask a^k is independent of k, namely if $a^k = a, k \in \mathbb{N}$, the subdivision scheme is called stationary otherwise it is called non-stationary. Sometimes, in computer graphics and geometric modeling, it is required to have schemes to construct circular parts or parts of conics. It seems that (linear) stationary schemes cannot generate conics and non-stationary schemes have such a capability to generate trigonometric polynomials, trigonometric splines and, in particular, circles, ellipses and so on. Such schemes are useful in computer graphics and geometric modeling. Successful efforts have been made to establish approximating and interpolating non-stationary schemes which can provide smooth curves and reproduce circle or some trigonometric curves.

The theoretical bases regarding non-stationary schemes are derived from the analysis of stationary schemes. Jena *et al.* [1] worked on 4-point binary non-stationary subdivision scheme for curve interpolation. Yoon [2] presented the analysis of binary non-stationary interpolating scheme based on exponential polynomials. Beccari *et al.* [3] worked on 4-point binary non-stationary uniform tension controlled interpolating scheme reproducing conics. Daniel and Shunmugaraj [4] presented 4-point ternary non-stationary interpolating subdivision scheme. In this paper, we present an algorithm to construct 4-point n-ary scheme. For simplicity, we have discussed and analyzed 4-point ternary scheme.

This paper is organized as follows. Section 2 presents the construction of 4-point n-ary non-stationary interpolating subdivision schemes. As an example, 4-point ternary scheme is presented in this section. Section 3 provides the smoothness of proposed schemes. In the last section conclusion and visual performance of proposed schemes are presented.

2. Construction of 4-Point n-Ary Scheme

Here we suggest the following algorithm to construct the non-stationary n-ary 4-point $(n = 2,3,4,\cdots)$, interpolating schemes for trigonometric, hyperbolic and polynomial cases.

- Choose interpolating function
  $f(x) = a_0 + a_1 x + a_2 \cos x + a_3 \sin x$, or
  $f(x) = a_0 + a_1 x + a_2 \cosh x + a_3 \sinh x$, or
  $f(x) = a_0 + a_1 x + a_2 x^2 + a_3 x^3$.
- Then define the points $\{p_i^k | i \in \mathbb{Z}\}$ at level k and get system of linear equations by interpolating.
- The data p_{i+h}^k corresponding to the abscissas
  $x = \frac{ht}{n^k}, h = -1, 0, 1, 2.$
- Solve the system of linear equations by any well known method to get the values of unknowns.
- Evaluate the interpolating function $f(x)$ at the grid points $\frac{(r-1)t}{n^{k+1}} : r = 2, 3, \cdots, n.$
- Define the new points $p_{ni}^{k+1} = p_i^k$.
- Define the new points
  $p_{ni+j}^{k+1} = f\left(\frac{(r-1)t}{n^{k+1}}\right), j = r-1, r = 2, 3, \cdots, n,$ as a linear combination of four consecutive points p_{i-1}^k, p_i^k, p_{i+1}^k and p_{i+2}^k.

Ternary 4-Point Interpolating Scheme

Given a set of control points P^k at level k, using above algorithm, we define a unified ternary 4-point interpolating scheme that makes a new set of control points p^{k+1} by the rule:

$$\begin{cases} p_{3i}^{k+1} = p_i^k, \\ p_{3i+1}^{k+1} = \lambda_1^k p_{i-1}^k + \lambda_2^k p_i^k + \lambda_3^k p_{i+1}^k + \lambda_4^k p_{i+2}^k, \\ p_{3i+2}^{k+1} = \lambda_4^k p_{i-1}^k + \lambda_3^k p_i^k + \lambda_2^k p_{i+1}^k + \lambda_1^k p_{i+2}^k, \end{cases} \tag{2.1}$$

where

$$\lambda_1^k = \{-2(4\upsilon_{k+1} + 1)\}/\lambda_5^k,$$

$$\lambda_2^k = \{-4(\upsilon_{k+1} + 1)(-16\upsilon_{k+1}^3 + 1)\}/\lambda_5^k,$$

$$\lambda_3^k = \{4(8\upsilon_{k+1}^4 + 8\upsilon_{k+1}^3 - 2\upsilon_{k+1} + 1)\}/\lambda_5^k,$$

$$\lambda_4^k = \{-4(\upsilon_{k+1} + 1)\}/\lambda_5^k,$$

with

$$\lambda_5^k = 6(4\upsilon_{k+1}^2 - 1)(2\upsilon_{k+1} + 1)^2,$$

where the parameter υ_{k+1} can easily be updated at each subdivision step through following equation

$$\upsilon_{k+1} = \sqrt{\frac{1 + \upsilon_k}{2}}, \quad k = 0, 1, 2, \cdots, \upsilon_0 \in (-1, \infty). \tag{2.2}$$

Therefore, given parameter υ_k, the subdivision rules are achieved by first computing υ_{k+1} using Equation (2.2) and then by substituting υ_{k+1} into (2.1). As a result, depending on the choice of the parameter, we get different schemes. For $\upsilon_{k+1} = \cos(t/3^{k+1}), \cosh(t/3^{k+1})$ and 1 in (2.1), we can generate following schemes exact for trigonometric (2.3), hyperbolic (2.4) and polynomial (2.5) respectively.

$$\begin{cases} p_{3i}^{k+1} = p_i^k, \\ p_{3i+1}^{k+1} = \eta_1^k p_{i-1}^k + \eta_2^k p_i^k + \eta_3^k p_{i+1}^k + \eta_4^k p_{i+2}^k, \\ p_{3i+2}^{k+1} = \eta_4^k p_{i-1}^k + \eta_3^k p_i^k + \eta_2^k p_{i+1}^k + \eta_1^k p_{i+2}^k, \end{cases} \tag{2.3}$$

where

$$\eta_1^k = \{-2\sin(t/3^k) + 3\sin(2t/3^{k+1})\}/\eta^k,$$

$$\eta_2^k = \{2\sin(2t/3^k) - \sin(t/3^k) - 6\sin(2t/3^{k+1}) + 3\sin(t/3^{k+1})\}/\eta^k,$$

$$\eta_3^k = \{3\sin(2t/3^{k+1}) + \sin(2t/3^k) - 2\sin(t/3^k) - 6\sin(t/3^{k+1})\}/\eta^k,$$

$$\eta_4^k = \{-\sin(t/3^k) + 3\sin(t/3^{k+1})\}/\eta^k,$$

$$\eta^k = 6\sin(t/3^k)\{\cos(t/3^k) - 1\}.$$

$$\begin{cases} p_{3i}^{k+1} = p_i^k, \\ p_{3i+1}^{k+1} = \chi_1^k p_{i-1}^k + \chi_2^k p_i^k + \chi_3^k p_{i+1}^k + \chi_4^k p_{i+2}^k, \\ p_{3i+2}^{k+1} = \chi_4^k p_{i-1}^k + \chi_3^k p_i^k + \chi_2^k p_{i+1}^k + \chi_1^k p_{i+2}^k, \end{cases} \tag{2.4}$$

where $\chi_1^k = \eta_1^k$, $\chi_2^k = \eta_2^k$, $\chi_3^k = \eta_3^k$, $\chi_4^k = \eta_4^k$, after replacing sin and cos functions by sinh and cosh functions in $\eta_1^k, \eta_2^k, \eta_3^k, \eta_4^k$.

$$\begin{cases} p_{3i}^{k+1} = p_i^k, \\ p_{3i+1}^{k+1} = -\frac{5}{81} p_{i-1}^k + \frac{60}{81} p_i^k + \frac{30}{81} p_{i+1}^k - \frac{4}{81} p_{i+2}^k, \\ p_{3i+2}^{k+1} = -\frac{4}{81} p_{i-1}^k + \frac{30}{81} p_i^k + \frac{60}{81} p_{i+1}^k - \frac{5}{81} p_{i+2}^k, \end{cases} \tag{2.5}$$

Remark 2.1. The scheme (2.3) and (2.4) can be considered as a non-stationary counterpart of the DD stationary scheme [5] *i.e.* scheme (2.5) because, the masks of the schemes (2.3) and (2.4) converge to the mask of scheme (2.5):

$$\eta_1^k = \chi_1^k \to -\frac{5}{81}, \eta_2^k = \chi_2^k \to \frac{60}{81}, \eta_3^k = \chi_3^k \to \frac{30}{81},$$

$$\eta_4^k = \chi_4^k \to -\frac{4}{81}, \quad \text{as } k \to \infty.$$

3. Smoothness Analysis

The subdivision scheme given in the previous subsection, the coefficients $\{\lambda_i^k\}_{i=1,2,3,4}$ in (2.1) may vary from one refinement level to another. Hence the scheme is non-stationary and its smoothness properties can be derived by asymptotical equivalence [6] with the corresponding stationary scheme. Two subdivision schemes S_{a^k} and S_a are said asymptotically equivalent if $\sum_{k\in\mathbb{Z}^+} \|S_{a^k} - S_a\| < +\infty$. In particular, our analysis is based on the generalization of Theorem 8 in [6] to ternary subdivision.

Since our schemes (2.3) and (2.4) are non-stationary then we can use the theory of asymptotic equivalence and generating function formalism [7] to investigate the smoothness of the schemes. First, we need some estimates of η_i^k and $\chi_i^k, i = 1,2,3,4,$ which are specified in subsequent lemmas.

Lemma 3.1.

The mask of scheme (2.3) *satisfies following inequalities for sufficiently large* k.

1) $-\frac{1}{3} \le \eta_1^k \le \frac{2}{3},$

2) $-\frac{4}{3} \le \eta_2^k \le 1,$

3) $-\frac{10}{27} \le \eta_3^k \le \frac{4}{3},$

4) $-\frac{1}{6} \le \eta_4^k \le \frac{1}{3}.$

Proof. We make use the inequalities $\frac{\sin a}{\sin b} \ge \frac{a}{b}$ for $0 \le a \le b < \frac{\pi}{2}$, $\csc\theta < t\csc t$ for $0 < \theta < t < \frac{\pi}{2}$ and $\cos x < \frac{\sin x}{x}$ (or $\csc x < \frac{1}{x\cos x}$) for $0 < x < \frac{\pi}{2}$ to prove the above inequalities:

Since

$$\eta_1^k = \frac{2\sin(t/3^k) - 6\sin(t/3^{k+1})\cos(t/3^{k+1})}{12\sin(t/3^k)\sin^2(t/2\cdot 3^k)}$$

$$= \frac{\sin(t/3^k)}{6\sin(t/3^k)\sin^2(t/2\cdot 3^k)} - \frac{\sin(t/3^k)\cos(t/3^{k+1})}{2\sin(t/3^k)\sin^2(t/2\cdot 3^k)}.$$

Then for $k \to \infty$

$$\eta_1^k = \frac{1-3\cos(t/3^{k+1})}{6\sin^2(t/2\cdot 3^k)} \ge \frac{1-3\cos(t/3^{k+1})}{6} = -\frac{1}{3}$$

and also for $k \to \infty$, we get

$$\eta_1^k \le \frac{-2\sin(t/3^k) + 6\sin(t/3^k)\cos(t/3^{k+1})}{-12\sin(t/3^k)\sin^2(t/2\cdot 3^k)}$$

$$= \frac{2(2 - 6\sin^2(t/2\cdot 3^{k+1}))}{-12\sin^2(t/2\cdot 3^k)} = \frac{2}{3}.$$

This proves 1). The proofs of 2), 3) and 4) are similar.

Lemma 3.2.

The coefficients in the scheme (2.4) *satisfy following inequalities when subdivision level* $k \to \infty$.

1) $-\frac{1}{3} \le \chi_1^k \le 0,$

2) $-\frac{7}{6} \le \chi_2^k \le \frac{7}{6},$

3) $-\frac{4}{3} \le \chi_3^k \le \frac{4}{3},$

4) $-\frac{1}{6} \le \chi_4^k \le 0.$

Proof. We make use of following inequality of [8]

$$\frac{1}{\cosh x} \le \frac{1+\cos x}{2} = \cos^2\left(\frac{x}{2}\right), x \in \left(-\frac{\pi}{2}, \frac{\pi}{2}\right).$$

This claim holds true if the function $f(x)$ is non-negative on $[0, \pi/2)$.

Some other inequalities for $0 < x < \frac{\pi}{2}$ are

$$\sinh\left(\frac{2x}{3^{k+1}}\right) \le \sinh\left(\frac{x}{3^k}\right), \sinh x \ge 0,$$

$$\frac{1}{\cosh x} < \frac{x}{\sinh x}, \frac{1}{1-\cosh x} \le 0.$$

Since

$$\chi_1^k = \frac{-2\sinh(t/3^k) + 3\sinh(2t/3^{k+1})}{6\sinh(t/3^k)\{\cosh(t/3^k) - 1\}}$$

$$\ge \frac{-2\sinh(t/3^k)}{6\sinh(t/3^k)\{\cosh(t/3^k) - 1\}}.$$

Then for $k \to \infty$

$$\chi_1^k = -\frac{1}{33\{\cosh(t/3^k) - 1\}} \ge -\frac{1+\cos(t/3^k)}{3(1-\cos(t/3^k))}$$

$$= -\frac{1+\cos(t/3^k)}{3(2\sin^2(t/2\cdot 3^k))} = -\frac{1}{3},$$

and similarly for $k \to \infty$

$$\chi_1^k = \frac{-2\sinh(t/3^k) + 3\sinh(2t/3^{k+1})}{6\sinh(t/3^k)\{\cosh(t/3^k) - 1\}} \le \frac{3\sinh(t/3^k)}{6\sinh(t/3^k)\{\cosh(t/3^k) - 1\}} = 0.$$

This proves 1). The proofs of 2), 3) and 4) are similar.

The following two Lemmas are the consequence of previous Lemmas.

Lemma 3.3.

1) $\left|\eta_1^k - \left(\frac{-5}{81}\right)\right| \le \frac{59}{81}$,

2) $\left|\eta_2^k - \frac{60}{81}\right| \le \frac{21}{81}$,

3) $\left|\eta_3^k - \frac{30}{81}\right| \le \frac{78}{81}$,

4) $\left|\eta_4^k - \left(\frac{-4}{81}\right)\right| \le \frac{31}{81}$.

Proof. Since $\eta_1^k \to -\frac{5}{81}$ as $k \to \infty$ and by 1) of Lemma 3.1, we have 1). Similarly, we get 2), 3) and 4).

Lemma 3.4.

1) $\left|\chi_1^k - \left(\frac{-5}{81}\right)\right| \le -\frac{5}{81}$,

2) $\left|\chi_2^k - \frac{60}{81}\right| \le \frac{23}{54}$,

3) $\left|\chi_3^k - \frac{30}{81}\right| \le \frac{78}{81}$,

4) $\left|\chi_4^k - \left(\frac{-4}{81}\right)\right| \le \frac{4}{81}$.

Proof. Since $\chi_1^k \to -\frac{5}{81}$ as $k \to \infty$ and by 1) of Lemma 3.2, we have 1). Similarly, we get 2), 3) and 4).

Lemma 3.5.

The Laurent polynomial $\lambda^k(z)$ of the kth level of the scheme S_{λ^k} defined by (2.1) can be written as $\lambda^k(z) = \left(\frac{1+z+z^2}{3}\right)a^k(z)$ where

$$a^k(z) = 3\lambda_4^k z^5 + 3(\lambda_1^k - \lambda_4^k)z^4 - 3\lambda_1^k z^3 + 3(\lambda_3^k + \lambda_4^k)z^2 + 3(\lambda_1^k + \lambda_2^k - \lambda_3^k - \lambda_4^k)z + 3(\lambda_3^k + \lambda_4^k) - 3\lambda_1^k z^{-1} + 3(\lambda_1^k - \lambda_4^k)z^{-2} + 3\lambda_4^k z^{-3}.$$

Proof. By (2.1), we have

$$\lambda^k(z) = \lambda_4^k z^5 + \lambda_1^k z^4 + \lambda_3^k z^2 + \lambda_2^k z^1 + 1 + \lambda_2^k z^{-1} + \lambda_3^k z^{-2} + \lambda_1^k z^{-4} + \lambda_4^k z^{-5}.$$

It can be easily verified that

$$\lambda^k(z) = \left(\frac{1+z+z^2}{3}\right)a^k(z).$$

Lemma 3.6.

The stationary scheme $\{S_a\}$ defined by (2.5) associated with the symbol

$$a(z) = \frac{1}{81}\{-4z^5 - 5z^4 + 30z^2 + 60z^1 + 81 + 60z^{-1} + 30z^{-2} - 5z^{-4} - 4z^{-5}\}$$

is C^1.

Proof. To prove that $\{S_a\}$ is C^1, consider

$$b(z) = \frac{3a(z)}{(1+z+z^2)^2} = \frac{1}{27}\{-4z^{-3} + 3z^{-2} + 6z^{-1} + 17 + 6z + 3z^2 - 4z^3\}.$$

Since

$$\|S_b\|_\infty = \max\left\{\sum_{j\in\mathbb{Z}}|b_{3j}|, \sum_{j\in\mathbb{Z}}|b_{3j+1}|, \sum_{j\in\mathbb{Z}}|b_{3j+2}|\right\} = \max\left\{\frac{25}{27}, \frac{9}{27}, \frac{9}{27}\right\} < 1$$

then by [[7], corollary 4.11] the scheme $\{S_a\}$ is C^1.

Lemma 3.7.

The scheme S_{λ^k} defined by (2.1) is C^1.

Proof. Since $\{S_a\}$ is C^1 by Lemma 3.6, in view of [[6], Theorem 8(a)], it is sufficient to show that

$$\sum_{k=0}^{\infty} 3^k \|S_{\lambda^k} - S_a\|_\infty < \infty$$

where

$$\|S_{\lambda^k} - S_a\|_\infty = \max\left\{\sum_{j\in\mathbb{Z}}|\lambda_{3j}^k - a_{3j}|, \sum_{j\in\mathbb{Z}}|\lambda_{3j+1}^k - a_{3j+1}|, \sum_{j\in\mathbb{Z}}|\lambda_{3j+2}^k - a_{3j+2}|\right\}$$
$$= \max\left\{\left|-3\lambda_1^k - \frac{5}{27}\right| + \left|3\lambda_4^k + \frac{4}{27}\right| + \left|3\lambda_3^k + 3\lambda_4^k - \frac{26}{27}\right|, 2\left|3\lambda_1^k - 3\lambda_4^k + \frac{1}{27}\right| + \left|3\lambda_1^k + 3\lambda_2^k - 3\lambda_3^k - 3\lambda_4^k - \frac{29}{27}\right|\right\}.$$

Note that

$$\left|3\lambda_1^k + 3\lambda_2^k - 3\lambda_3^k - 3\lambda_4^k - \frac{29}{27}\right| \le \left|\lambda_1^k - \frac{-5}{81}\right| + \left|\lambda_2^k - \frac{60}{81}\right| + \left|\lambda_3^k - \frac{30}{81}\right| + \left|\lambda_4^k - \frac{-4}{81}\right|.$$

Since $\sum_{k=0}^{\infty} 3^k \left|\lambda_1^k - \left(\frac{-5}{81}\right)\right| < \infty$,

$\sum_{k=0}^{\infty} 3^k \left| \lambda_2^k - \left(\frac{60}{81}\right) \right| < \infty$, $\sum_{k=0}^{\infty} 3^k \left| \lambda_3^k - \left(\frac{30}{81}\right) \right| < \infty$ and $\sum_{k=0}^{\infty} 3^k \left| \lambda_1^k + \left(\frac{4}{81}\right) \right| < \infty$, then by Lemma 3.3, it follows that

$$\sum_{k=0}^{\infty} 3^{k+1} \left| \lambda_1^k + \lambda_2^k - \lambda_3^k - \lambda_4^k - \frac{29}{81} \right| < \infty.$$

By Lemma 3.3, we can also show that

$\sum_{k=0}^{\infty} 3^k \left| -3\lambda_1^k - \frac{5}{27} \right| < \infty$, $\sum_{k=0}^{\infty} 3^k \left| 3\lambda_4^k + \frac{4}{27} \right| < \infty$,

$\sum_{k=0}^{\infty} 3^k \left| 3\lambda_3^k + 3\lambda_4^k - \frac{26}{27} \right| < \infty$ and

$\sum_{k=0}^{\infty} 3^k \left| 6\lambda_1^k - 6\lambda_4^k + \frac{2}{27} \right| < \infty.$

Hence $\sum_{k=0}^{\infty} 3^k \left\| S_{\lambda^k} - S_a \right\|_{\infty}$.

4. Conclusions

To the aim of reproducing conic sections, we introduce an algorithm for generation of 4-point *n*-ary interpolating scheme. In particular, we define 4-point interpolating scheme that unifies three different curves schemes which are capable of representing trigonometric, hyperbolic and polynomial functions.

The resulting ternary algorithm allows us to efficiently define limit curves that combine all the ingredients of locality, C^1 smoothness, user-independence, local tension control and reproduction of conics, see **Figure 1**.

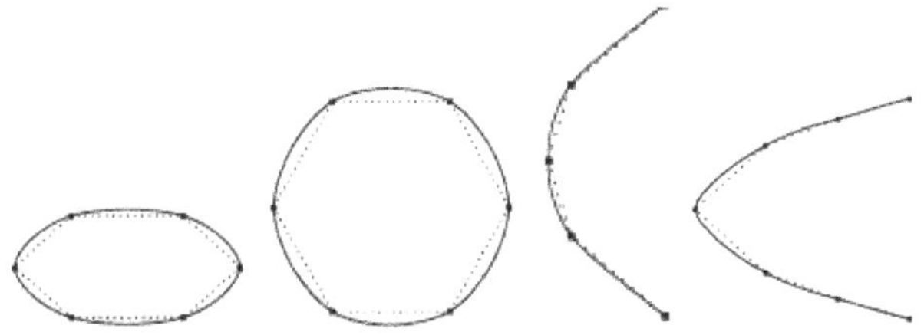

Figure 1. Dotted lines indicate the initial closed and open polygons. Solid continuous curves are generated by proposed ternary interpolating scheme for trigonometric case.

5. Acknowledgements

This work is supported by the Indigenous Ph.D Scholarship Scheme of Higher Education Commission (HEC) Pakistan.

REFERENCES

[1] M. K. Jena, P. Shunmugaraj and P. C. Das, "A Non-Stationary Subdivision Scheme for Curve Interpolation," *ANZIAM Journal*, Vol. 44, No. E, 2003, pp. 216-235.

[2] J. Yoon, "Analysis of Non-Stationary Interpolatory Subdivision Schemes Based on Exponential Polynomials," *Geometric Modeling and Processing*, Vol. 4077, 2006, pp. 563-570.

[3] C. Beccari, G. Casciola and L. Romani, "A Non-Stationary Uniform Tension Controlled Interpolating 4-Point Scheme Reproducing Conics," *Computer Aided Geometric Design*, Vol. 24, No. 1, 2007, pp. 1-9.

[4] S. Daniel and P. Shunmugaraj, "Some Interpolating Non-Stationary Subdivision Schemes," *International Symposium on Computer Science and Society*, Kota Kinabalu, 16-17 July 2011, pp. 400-403.

[5] G. Deslauriers and S. Dubuc, "Symmetric Iterative Interpolation Processes," *Constructive Approximation*, Vol. 5, No. 1, 1989, pp. 49-68.

[6] N. Dyn and D. Levin, "Analysis of Asymptotically Equivalent Binary Subdivision Schemes," *Journal of Mathematical Analysis and Applications*, Vol. 193, No. 2, 1995, pp. 594-621.

[7] N. Dyn and D. Levin, "Subdivision Schemes in Geometric Modelling," *Acta Numerica*, Vol. 11, 2002, pp. 73-144.

[8] R. Klen, M. Lehtonen and M. Vuorinen, "On Jordan Type Inequalities for Hyperbolic Functions," *Journal of Inequalities and Applications*, Vol. 2010, 2010, Article ID: 362548.

Enhanced Frequency Resolution in Data Analysis

Luca Perotti, Daniel Vrinceanu, Daniel Bessis
Department of Physics, Texas Southern University, Houston, USA

ABSTRACT

We present a numerical study of the resolution power of Padé Approximations to the *Z*-transform, compared to the Fourier transform. As signals are represented as isolated poles of the Padé Approximant to the *Z*-transform, resolution depends on the relative position of signal poles *in the complex plane i.e.* not only the difference in frequency (separation in angular position) but also the difference in the decay constant (separation in radial position) contributes to the resolution. The frequency resolution increase reported by other authors is therefore an upper limit: in the case of signals with different decay rates frequency resolution can be further increased.

Keywords: Frequency Resolution; *Z*-Transform; Padé Approximations

1. Introduction

It is known that Padé approximants to the *Z*-transform of a time series allow "super resolution" of signals in low noise: for instance, when the signal to noise ratio (SNR) is high enough (more than 10^5) we can resolve frequencies separated by $10^{-4} f_{\max}/\pi$ with less than 100 data points: 10^2 better than discrete Fourier Transform (DFT) in the same noise conditions. This property has been observed both by P. Barone [1,2] and D. Belkich [3].

Up to now no extensive study has been made of this remarkable property of Padé approximants, not only concerning the limits of super resolution but also of the implications of why it is so.

The key point we want to address in the present note is that resolution for Padé approximants is in the complex plane, not in frequency alone as is in the case of DFT. This allows not only even better frequency resolutions of those reported above, when the decay rates of neighbouring peaks are different, but also resolution of wide overlapping peaks.

Given a data series $s_0, s_1, s_2, \cdots, s_N$, its "Z-transform" is:

$$Z(z) = \sum_{i=0}^{N} s_i z^{-i}; \qquad (1)$$

DFT is clearly the "Z-transform" calculated on the $N+1$ roots of unity. This means that DFT has two intrinsic limitations.

1) There is a resolution limit, due to the time of observation:

Let T be the total sampling time and N the number of sampled points; the time step will be $\Delta = T/N$; the maximum detectable frequency will therefore be $f_{\max} = 1/2\Delta = N/2T$ and the frequency step (resolution) $\Delta_f = 1/N\Delta = 1/T$.

Scaling to the unit circle where the roots of unity reside, $f_{\max}$ becomes $1/2$ the frequency step is therefore $\Delta_{f_{U.C.}} = \Delta_f / 2 f_{\max} = 1/N$.

This is the well known Nyquist limit [4]. The Padé approximation to the "Z-transform" is supperior to it through nonlinearity: because of its discreteness, the average density of peaks is the same as for the DFT; but while the local density in DFT is the same as the average density, it can be very different when a Padé analysis is used, since its peaks are not bound to fall on the regular lattice of the roots of unity. This is the source of "super resolution" for constant amplitude signals reported in the papers mentioned above.

2) Damped signals have a natural width on the unit circle.

The peak for a damped signal is a Lorentzian of the form

$$F(\omega) = \frac{I\gamma^2}{\gamma^2 + (\omega - \omega_0)^2} \qquad (2)$$

where I is the height of the peak, γ is the decay factor and $\omega = 2\pi f$. The half height width is therefore $W_\omega = 2\gamma$ or $W_f = \gamma/\pi$. Scaling to the unit circle, $f_{\max}$

becomes $1/2$. The width is therefore $W_{U.C.} = W_f / 2f_{\max} = \gamma\Delta/\pi = \Gamma/\pi$ where $\Gamma = \gamma\Delta$ is the dimensionless decay factor.

This means that, when using the Fourier Transform, increasing numerical resolution (sampling rate) beyond the natural linewidth of the signals involved is not much use in separating neighbouring peaks.

When using the Padé Approximant approach, things are very different: signals are represented by poles in the complex plane and all poles are by definition singularities of *Z*(*z*) and therefore sharp. The basic point is that poles corresponding to damped signals are off the unit circle and are sharp only if we look at them in the complex plane; if we only look on the unit circle, as it is the case when using DFT, we do not see the singularity itself but the profile of its tail as the intersection of *Z*(*z*) with the unit circle.

This has two consequences:

1) as we are looking at the poles themselves, there are no tails of strong wide peaks that can hide nearby peaks.

2) Since what counts is the distance of neighbouring peaks *in the complex plane*, damped signals can be even closer in frequency than reported above if their damping constants (radial positions) are different.

2. Summary of the Method

Given a data series $s_0, s_1, s_2, \cdots, s_N$, we build its generating function, or "*Z*-transform" Equation 1 and construct its diagonal Padé Approximant, *i.e.* a rational function with the numerator and denominator having the same degree and whose Taylor expansion equals the *Z*-transform up to order N. The aim is to try and predict the "*Z*-transform" for $N \to \infty$.

The choice of a diagonal rational approximation is the best for both signal and noise because of the following considerations.

For a finite ensemble of damped oscillators, the *Z*-transform tends, when the number of data points goes to infinity, to a $[n/n]$ rational function in *z*, with $n = N/2$ equal to twice the number of oscillators [5]. A diagonal Padé Approximant therefore has the right structure for the signal.

For pure noise, the organization of poles and zeros in Froissart doublets [6-9] is again best approximated by a $[n/n]$ rational function in z.

Most data analysts stopped using Padé Approximants because of instabilities due to the fact that for a pure signal, singularities appear when one tries to construct a Padé Approximant of order higher than $[n/n]$. The problem is conveniently solved by the presence of noise whose Froissart doublets act as additional "signals".

To numerically calculate poles and zeros of the Padé Approximant of the *Z*-transform of a finite time series, we build directly from the time series two tridiagonal Hilbert space operators, called *J*-Matrices, one for the numerator and one for the denominator. The eigenvalues of these matrices readily provide the desired zeros and poles. Details of our method can be found in [5]. Knowledge of the positions of all poles and zeros also gives us the residues for all poles and therefore the amplitudes and phases of the signal oscillations.

3. Results and Sensitivity of the Method

When dealing with resolution, key parameters for both Padé Approximant and Fourier Transform are:

1) the angular distance of the two signal poles, *i.e.* the frequency difference scaled to the maximum detectable frequency: this is the *resolution* itself.

2) The radial position ρ of the two signal poles, *i.e.* the decay factor $\Gamma = -\ln \rho$ of each of the signals.

3) The relative amplitude of the two signals.

4) The signal to noise ratio of the smaller of the two signals, or equivalently, its precision in number of digits.

5) The number of data points.

These are the factors that in practice can limit resolution, which assuming infinite data precision and arbitrarily small noise has no limitation for Padé Approximants.

We now pass to look at a few cases that can help clarify how parameters 2, 3, 4 and 5 affect resolution.

3.1. Two Equal Peaks on the Unit Circle

As a first example, let us consider two peaks on the unit circle separated by $\delta = 4\times10^{-3}$ along the circumference, *i.e.* two constant amplitude waves whose frequency difference is $4\times10^{-3} 2f_{\max}$. Using the DFT to resolve them we need a frequency step at least half the distance, *i.e.* $\Delta_{f_{U.C.}} < 4\times10^{-3}/2$, which means a number of data points $N > \pi\times10^3$.

Figure 1 shows what can be seen with $N = 256$ data points. By increasing *N* to $N = 8192$ (**Figure 2**) a decent resolution can be obtained.

Assuming the noise average amplitude to be 10^{-5} times the amplitude of each of the two signals, and assuming the data to be in double precision, the diagonal Padé Approximant instead needs only 40 data points to resolve the two signals with reasonable accuracy, a reduction by two orders of magnitude. **Figure 3** shows the positions and residues of the reconstructed signal poles for 8 different realizations of the noise: there is some spread in position and amplitude (pole residue) which completely disappears by doubling the number of data points, but the 16 poles are clearly grouped around the positions of the two input poles marked by large red dots. 8 zeros fall between the two groups of poles. The resolution transition between 1 and 2 signals takes place as follows. For low N only 8 poles with sizable residues are visible, one for each noise realization, clustered halfway between the positions of the two signal poles; **Figure 4** shows the

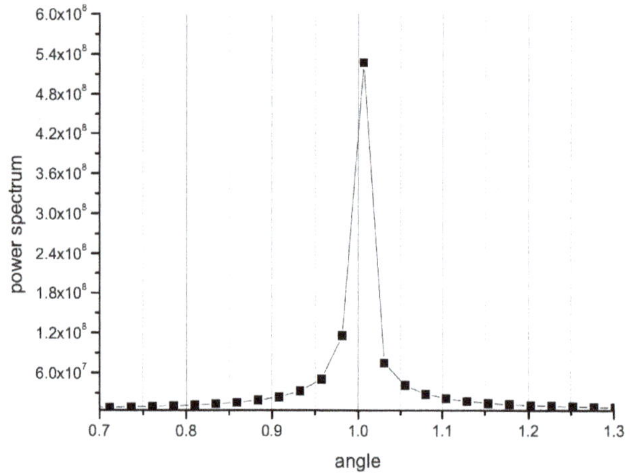

Figure 1. FFT for a signal corresponding to two peaks on the unit circle distant $\delta = 4\times10^{-3}$ using 256 data points. The noise average amplitude is 10^{-5} times the amplitude of each of the two signals.

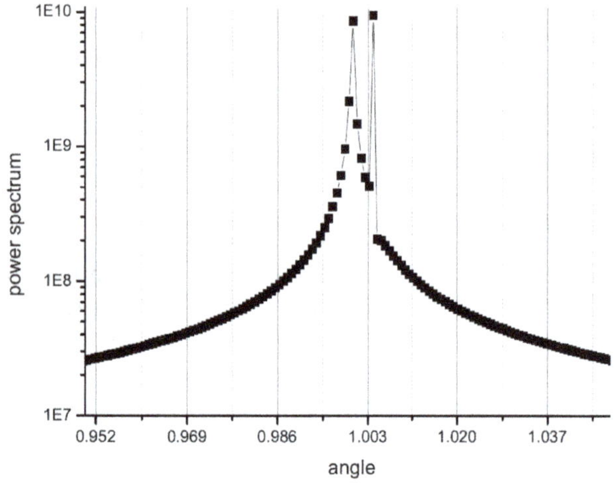

Figure 2. FFT for the same signal as in Figure 1 using 8192 data points.

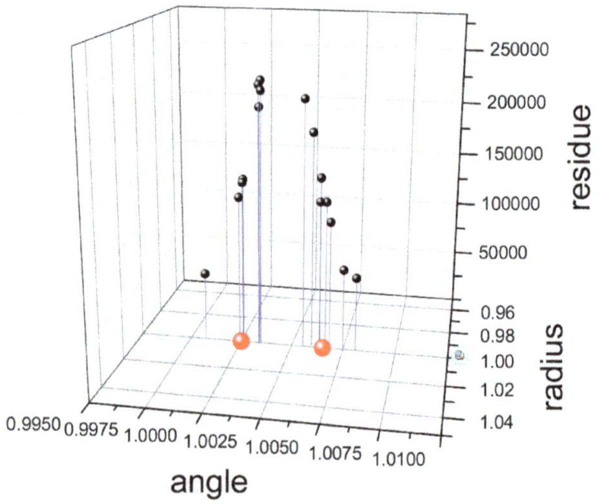

Figure 3. Residues of the poles of the Padé Approximant to the *Z*-transform of the same signal as in Figure 1 using 40 data points. Large red dots indicate the position of the signal poles.

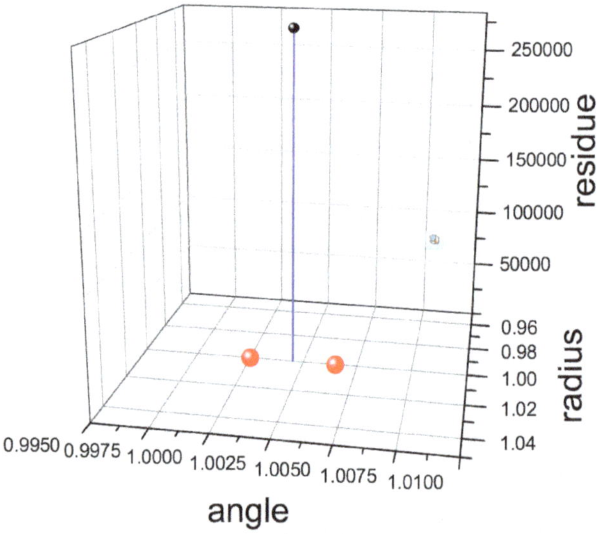

Figure 4. Residues of the poles of the Padé Approximant to the *Z*-transform of the same signal as in Figure 1 using only 10 data points. Large red dots indicate the position of the signal poles.

case for $N = 10$. The noise doublets are spread around the unit circle away from the signal poles; we discussed this repulsion in [10]. By increasing N, we see that 8 doublets, one for each data sequence, approach the signal poles, see **Figure 5**.

Close to the signal poles the doublets split while the central cluster of 8 poles also breaks up; finally the poles regroup in the two clusters visible in **Figure 3** with the zeros halfway between them. **Figure 6** shows this sequence of events. Already for $N = 30$ all 16 poles are visible, even if the two clusters visible in **Figure 3** are not yet fully formed: see **Figure 7**.

The case we have shown is that of equal phases of the two signals; for different phases the zeros are on the straight line perpendicular to the line connecting the two poles and crossing it halfway between the two poles. **Figure 8** shows the case when the signal residues have opposite sign: no zero is present near the signal poles as in **Figure 8(a)**; all the 8 zeros are clustered at the origin as in **Figure 8(b)**.

Noise amplitude determines the number of data points necessary to resolve two signals at a given distance: if we reduce the noise level from 10^{-5} to 10^{-7} then $N = 30$ data points are again sufficient to get a very good resolution. **Figure 9** shows the positions of the poles (black dots) and zeros (red dots) for 8 different realizations of the noise: the spread of each cluster of poles is minimal and all the zeros are located halfway between them because the two signals have again the same amplitude

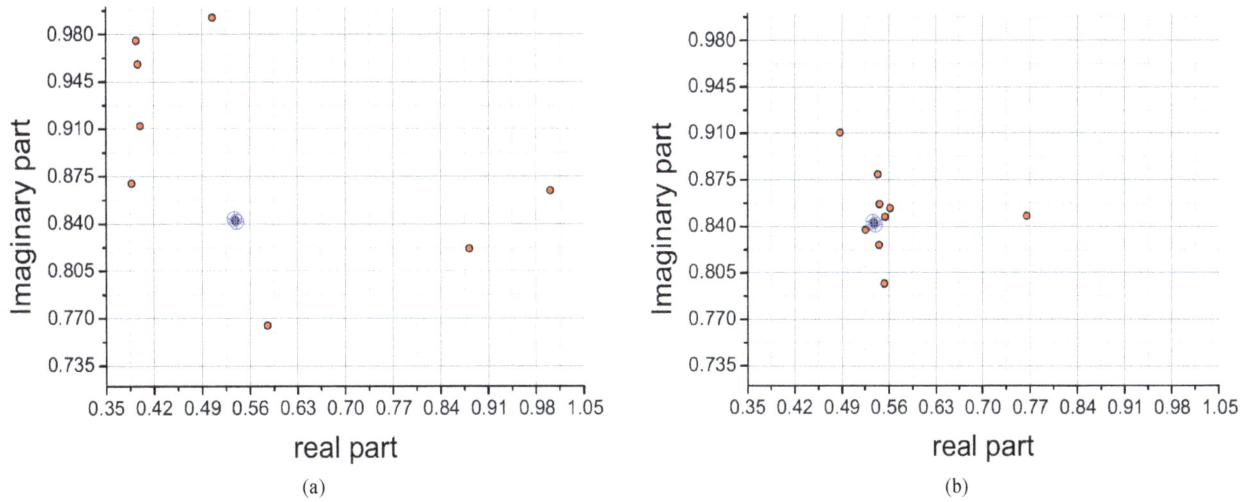

Figure 5. Poles (black dots) and zeros (red dots) of the Padé Approximant to the *Z*-transform of the same signal as in Figure 1 using (a) 10 and (b) 20 data points. Crosses inscribed in circles indicate the position of the signal poles.

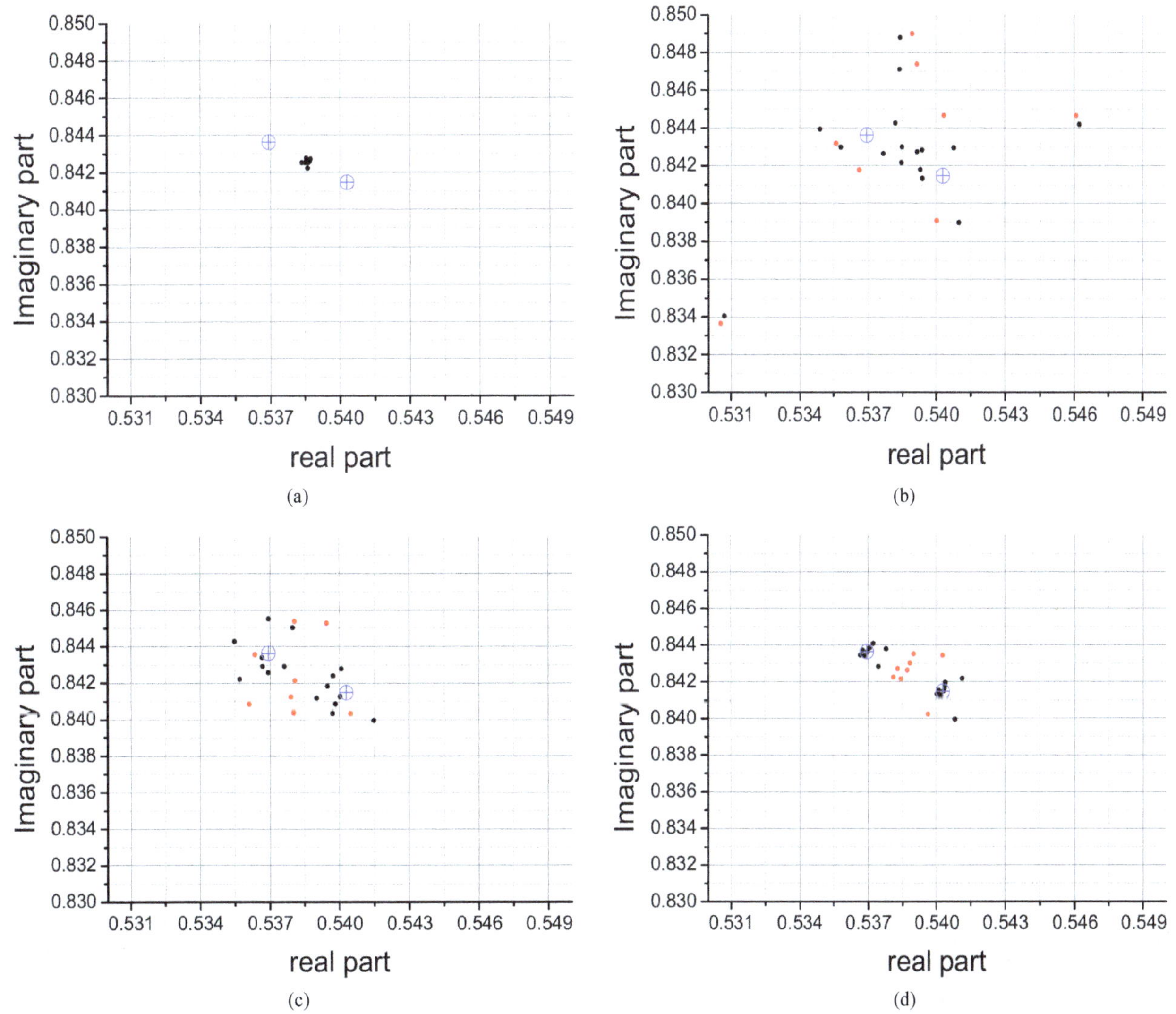

Figure 6. Poles (black dots) and zeros (red dots) of the Padé Approximant to the *Z*-transform of the same signal as in Figure 1 using (a) 20; (b) 30; (c) 40; and (d) 50 data points. Crosses inscribed in circles indicate the position of the signal poles.

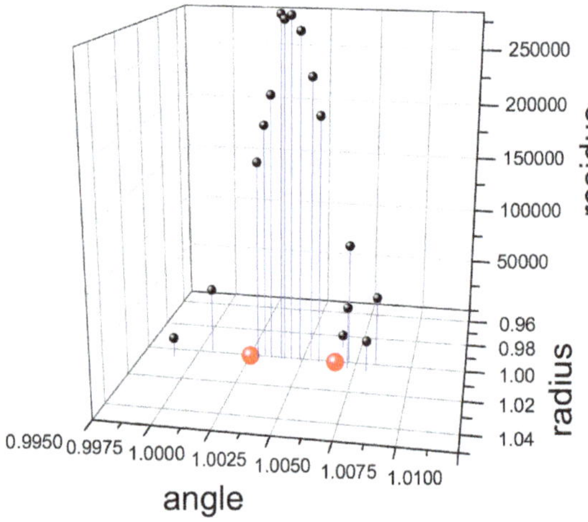

Figure 7. Residues of the poles of the Padé Approximant to the *Z*-transform of the same signal as in Figure 1 using 30 data points. Large red dots indicate the position of the signal poles.

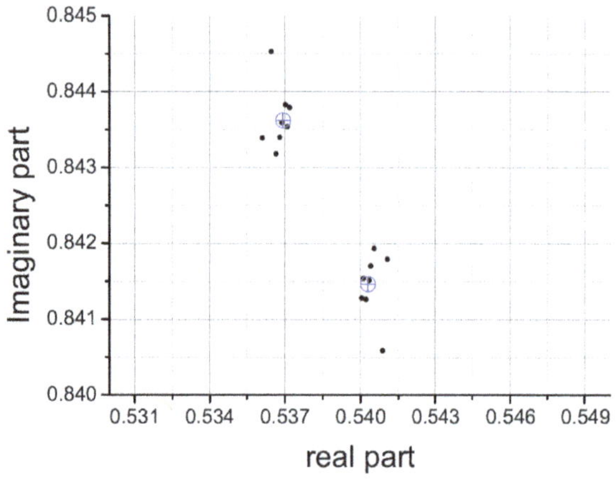

(a)

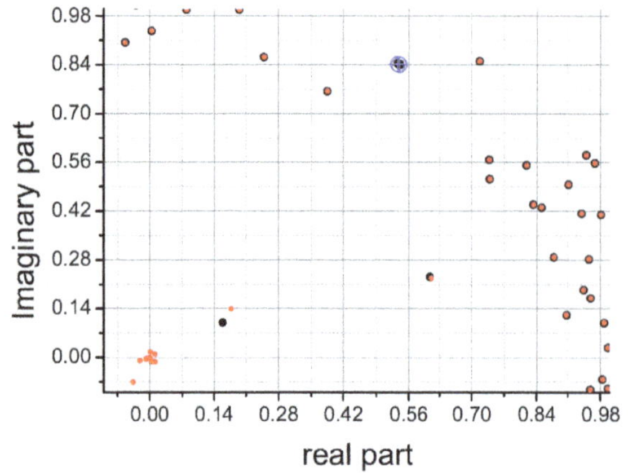

(b)

Figure 8. Poles (black dots) and zeros (red dots) of the Padé Approximant to the Z-transform of the same signal as in Figure 1 using 50 data points. Crosses inscribed in circles indicate the position of the signal poles; (a) shows the region close to the two poles and (b) a more extended area so that the zeros can be seen.

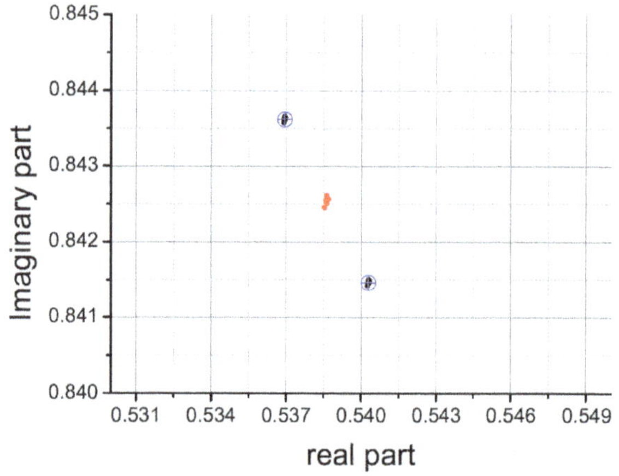

Figure 9. Poles (black dots) and zeros (red dots) of the Padé Approximant to the *Z*-transform of the same signal as in Figure 1 using 30 data points; the noise average amplitude is now reduced to 10^{-7} times the amplitude of each of the two signals. Crosses inscribed in circles indicate the position of the signal poles.

and phase. In this case, we have early indications of a second pole since even with $N = 10$, we can see all the 16 poles: as in **Figure 10**.

3.2. Two Unequal Peaks on the Unit Circle

If the residues of the two poles are unequal, there is an obvious migration of the intervening zero from equidistant to the two poles toward the weaker of the two poles. When the signals also have different phases, the zero is on a circle of radius $2k/\left|1-k^2\right|$ whose center lies on the line connecting the two poles at a distance $\left(1+k^2\right)/\left|1-k^2\right|$ from their middle point, where k is the ratio of the magnitudes of the two poles.

Resolution in this case depends on the SNR of the smaller of the two signals only: increasing the residue of the larger of the two poles does not alter the spread of the poles of the weaker one.

Figure 11 shows an example where only the residue of larger of the two signals is increased while all the other parameters are kept fixed. We keep the same noise realization, so as to do not move the poles of the weaker of the two signals. Two effects are clearly visible: the reduction of the spread of the poles of the stronger of the two signals and the migration of the zeros of toward the weaker of the two signals.

3.3. Two Equal Peaks off the Unit Circle

The case of signal poles on the unit circle is a special one: each new data point gives information with the same precision (assuming a constant noise level). Increasing the number of data points (at constant sampling rate) will therefore always improve resolution, even if more and more slowly.

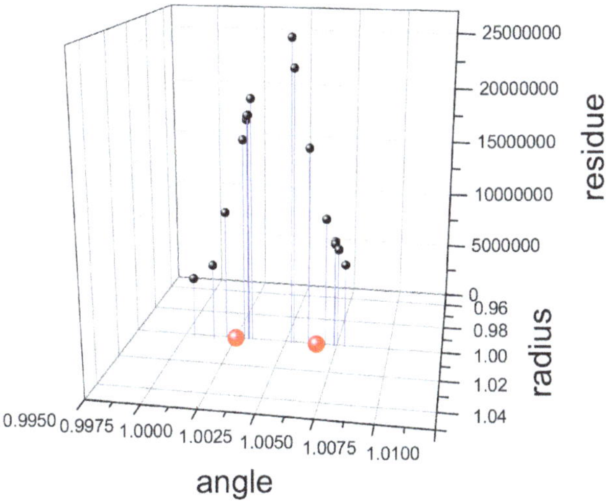

Figure 10. Residues of the poles of the Padé Approximant to the *Z*-transform of the same signal as in Figure 9 using 10 data points. Large red dots indicate the position of the signal poles.

If it is possible to increase the sampling rate at will, then—for any given decay time—the signal poles can be moved as close as we want to the unit circle and we are back to the unit circle case.

If instead the sampling rate is limited, then—for any given distance between signal poles—we'll have to search for the optimal number of data points as a function not only of the SNR but also of the data poles distance from the unit circle.

Signal poles off the unit circle correspond to damped signals; again assuming constant noise amplitude (and data precision) each new point will have lower and lower precision. One might therefore expect (for a given noise level and data precision) the resolution to first increase and then decrease when increasing the number of data points. We do not have evidence of this kind of behaviour.

What we instead see is:

1) Compared to the unit circle case, a very slow resolution increase with the number of points: for $\text{noise} = 10^{-5}$, two peaks, radial position $\rho = 0.95$, angle $\varphi = 1.0$, and separation $\delta = 4\times10^{-3}$, not much difference is seen going from 100 (**Figure 12(a)**) to 200 (**Figure 12(b)**) data points.

2) A decrease of resolution when moving off the unit circle: in the case of two peaks, angle $\varphi = 1.0$, and separation $\delta = 4\times10^{-3}$, going from the unit circle to a radial position $\rho = 0.95$, noise has to be reduced from 10^{-5} to 10^{-7} to get a comparable resolution with 300 data points: see **Figure 13**.

3) The relevant distance is not the frequency one, but the one in the complex plane. For example, for $\text{noise} = 10^{-5}$, two peaks, radial position ρ = 0.950, angle φ = 1.000, separation $\delta = 4\times10^{-3}$, and N = 300, there is no relevant difference in the spread of the signal poles

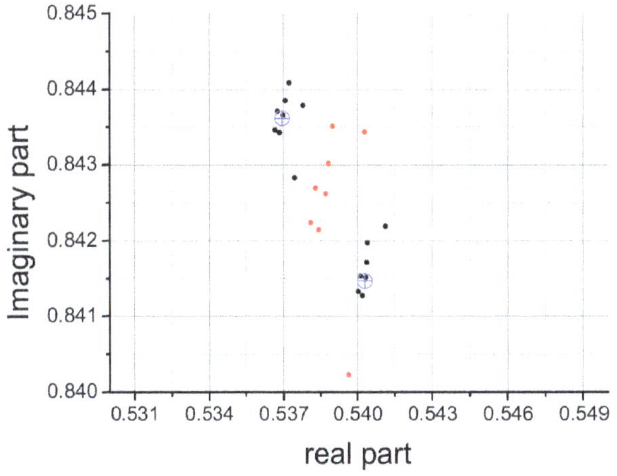

(a)

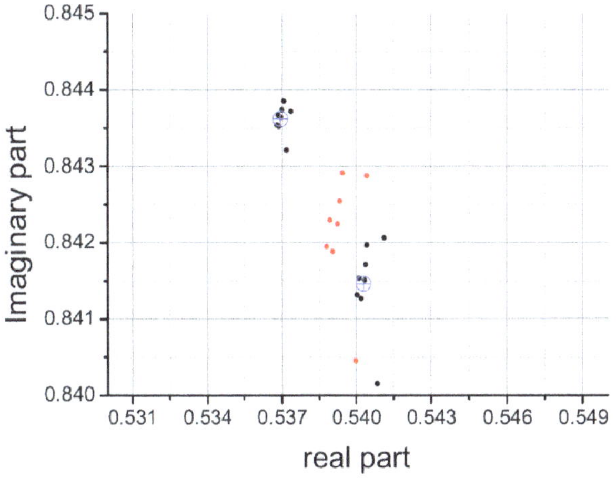

(b)

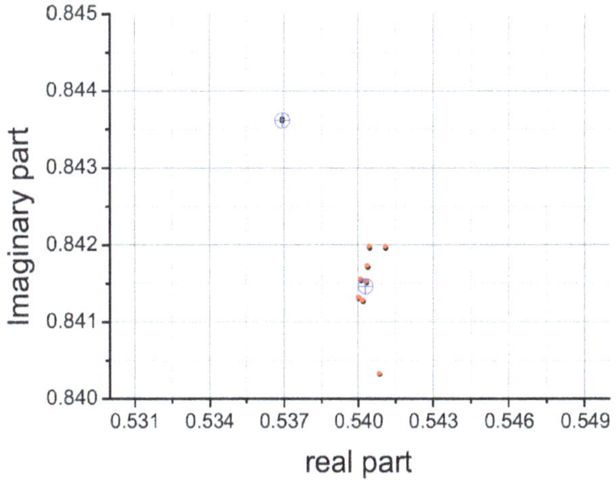

(c)

Figure 11. Poles (black dots) and zeros (red dots) of the Padé Approximant to the Z-transform of two signals on the unit circle distant $\delta = 4\times10^{-3}$ using 50 data points. The noise average amplitude is 10^{-5} times the amplitude of the residue of the weaker pole. (a) Equal residues; (b) Second pole twice as big as the first one; (c) Second pole 100 times bigger than the first one; Crosses inscribed in circles indicate the position of the signal poles.

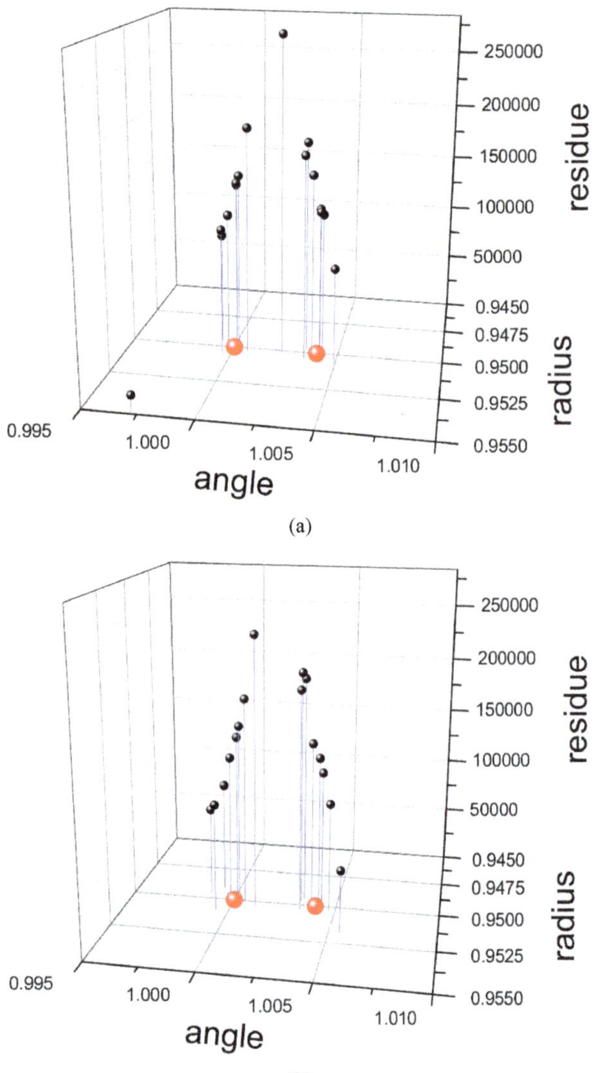

Figure 12. Residues of the poles of the Padé Approximant to the *Z*-transform of the signal generated by 2 poles in (ρ, φ) = (0.950, 1.000) and (0.950, 1.004) with residues $r = 10^5$. Noise amplitude is unitary. We use 8 data samples of (a) 100 and (b) 200 points each. Large red dots indicate the position of the signal poles.

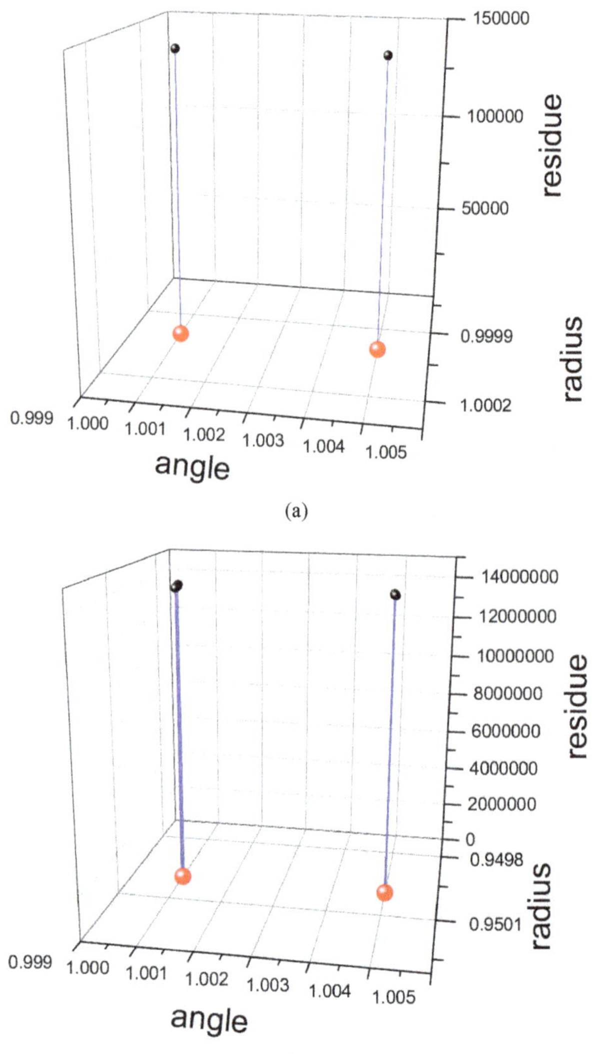

Figure 13. Residues of the poles of the Padé Approximant to the *Z*-transform of the signal generated by 2 poles in (a) (ρ, φ) = (1.000, 1.000) and (1.000, 1.004) with residues $r = 10^5$ and in (b) (ρ, φ) = (0.950, 1.000) and (0.950, 1.004) with residues $r = 10^7$. Noise amplitude is unitary. We use 8 data samples of 300 points each. Large red dots indicate the position of the signal poles.

between the case of two poles having the same radial position (decay rate) and different angles (frequencies), **Figure 14(a)** and the case where the two poles have different radial positions and the same angle, **Figure 14(b)**.

3.4. Four Unequal Peaks

To show that the presence of more signals does not alter the above picture, we now present results for a tight cluster of 4 poles with different residues and decay rates.

Figure 15 shows poles and zeros for an example where the four signal poles are at (ρ, φ) = (0.95, 1.00), (0.90, 1.01), (0.85, 1.02), (0.80, 1.05) with residues $r = 10^5, 10^5, 10^5, 2\times10^5$ respectively. Noise amplitude is unity. We use 8 samples of 300 points each. The picture is quite clear (large red dots indicate the positions of the signal poles): only one reconstructed pole appears to fall on the signal pole in the foreground (the one with the longer decay time); in effect it's 8 poles superimposed, as they are almost identical. When we look at the position of poles and zeros (**Figure 16**), we see that again zeros appear between the signal poles, as the phases of the residues of the poles are equal.

In this case we did not plot the residues of the poles; to make the picture more evident and distinguish signal poles (poles from different data sequences form clusters)

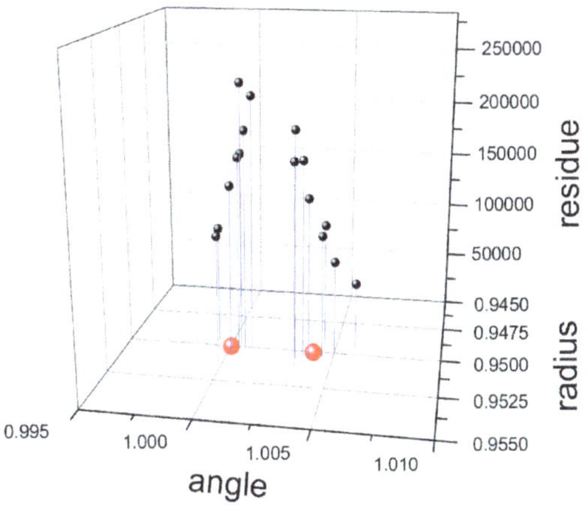

(a)

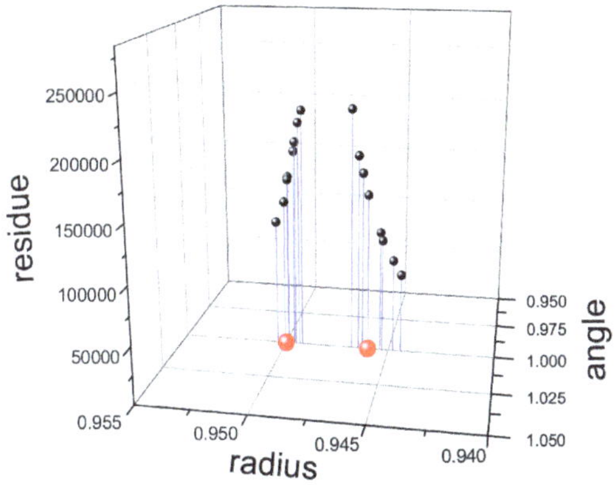

(b)

Figure 14. Residues of the poles of the Padé Approximant to the Z-transform of the signal generated by 2 poles in (a) (ρ, φ) = (0.950, 1.000) and (0.950, 1.004) and in (b) (ρ, φ) = (0.950, 1.000) and (0.946, 1.000) with residues $r = 10^5$. Noise amplitude is unitary. We use 8 data samples of 300 points each. Large red dots indicate the position of the signal poles.

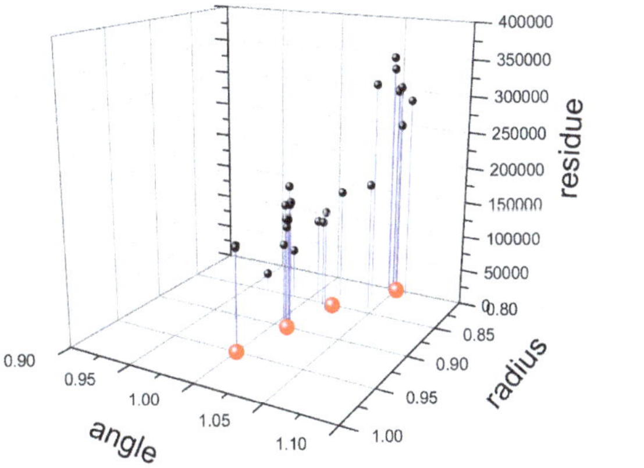

Figure 15. Residues of the poles of the Padé Approximant to the Z-transform of the signal generated by 4 poles in (ρ, φ) = (0.95, 1.00), (0.90, 1.01), (0.85, 1.02), (0.80, 1.05) with residues r = 10^5, 10^5, 10^5, 2 × 10^5 respectively. Noise amplitude is unitary. We use 8 data samples of 300 points each. Large red dots indicate the position of the signal poles.

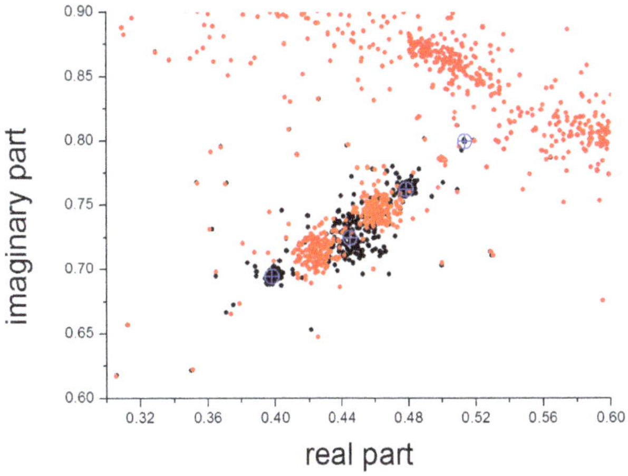

Figure 16. Poles (black dots) and zeros (red dots) of the Padé Approximant to the *Z*-transform of the same signal as in Figure 15. We use the 163 combinations of 4 or more of the 8 samples available, each consisting of 300 data points. Crosses inscribed in circles mark the positions of the signal poles.

from noise ones (which do not form clusters), we took advantage of the nonlinearity of the Padé Approximants and calculated them for a number of linear combinations of the available data sequences [10], as in **Figure 16** where we used the 163 combinations of 4 or more of the 8 samples available.

None of this structure is visible in the DFT. **Figure 17(a)** shows the relevant section of the DFT over 256 points: the red lines mark the four signals but only a single wide peak is visible.

Extending the samples to 8192 points does not reveal any additional structure as can be seen in **Figure 17(b)**: here the vertical scale is logarithmic and only the region around the center of the peak has been plotted to better check for the presence of structures.

Figures 18 and **19** show a less extreme case where the average separation of the poles is increased so that the four signal poles are at (ρ, φ) = (0.95, 0.85), (0.90, 0.90), (0.95, 1.00), (0.90, 1.05) with residues r = 10^3, 5 × 10^3, 10^3, 5 × 10^3 respectively. Again, noise amplitude is unitary and we use 8 samples of 300 points each. The residue picture, **Figure 18**, is very clear.

Of course, the DFT performance is also somehow improved but only two of the four peaks are now clearly visible in **Figure 19**.

4. Conclusions

We have thus extended and generalized the remarks previously made in the literature [1-3] about the super-resolution properties of the Padé Approximations to the *Z*-transform of a signal, stressing the point that resolution needs to be considered in the complex plane and not only in the frequency domain. We also investigated the effect

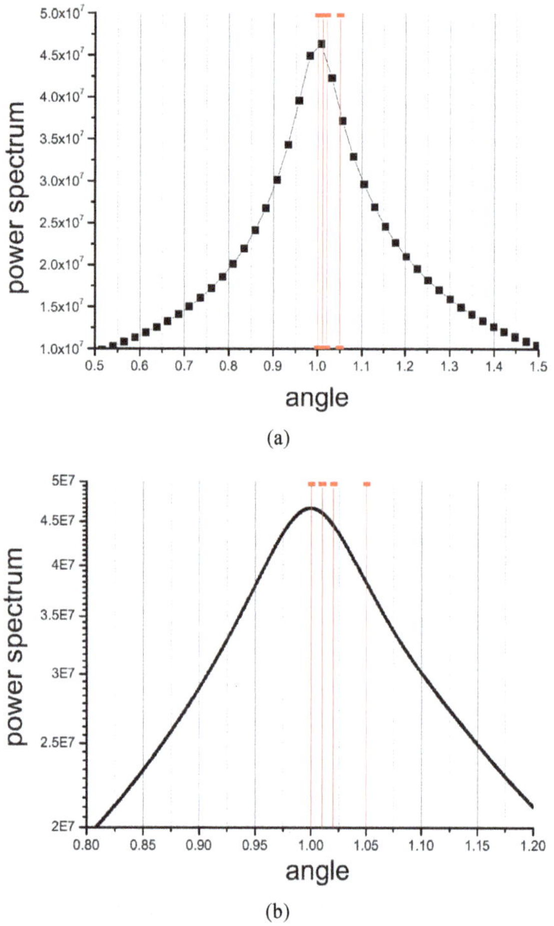

Figure 17. FFT for the same signal as in Figure 15 using (a) 256 data points and (b) 8192 data points.

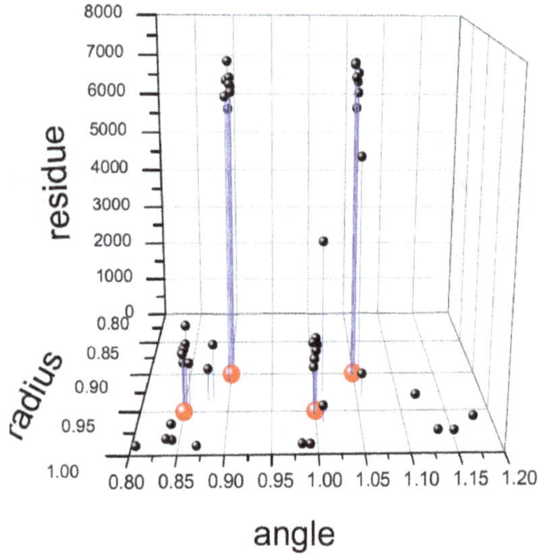

Figure 18. Residues of the poles of the Padé Approximant to the Z-transform of the signal generated by 4 poles in (ρ, φ) = (0.95, 0.85), (0.90, 0.90), (0.95, 1.00), (0.90, 1.05) with residues r = 10^3, 5 × 10^3, 10^3, 5 × 10^3 respectively. Noise amplitude is unitary. We use 8 samples of 300 points each. Large red dots indicate the position of the signal poles.

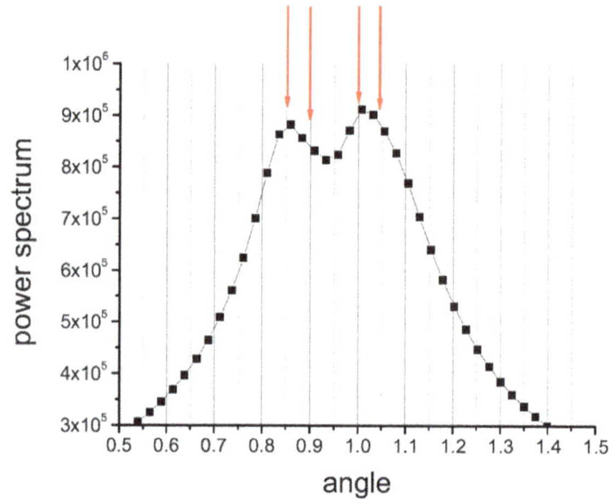

Figure 19. FFT for the same signal as in Figure 18 using 256 data points.

of noise, or equivalently of the number of significant digits of the input data.

In passing, let us note that super resolution is not unique to Padé approximants to the *Z*-transform: see e.g. [11] and references therein. The advantages of Padé approximants are that 1) they only require knowledge that the spectrum is made up of a finite number of damped oscillators; 2) they are stable with respect to the presence of small amounts of noise [5,10].

5. Acknowledgements

We thank Professor Marcel Froissart, from College de France, for discussions and suggestions.

We thank Professor Carlos Handy, Head of the Physics Department at Texas Southern University, for his support.

Special thanks to Professor Mario Diaz, Director of the Center for Gravitational Waves at the University of Texas at Brownsville: without his constant support this work would have never been possible.

This work has been supported by NASA through an award made to the Center for Gravitational Waves at the University of Texas at Brownsville.

REFERENCES

[1] P. Barone and R. March, "On the Super-Resolution Properties of Prony's Method," *ZAMM: Zeitschrift Fur Angewandte Mathematik Und Mechanik*, Vol. 76, Suppl. 2, 1996, pp. 177-180.

[2] P. Barone and R. March, "Some Properties of the Asymptotic Location of Poles of Pade Approximants to Noisy Rational Functions, Relevant for Modal Analysis," *IEEE Transactions on Signal Processing*, Vol. 46, No. 9, 1998, pp. 2448-2457.

[3] D. Belkic and K. Belkic, "Optimized Molecular Imaging through Magnetic Resonance for Improved Target Defi-

nition in Radiation Oncology," In: G. Garca Gomez-Tejedor and M. C. Fuss, Eds., *Radiation Damage in Biomolecular Systems*, Springer Netherlands, Dordrecht, 2012, pp. 411-430.

[4] C. E. Shannon, "Communication in the Presence of Noise," *Proceedings of IEEE*, Vol. 86, No. 2, 1998, pp. 447-457.

[5] D. Bessis and L. Perotti, "Universal Analytic Properties of Noise: Introducing the J-Matrix Formalism," *Journal of Physics A*, Vol. 42, No. 36, 2009, Article ID: 365202.

[6] J. Gilewicz and B. Truong-Van, "Froissart Doublets in Padé Approximants and Noise," Constructive Theory of Functions 1987, Bulgarian Academy of Sciences, Sofia, 1988, pp. 145-151.

[7] J.-D. Fournier, G. Mantica, A. Mezincescu and D. Bessis, "Universal Statistical Behavior of the Complex Zeros of Wiener Transfer Functions," *Europhysics Letters*, Vol. 22, No. 5, 1993, pp. 325-331.

[8] J.-D. Fournier, G. Mantica, A. Mezincescu and D. Bessis, "Statistical Properties of the Zeros of the Transfer Functions in Signal Processing," In: D. Benest and C. Froeschle, Eds., *Chaos and Diffusion in Hamiltonian Systems*, Editions Frontières, Paris, 1995.

[9] D. Bessis, "Padé Approximations in Noise Filtering," *Journal of Computational and Applied Mathematics*, Vol. 66, No. 1-2, 1996, pp. 85-88.

[10] L. Perotti, D. Vrinceanu and D. Bessis, "Beyond the Fourier Transform: Signal Symmetry Breaking in the Complex Plane," *IEEE Signal Processing Letters*, Vol. 19, No. 12, 2012, pp. 865-867.

[11] V. F. Pisarenko, "The Retrieval of Harmonics from a Covariance Function," *Geophysical Journal of the Royal Astronomical Society*, Vol. 33, No. 3, 1973, pp. 347-366.

Half-Step Continuous Block Method for the Solutions of Modeled Problems of Ordinary Differential Equations

A. A. James[1], A. O. Adesanya[2], J. Sunday[3], D. G. Yakubu[4]
[1]Department of Mathematics, American University of Nigeria, Yola, Nigeria
[2]Department of Mathematics, Modibbo University of Technology, Yola, Nigeria
[3]Department of Mathematical Sciences, Adamawa State University, Mubi, Nigeria
[4]Department of Mathematics, Tafawa Balewa Federal University of Bauchi, Bauch State, Nigeria

ABSTRACT

In this paper, we developed a new continuous block method using the approach of collocation of the differential system and interpolation of the power series approximate solution. A constant step length within a half step interval of integration was adopted. We evaluated at grid and off grid points to get a continuous linear multistep method. The continuous linear multistep method is solved for the independent solution to yield a continuous block method which is evaluated at selected points to yield a discrete block method. The basic properties of the block method were investigated and found to be consistent and zero stable hence convergent. The new method was tested on real life problems namely: SIR model, Growth model and Mixture Model. The results were found to compete favorably with the existing methods in terms of accuracy and error bound.

Keywords: Approximate Solution; Interpolation; Collocation; Half Step; Converges; Block Method

1. Introduction

We consider the numerical solution of first order initial value problems of the form:

$$y' = f(x, y), y(x_0) = y_0 \tag{1}$$

where f is continuous and satisfies Lipchitz's condition that guarantees the uniqueness and existence of a solution.

Problem in the form (1) has wide application in physical science, engineering, economics, etc. Very often, these problems do not have an analytical solution, and this has necessitated the deviation of numerical schemes to approximate their solutions.

In the past, scholars have developed a continuous linear multistep in solving (1). These authors proposed methods with different basis functions and among them were [1-6] to mention a few.

These authors proposed methods ranging from predictor corrector method to discrete block method.

Scholars later proposed block method. This block method has the properties of Runge-kutta method for being self-starting and does not require development of separate predictors or starting values. Among these authors are [7-12]. Block method was found to be cost effective and gave better approximation.

This paper is divided into sections as follows: Section 1 is the introduction and background of the study; Section 2 contains the discussion about the methodology involved in deriving the continuous multistep method and the continuous block method. Section 3 considers the analysis of the block method in terms of the order, zero stability and the region of absolute stability. Section 4 focuses on the application of the new method on some numeric examples and Section 5 is on the discussion of result. We tested our method on first order ordinary differential equations and compared our result with existing methods.

2. Methodology

Consider power series approximate solution in the form

$$y(x) = \sum_{j=0}^{s+r-1} a_j x^j \tag{2}$$

where S and r are the number of interpolation and collocation points respectively.

The first derivative of (2) gives

$$y'(x) = \sum_{j=0}^{s+r-1} j a_j x^{j-1} \qquad (3)$$

Substituting (3) into (2) gives

$$f(x,y) = \sum_{j=0}^{s+r-1} j a_j x^{j-1} \qquad (4)$$

Collocating (3) at $x_{n+s}, s = 0\left(\frac{1}{12}\right)\frac{1}{2}$ and interpolating (2) at x_n gives a system of non-linear equation in the form

$$AX = U \qquad (5)$$

where

$$A = \begin{bmatrix} 1 & x_n & x_n^2 & x_n^3 & x_n^4 & x_n^5 & x_n^6 & x_n^7 \\ 0 & 1 & 2x_n & 3x_n^2 & 4x_n^3 & 5x_n^4 & 6x_n^5 & 7x_n^6 \\ 0 & 1 & 2x_{n+\frac{1}{12}} & 3x^2_{n+\frac{1}{12}} & 4x^3_{n+\frac{1}{12}} & 5x^4_{n+\frac{1}{12}} & 6x^5_{n+\frac{1}{12}} & 7x^6_{n+\frac{1}{12}} \\ 0 & 1 & 2x_{n+\frac{1}{6}} & 3x^2_{n+\frac{1}{6}} & 4x^3_{n+\frac{1}{6}} & 5x^4_{n+\frac{1}{6}} & 6x^5_{n+\frac{1}{6}} & 7x^6_{n+\frac{1}{6}} \\ 0 & 1 & 2x_{n+\frac{1}{4}} & 3x^2_{n+\frac{1}{4}} & 4x^3_{n+\frac{1}{4}} & 5x^4_{n+\frac{1}{4}} & 6x^5_{n+\frac{1}{4}} & 7x^6_{n+\frac{1}{4}} \\ 0 & 1 & 2x_{n+\frac{1}{3}} & 3x^2_{n+\frac{1}{3}} & 4x^3_{n+\frac{1}{3}} & 5x^4_{n+\frac{1}{3}} & 6x^5_{n+\frac{1}{3}} & 7x^6_{n+\frac{1}{3}} \\ 0 & 1 & 2x_{n+\frac{5}{12}} & 3x^2_{n+\frac{5}{12}} & 4x^3_{n+\frac{5}{12}} & 5x^4_{n+\frac{5}{12}} & 6x^5_{n+\frac{5}{12}} & 7x^6_{n+\frac{5}{12}} \\ 0 & 1 & 2x_{n+\frac{1}{2}} & 3x^2_{n+\frac{1}{2}} & 4x^3_{n+\frac{1}{2}} & 5x^4_{n+\frac{1}{2}} & 6x^5_{n+\frac{1}{2}} & 7x^6_{n+\frac{1}{2}} \end{bmatrix}, \quad \begin{bmatrix} a_0 \\ a_1 \\ a_2 \\ a_3 \\ a_4 \\ a_5 \\ a_6 \\ a_7 \end{bmatrix}, U = \begin{bmatrix} y_n \\ f_n \\ f_{n+\frac{1}{12}} \\ f_{n+\frac{1}{6}} \\ f_{n+\frac{1}{4}} \\ f_{n+\frac{1}{3}} \\ f_{n+\frac{5}{12}} \\ f_{n+\frac{1}{2}} \end{bmatrix}$$

Solving (5) for the a_js and substituting back into (4) gives a continuous multistep method in the form

$$y(x) = \alpha_0 y_n + h\sum_{j=0}^{\frac{1}{2}} \beta_j(x) f_{n+j}, j = 0\left(\frac{1}{12}\right)\frac{1}{2} \qquad (6)$$

where $a_0 = 1$ and the coefficients of f_{n+j} gives

$$\beta_0 = \left(\frac{1}{120}\right)\left(124416t^7 - 254016t^6 + 211680t^5 - 92610t^4 + 22736t^3 - 3087t^2 + 210t\right)$$

$$\beta_{\frac{1}{12}} = \left(-\frac{1}{35}\right)\left(-124416t^7 - 241920t^6 - 187488t^5 + 73080t^4 - 14616t^3 + 1260t^2\right)$$

$$\beta_{\frac{1}{6}} = \left(\frac{1}{35}\right)\left(311040t^7 - 57456t^6 + 414288t^5 - 145215t^4 + 24570t^3 - 1575t^2\right)$$

$$\beta_{\frac{1}{4}} - \left(-\frac{1}{105}\right)\left(1244160t^7 - 2177280t^6 - 1463616t^5 + 468720t^4 - 71120t^3 + 4200t^2\right)$$

$$\beta_{\frac{1}{3}} = \left(\frac{1}{70}\right)\left(622080t^7 - 1028160t^6 + 647136t^5 - 193410t^4 + 27720t^3 - 1575t^2\right)$$

$$\beta_{\frac{5}{12}} = \left(-\frac{1}{35}\right)\left(-124416t^7 + 193536t^6 - 114912t^5 + 32760t^4 - 4536t^3 + 252t^2\right)$$

$$\beta_{\frac{1}{2}} = \left(\frac{1}{105}\right)\left(62208t^7 - 90720t^6 + 51408t^5 - 14175t^4 + 1918t^3 - 105t^2\right)$$

where $t = \frac{x - x_n}{h}$. Solving (6) for the independent solution gives a continuous block method in the form

$$y_{n+k} = \sum_{j=0}^{\mu-1} \frac{(jh)^m}{m!} y_n^{(m)} + h^{\mu} \sum_{j=0}^{s} \sigma_j(x) f_{n+j} \quad (7)$$

where μ is the order of the differential equation s is the collocation points. Hence the coefficient of f_{n+j} in (7)

$$\sigma_0 = \left(\frac{1}{120}\right)\left(124416t^7 - 254016t^6 + 211680t^5 - 92610t^4 + 22736t^3 - 3087t^2 + 210t\right)$$

$$\sigma_{\frac{1}{12}} = \left(-\frac{1}{35}\right)\left(-124416t^7 - 241920t^6 - 187488t^5 + 73080t^4 - 14616t^3 + 1260t^2\right)$$

$$\sigma_{\frac{1}{6}} = \left(\frac{1}{35}\right)\left(311040t^7 - 57456t^6 + 414288t^5 - 145215t^4 + 24570t^3 - 1575t^2\right)$$

$$\sigma_{\frac{1}{4}} = \left(-\frac{1}{105}\right)\left(-1244160t^7 - 2177280t^6 - 1463616t^5 + 468720t^4 - 71120t^3 + 4200t^2\right)$$

$$\sigma_{\frac{1}{3}} = \left(\frac{1}{70}\right)\left(622080t^7 - 1028160t^6 + 647136t^5 - 193410t^4 + 27720t^3 - 1575t^2\right)$$

$$\sigma_{\frac{5}{12}} = \left(-\frac{1}{35}\right)\left(-124416t^7 + 193536t^6 - 114912t^5 + 32760t^4 - 4536t^3 + 252t^2\right)$$

$$\sigma_{\frac{1}{2}} = \left(\frac{1}{105}\right)\left(62208t^7 - 90720t^6 + 51408t^5 - 14175t^4 + 1918t^3 - 105t^2\right)$$

where $t = \frac{x - x_n}{h}$.

Evaluating (2.5) at $t = \frac{1}{12}\left(\frac{1}{12}\right)\frac{1}{2}$ gives a discrete block formula of the form

$$Y_m = ey_n + hdf(y_n) + hdf(Y_m) \quad (8)$$

where $e, d,$ are $r \times r$ matrix
where

$$d = \begin{bmatrix} \frac{19087}{725760} & \frac{1139}{45360} & \frac{137}{5376} & \frac{143}{5670} & \frac{3715}{145152} & \frac{41}{1680} \end{bmatrix}^T$$

$$Y_m = \begin{bmatrix} y_{n+\frac{1}{12}}, y_{n+\frac{1}{6}}, y_{n+\frac{1}{4}}, y_{n+\frac{1}{3}}, y_{n+\frac{5}{12}}, y_{n+\frac{1}{2}} \end{bmatrix}^T$$

$$e = \begin{bmatrix} 0 & 0 & 0 & 0 & 0 & 1 \\ 0 & 0 & 0 & 0 & 0 & 1 \\ 0 & 0 & 0 & 0 & 0 & 1 \\ 0 & 0 & 0 & 0 & 0 & 1 \\ 0 & 0 & 0 & 0 & 0 & 1 \\ 0 & 0 & 0 & 0 & 0 & 1 \end{bmatrix}$$

$$b = \begin{bmatrix} \frac{2713}{30240} & \frac{-15487}{241920} & \frac{293}{5670} & \frac{-6737}{241920} & \frac{263}{30240} & \frac{-863}{725760} \\ \frac{47}{378} & \frac{11}{15120} & \frac{83}{2835} & \frac{-269}{15120} & \frac{11}{1890} & \frac{-37}{45360} \\ \frac{27}{224} & \frac{387}{8960} & \frac{17}{210} & \frac{-243}{8960} & \frac{9}{1120} & \frac{-29}{26880} \\ \frac{116}{945} & \frac{32}{945} & \frac{376}{2835} & \frac{29}{1890} & \frac{4}{945} & \frac{-2}{2835} \\ \frac{725}{6048} & \frac{2125}{48384} & \frac{125}{1134} & \frac{3875}{48384} & \frac{235}{6048} & \frac{-275}{145152} \\ \frac{9}{70} & \frac{9}{560} & \frac{17}{105} & \frac{9}{560} & \frac{9}{70} & \frac{41}{1680} \end{bmatrix}$$

3. Analysis of the Basic Properties of Our New Method

3.1. Order of the Method

Let the linear operator $L\{y(x):h\}$ associated with the block formular be defined as

$$L\{y(x):h\} = A^{(0)}Y_m - ey_n - h^{\mu}df(y_n) - h^{\mu}bF(Y_m) \quad (9)$$

expanding in Taylor series and comparing the coefficient of h gives

$$L\{y(x);h\} = c_0 y(x) + c_1 hy'^{(x)} + c_1 hy''^{(x)} \cdots c_p h^p y^p(x) + c_{p+1}h^{p+1}y^{p+1}(x) + c_{p+2}h^{p+2}y^{p+2} \quad (10)$$

Definition:-The linear operator L and the associated continuous linear multistep method (3.1) are said to be of order p if $c_0 = c_1 = c_2 = \cdots = c_p = 0$ and $c_{p+1} \neq 0$ is called the error constant and implies that the local truncation error is given by $t_{n+k} = C_{p+2}h^{p+1}y^{p+1}(x_n) + 0(h^{p+2})$.

For our method;

$$L\{y(x);h\} = \begin{bmatrix} y_{n+\frac{1}{12}} \\ y_{n+\frac{1}{6}} \\ y_{n+\frac{1}{4}} \\ y_{n+\frac{1}{3}} \\ y_{n+\frac{5}{12}} \\ y_{n+\frac{1}{2}} \end{bmatrix} - \begin{bmatrix} 1 & \frac{1}{12}h \\ 1 & \frac{1}{6}h \\ 1 & \frac{1}{4}h \\ 1 & \frac{1}{3}h \\ 1 & \frac{5}{12}h \\ 1 & \frac{1}{2}h \end{bmatrix} [y_n] - \begin{bmatrix} f_n \\ f_{n+\frac{1}{12}} \\ f_{n+\frac{1}{6}} \\ f_{n+\frac{1}{4}} \\ f_{n+\frac{1}{3}} \\ f_{n+\frac{5}{12}} \\ f_{n+\frac{1}{2}} \end{bmatrix} = 0$$

Expanding in Taylor series expansion gives

$$\sum_{j=0}^{\infty}\frac{\left(\frac{1}{12}\right)^j}{j!}y_n^j - y_n - \frac{19087}{725760}hy_n' - \sum_{j=0}^{\infty}\frac{h^{j+1}}{j!}y_n^{j+1}\left\{\frac{2713}{30240}\left(\frac{1}{12}\right)^j + \frac{15487}{241920}\left(\frac{1}{6}\right)^j - \frac{293}{5670}\left(\frac{1}{4}\right)^j + \frac{6737}{241920}\left(\frac{1}{3}\right)^j + \frac{263}{30240}\left(\frac{5}{6}\right)^j - \frac{863}{725760}\left(\frac{1}{2}\right)^j\right\}$$

$$\sum_{j=0}^{\infty}\frac{\left(\frac{1}{6}\right)^j}{j!}y_n^j - y_n - \frac{1139}{45360}hy_n' - \sum_{j=0}^{\infty}\frac{h^{j+1}}{j!}y_n^{j+1}\left\{\frac{47}{378}\left(\frac{1}{12}\right)^j + \frac{11}{15120}\left(\frac{1}{6}\right)^j + \frac{83}{2835}\left(\frac{1}{4}\right)^j - \frac{269}{15120}\left(\frac{1}{3}\right)^j - \frac{11}{1890}\left(\frac{5}{12}\right)^j - \frac{37}{45360}\left(\frac{1}{2}\right)^j\right\}$$

$$\sum_{j=0}^{\infty}\frac{\left(\frac{1}{4}\right)^j}{j!}y_n^j - y_n - \frac{137}{5376}hy_n' - \sum_{j=0}^{\infty}\frac{h^{j+1}}{j!}y_n^{j+1}\left\{\frac{27}{224}\left(\frac{1}{12}\right)^j + \frac{387}{8960}\left(\frac{1}{6}\right)^j + \frac{17}{210}\left(\frac{1}{4}\right)^j - \frac{243}{8960}\left(\frac{1}{3}\right)^j + \frac{9}{1120}\left(\frac{5}{12}\right)^j - \frac{29}{26880}\left(\frac{1}{2}\right)^j\right\}$$

$$\sum_{j=0}^{\infty}\frac{\left(\frac{1}{3}\right)^j}{j!}y_n^j - y_n - \frac{143}{5670}hy_n' - \sum_{j=0}^{\infty}\frac{h^{j+1}}{j!}y_n^{j+1}\left\{\frac{116}{945}\left(\frac{1}{12}\right)^j + \frac{32}{945}\left(\frac{1}{6}\right)^j + \frac{376}{2835}\left(\frac{1}{4}\right)^j + \frac{29}{1890}\left(\frac{1}{3}\right)^j + \frac{4}{945}\left(\frac{5}{12}\right)^j - \frac{2}{2835}\left(\frac{1}{2}\right)^j\right\}$$

$$\sum_{j=0}^{\infty}\frac{\left(\frac{5}{12}\right)^j}{j!}y_n^j - y_n - \frac{3715}{145152}hy_n' - \sum_{j=0}^{\infty}\frac{h^{j+1}}{j!}y_n^{j+1}\left\{\frac{725}{6048}\left(\frac{1}{12}\right)^j + \frac{2125}{48384}\left(\frac{1}{6}\right)^j + \frac{125}{1134}\left(\frac{1}{4}\right)^j + \frac{3875}{48384}\left(\frac{1}{3}\right)^j + \frac{235}{6048}\left(\frac{5}{12}\right)^j - \frac{275}{145152}\left(\frac{1}{2}\right)^j\right\}$$

$$\sum_{j=0}^{\infty}\frac{\left(\frac{1}{2}\right)^j}{j!}y_n^j - y_n - \frac{41}{1680}hy_n' - \sum_{j=0}^{\infty}\frac{h^{j+1}}{j!}y_n^{j+1}\left\{\frac{9}{70}\left(\frac{1}{12}\right)^j + \frac{9}{560}\left(\frac{1}{6}\right)^j + \frac{17}{105}\left(\frac{1}{4}\right)^j + \frac{9}{560}\left(\frac{1}{3}\right)^j + \frac{9}{70}\left(\frac{5}{12}\right)^j - \frac{41}{1680}\left(\frac{1}{2}\right)^j\right\}$$

Equating coefficients of the Taylor series expansion to zero yield

$$c_0 = c_1 = \cdots = c_6 = 0$$

Hence we arrived at a uniform order 6 for our method with error constants

$$c_7 = \left[3.78(-09)2.53(-09)3.49(-09)1.85(-09)1.68(-08)5.56(-08)\right]$$

3.2. Zero Stability

Definition: The block (8) is said to be zero stable, if the roots Zs, $s = 1, 2, \cdots, N$ of the characteristic polynomial $\rho(z)$ defined by $\rho(z)\det\left(zA^{(0)} - E\right)$ satisfies $|z_s| \le 1$ and every root satisfying $|z_s| \le 1$ have multiplicity not exceeding the order of the differential equation. Moreover as $h \to 0, \rho(z) = z^{r-\mu}(z-1)^{\mu}$ where μ is the order of the differential equation, r is the order of the matrix $A^{(0)}$ and E.

For our method

$$\rho(z) = \left[z\begin{bmatrix} 1 & 0 & 0 & 0 & 0 & 0 \\ 0 & 1 & 0 & 0 & 0 & 0 \\ 0 & 0 & 1 & 0 & 0 & 0 \\ 0 & 0 & 0 & 1 & 0 & 0 \\ 0 & 0 & 0 & 0 & 1 & 0 \\ 0 & 0 & 0 & 0 & 0 & 1 \end{bmatrix} - \begin{bmatrix} 0 & 0 & 0 & 0 & 0 & 1 \\ 0 & 0 & 0 & 0 & 0 & 1 \\ 0 & 0 & 0 & 0 & 0 & 1 \\ 0 & 0 & 0 & 0 & 0 & 1 \\ 0 & 0 & 0 & 0 & 0 & 1 \\ 0 & 0 & 0 & 0 & 0 & 1 \end{bmatrix} \right] = 0$$

$\rho(z) = z^5(z-1)$. Hence our method is zero stable.

3.3 Region of Absolute Stability

The block formulated as a general linear method where it is partition in the form

$$\begin{bmatrix} Y \\ Y_{n-i} \end{bmatrix} = \begin{bmatrix} A_1 & B_1 \\ A_2 & B_2 \end{bmatrix}\begin{bmatrix} hf(y) \\ y_n \end{bmatrix}$$

The elements of A_1 and A_2 are obtained from the coefficients of the collocation points, B_1 and B_2 are obtained from the interpolation points.

Applying the test equation $y' = \lambda y$ leads to the recurrence equation

$$y^{i+1} = M(Z)y^i,\ Z = \lambda h,\ i = 1, 2, \cdots, \mu - 1$$

The stability function is given by

$$M(Z) = B_2 + ZA_2(I - ZA_1)^{-B_1}$$

and the stability polynomial of the method is given as

$$\rho(\lambda, Z) = \det(\lambda I - M(Z))$$

The region of absolute stability of the method is defined as $\rho(\lambda, Z) = 1,\ |\lambda| \le 1$.

For our method, writing the block in partition form gives

$$\begin{bmatrix} y_n y_{n+\frac{1}{12}} y_{n+\frac{1}{6}} y_{n+\frac{1}{4}} y_{n+\frac{1}{3}} y_{n+\frac{5}{12}} y_{n+1} \end{bmatrix}^{\mathrm{T}}$$

$$= \begin{bmatrix} 0 & 0 & 0 & 0 & 0 & 0 & 0 & \cdots & 0 & 1 \\ \frac{19087}{725760} & \frac{2713}{30240} & \frac{-15487}{241920} & \frac{293}{5670} & \frac{-6737}{241920} & \frac{263}{30240} & \frac{-863}{725760} & \cdots & 0 & 1 \\ \frac{1139}{45360} & \frac{47}{378} & \frac{11}{15120} & \frac{83}{2835} & \frac{-269}{15120} & \frac{11}{1890} & \frac{-37}{45360} & \cdots & 0 & 1 \\ \frac{137}{5376} & \frac{27}{224} & \frac{387}{8960} & \frac{17}{210} & \frac{-243}{8960} & \frac{9}{1120} & \frac{-29}{26880} & \cdots & 0 & 1 \\ \frac{143}{5670} & \frac{116}{945} & \frac{32}{945} & \frac{376}{2835} & \frac{29}{1890} & \frac{4}{945} & \frac{-2}{2835} & \cdots & 0 & 1 \\ \frac{3715}{145152} & \frac{725}{6048} & \frac{2125}{48384} & \frac{125}{1134} & \frac{3875}{48384} & \frac{235}{6048} & \frac{-275}{145152} & \cdots & 0 & 1 \\ \frac{41}{1680} & \frac{9}{70} & \frac{9}{560} & \frac{17}{105} & \frac{9}{560} & \frac{9}{70} & \frac{41}{1680} & \cdots & 0 & 1 \\ \vdots & \vdots & \vdots & \vdots & \vdots & \vdots & \vdots & \ddots & \vdots & \vdots \\ \frac{19087}{725760} & \frac{2713}{30240} & \frac{-15487}{241920} & \frac{293}{5670} & \frac{-6737}{241920} & \frac{263}{30240} & \frac{-863}{725760} & \cdots & 0 & 1 \\ \frac{41}{1680} & \frac{9}{70} & \frac{9}{560} & \frac{17}{105} & \frac{9}{560} & \frac{9}{70} & \frac{41}{1680} & \cdots & 1 & 1 \end{bmatrix} \begin{bmatrix} hf_n \\ hf_{n+\frac{1}{12}} \\ hf_{n+\frac{1}{6}} \\ hf_{n+\frac{1}{4}} \\ hf_{n+\frac{1}{3}} \\ hf_{n+\frac{5}{12}} \\ hf_{n+\frac{1}{2}} \\ \vdots \\ hf_{n+\frac{1}{12}} \\ hf_{n+\frac{1}{2}} \end{bmatrix}$$

4. Real Life Problems

4.1. Problem 1: (SIR MODEL)

The SIR model is an epidemiological model that computes the theoretical numbers of people infected with a contagious illness in a closed population over time. The name of this class of models derives from the fact that they involves coupled equations relating the number of susceptible people S(t), number of people infected I(t) and the number of people who have recovered R(t). This is a good and simple model for many infectious diseases including measles, mumps and rubella [13-15]. The SIR model is described by the three coupled equations.

$$\frac{\mathrm{d}s}{\mathrm{d}t} = \mu(1-S) - \beta IS \tag{11}$$

$$\frac{\mathrm{d}I}{\mathrm{d}t} = -\mu I - \gamma I + \beta IS \tag{12}$$

$$\frac{\mathrm{d}s}{\mathrm{d}t} = -\mu R + \gamma I \tag{13}$$

where μ, γ and β are positive parameters. Define y to be

$$y = S + I + R \tag{14}$$

Adding Equations (11)-(13), we obtain the following evolution equations for y

$$y' = \mu(1-y) \tag{15}$$

Taking $\mu = 0.5$ and attaching an initial condition $y(0) = 0.5$ (for a particular closed population), we obtain,

$$y'(t) = 0.5(1-y), y(0) = 0.5 \tag{16}$$

whose exact solution is,

$$y(t) = 1 - 0.5\mathrm{e}^{-0.5t} \tag{17}$$

Applying our new half step numerical scheme (8) to solve SIR model simplified as (17) gives results as shown in **Table 1**.

4.2. Problem 2 (Growth Model)

Let us consider the differential equation of the form;

$$\frac{\mathrm{d}N}{\mathrm{d}t} = \alpha N,\ N(0) = 1000,\ t \in [0,1] \tag{18}$$

Equation (18) represents the rate of growth of bacteria in a colony. We shall assume that the model grows continuously without restriction. One may ask; how many bacteria are in the colony after some minutes if an individual produces an offspring at an average growth rate of 0.2? We also assume that $N(t)$ is the population size at time t (**Table 2**).

The theoretical solution of (18) is given by;

s (19)

Note that, growth rate $\alpha = 0.2$ in (18).

Table 1. Showing results for SIR model problem.

X	Exact Result	Computed Solution	Error in half step method	Error in Sunday *et al.*
0.1	0.5243852877496430	0.5243852877496429	1.110223e−016	5.574430e−012
0.2	0.5475812909820202	0.5475812909820202	0.000000e+000	3.946177e−012
0.3	0.5696460117874711	0.5696460117874710	1.110223e−016	8.183232e−012
0.4	0.5906346234610092	0.5906346234610089	2.220446e−016	3.436118e−011
0.5	0.6105996084642976	0.6105996084642973	3.330669e−016	1.929743e−010
0.6	0.6295908896591411	0.6295908896591410	1.110223e−016	1.879040e−010
0.7	0.6476559551406433	0.6476559551406429	4.440892e−016	1.776835e−010
0.8	0.6648399769821805	0.6648399769821799	5.551115e−016	1.724676e−010
0.9	0.6811859241891134	0.6811859241891132	2.220446e−016	1.847545e−010
1.0	0.6967346701436834	0.6967346701436828	5.551115e−016	3.005770e−010

Table 2. Showing results for growth model problem.

X	Exact Result	Computed Solution	Error in half step method	Error in Sunday *et al.*
0.1	1020.2013400267558	1020.201340026755	0.000000e+000	1.830358e−011
0.2	1040.8107741923882	1040.8107741923882	0.000000e+000	1.250555e−011
0.3	1061.8365465453596	1061.8365465453596	0.000000e+000	1.227818e−011
0.4	1083.2870676749587	1083.2870676749585	2.273737e−013	3.137757e−011
0.5	1105.1709180756477	1105.1709180756475	2.273737e−013	2.216893e−010
0.6	1105.1709180756477	1127.4968515793755	2.273737e−013	2.060005e−010
0.7	1150.2737988572273	1150.2737988572271	2.273737e−013	2.171419e−010
0.8	1173.5108709918102	1173.5108709918102	0.000000e+000	2.216893e−010
0.9	1197.2173631218102	1197.2173631218102	0.000000e+000	2.744400e−010
1.0	1221.4027581601699	1221.4027581601699	0.000000e+000	4.899903e−010

Applying our new half step numerical scheme (8) to solve the Growth model (17) gives results as shown in **Table 2** [16].

4.3. Problem 3 (Decay Model)

A certain radioactive substance is known to decay at the rate proportional to the amount present. A block of this substance having a mass of 100 g originally is observed. After 40 mins, its mass reduced to 90 g. Find an expression for the mass of the substance at any time and test for the consistency of the block integrator on this problem for $t \in [0,1]$.

The problem has a differential equation of the form;

$$\frac{dN}{dt} = -\alpha N,\ N(0) = 100,\ t \in [0,1] \tag{21}$$

where N represents the mass of the substance at any time t and α is a constant which specifies the rate at which this particular substance decays. Note that,

$$f(0) = 100 \text{ g},\ t = 40 \text{ mins},\ f(40) = 90 \text{ g}$$

Since for any growth/decay problem,

$$f(t) = f(0)e^{\alpha t}$$

$$90 = 100e^{40\alpha}$$

$$\alpha = \frac{(\ln 9 - \ln 10)}{40} = -0.0026$$

Thus, the theoretical solution to (20) is given by,

$$f(t) = 100e^{-0.0026t} \tag{22}$$

which is also the expression for the mass of the substance at any time t.

Applying our new half step numerical scheme (8) to solve the Growth model (22) gives results as shown in **Table 3** [16,17] (**Table 3**).

5. Discussion of the Result

We have considered three real-life model problems to test the efficiency of our method. Problems 1 and 2 and 3 were solved by Sunday *et al.* [17]. They proposed an order six block integrator for the solution of first-order ordinary differential equations. Our half-step block method gave better approximation as shown in **Tables**

Table 3. Showing results for decay model problem.

X	Exact Result	Computed Solution	Error in half step method	Error in Sunday *et al.*
0.1	99.9740033797070850	99.9740033797070850	0.000000e+000	0.000000e+000
0.2	99.9480135176568470	99.9480135176568330	1.421085e−014	1.421085e−014
0.3	99.9220304120923400	99.9220304120923400	0.000000e+000	0.000000e+000
0.4	99.8960540612571460	99.8960540612571460	0.000000e+000	0.000000e+000
0.5	99.8700844633952300	99.8700844633952440	1.421085e−014	0.000000e+000
0.6	99.8441216167510670	99.8441216167510820	1.421085e−014	0.000000e+000
0.7	99.8181655195695610	99.8181655195695750	1.421085e−014	0.000000e+000
0.8	99.7922161700960970	99.7922161700960970	0.000000e+000	0.000000e+000
0.9	99.7662735665764730	99.7662735665764730	0.000000e+000	0.000000e+000
1.0	99.7403377072569700	99.7403377072569700	0.000000e+000	0.000000e+000

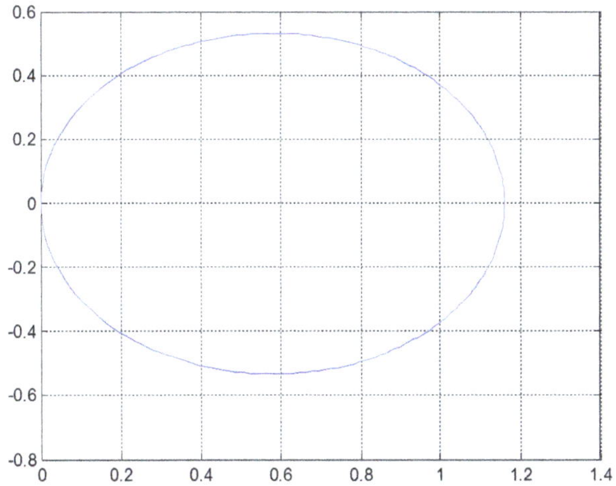

Figure 1. Showing region of absolute stability of our method.

1-3 because the iteration per step in the new method was lower than the method proposed by [17]. Our method was found to be zero stable, consistent and converges. **Figure 1** shows the region of absolute stability. From the numerical examples, we could safely conclude that our method gave better accuracy than the existing methods.

REFERENCES

[1] A. O. Adesanya, M. R. Odekunle, and A. A. James, "Order Seven Continuous Hybrid Methods for the Solution of First Order Ordinary Differential Equations," *Canadian Journal on Science and Engineering Mathematics*, Vol. 3 No. 4, 2012, pp. 154-158.

[2] A. M. Badmus and D. W. Mishehia, "Some Uniform Order Block Methods for the Solution of First Ordinary Differential Equation," J. N.A.M. P, Vol. 19, 2011, pp 149-154.

[3] J. Fatokun, P. Onumanyi and U. W. Sirisena, "Solution of First Order System of Ordering Differential Equation by Finite Difference Methods with Arbitrary," J.N.A.M.P, 2011, pp. 30-40.

[4] B. I. Zarina and S. I. Kharil, "Block Method for Generalized Multistep Method Adams and Backward Differential Formulae in Solving First Order ODEs," MATHEMATIKA, 2005, pp. 25-33.

[5] G. Dahlquist, "A Special Stability Problem for Linear Multistep Methods," BIT, Vol. 3, 1963, pp. 27-43.

[6] P. Onumanyi, U. W. Sirisena and S. A. Jator, "Solving Difference Equation," *International Journal of Computing Mathematics*, Vol. 72, 1999, pp. 15-27.

[7] E. A. Areo, R. A. Ademiluyi and P. O. Babatola, "Three-Step Hybrid Linear Multistep Method for the Solution of First Order Initial Value Problems in Ordinary Differential Equations," J.N.A.M.P, Vol. 19, 2011, pp. 261-266.

[8] E. A. Ibijola, Y. Skwame and G. Kumleng, "Formation of Hybrid of Higher Step-Size, through the Continuous Multistep Collocation," *American Journal of Scientific and Industrial Research*, Vol. 2, 2011, pp. 161-173.

[9] H. S. Abbas, "Derivation of a New Block Method Similar to the Block Trapezoidal Rule for the Numerical Solution of First Order IVPs," *Science Echoes*, Vol. 2, 2006, pp. 10-24.

[10] Y. A. Yahaya and G. M. Kimleng, "Continuous of Two-Step Method with Large Region of Absolute Stability," J.N.A.M.P, 2007, pp. 261-268.

[11] A. O. Adesanya, M. R. Odekunle and A. A. James, "Starting Hybrid Stomer-Cowell More Accurately by Hybrid Adams Method for the Solution of First Order Ordinary Differential Equation," *European Journal of Scientific Research*, Vol. 77 No. 4, 2012, pp. 580-588.

[12] S. Abbas, "Derivations of New Block Method for the Numerical Solution of First Order IVPs," *International Journal of Computer Mathematics*, Vol. 64, 1997, p. 325.

[13] W. B. Gragg and H. J. Stetter, "Generalized Multistep Predictor-Corrector Methods," *Journal of Association of Computing Machines*, Vol. 11, No. 2, 1964, pp. 188-209.

[14] J. B. Rosser, "A Runge-Kutta Method for All Seasons," *SIAM Review*, Vol. 9, No. 3, 1967, pp. 417-452.

[15] R. E. Mickens, "Nonstandard Finite Difference Models of Differential Equations," World Scientific, Singapore, 1994.

[16] O. D. Ogunwale, O. Y. Halid and J. Sunday, "On an Alternative Method of Estimation of Polynomial Regression Models," *Australian Journal of Basic and Applied Sciences*, Vol. 4, No. 8, 2010, pp. 3585-3590.

[17] J. Sunday, M. R. Odekunleans and A. O. Adeyanya, "Order Six Block Integrator for the Solution of First-Order Ordinary Differential Equations," *IJMS*, Vol. 3, No. 1 2013, pp. 87-96.

5

Simplified Optimization Routine for Tuning Robust Fractional Order Controllers

Cristina I. Muresan
Department of Automation, Faculty of Automation and Computer Science, Technical, University of Cluj-Napoca, Romania

ABSTRACT

Fractional order controllers have been used intensively over the last decades in controlling different types of processes. The main methods for tuning such controllers are based on a frequency domain approach followed by optimization routine, generally in the form of the Matlab *fminsearch*, but also evolving to more complex routines, such as the genetic algorithms. An alternative to these time consuming optimization routines, a simple graphical method has been proposed. However, these graphical methods are not suitable for all combinations of the imposed performance specifications. To preserve their simplicity, but also to make these graphical methods generally applicable, a modified graphical method using a very straightforward and simple optimization routine is proposed within the paper. Two case studies are presented, for tuning fractional order PI and PD controllers.

Keywords: Fractional Order Controllers; Graphical Tuning; Simplified Optimization Routine

1. Introduction

Fractional order PIDs (FO-PIDs) have been employed in various engineering fields ranging applications in a wide variety of domains. The fractional order PID controller is in fact a generalization of the classical integer order PID. In the fractional order PID control algorithm, the error signal is integrated and differentiated to any order, rather than to an integer order as with the traditional PIDs. The fractional order PIDs have two supplementary parameters compared to the traditional PIDs. It is for this reason, that the fractional order PIDs have the potential to meet more design specifications than the traditional PIDs and hence to increase the performance and robustness of closed loop systems [1-4]. A couple of interesting methods have been proposed for tuning such FO-PIDs with the great majority centered upon Matlab's *fminsearch* or graphical approaches [1, 5-7]. The current trend nowadays is directed to the latter methods, since they require less computational and time resources. Nevertheless, if no exact solution exists, the current graphical methods fail at the tuning of the FO-PID controller.

The purpose of this paper is to design an improved graphical method for tuning FO-PI and FO-PD controllers, based upon an optimization routine that selects the best possible tuning option even in the case of no exact solution. For exemplification, two case studies are considered. The first case study implies the design of FO-PI control for a simple first order process. The second case study consists in the design of a FO-PD controller for a second order process with integrator effect. Simulation results in both case studies show that the fractional order controllers tuned using the proposed algorithm can meet all performance specifications. To exemplify the optimized graphical methods for tuning fractional order controllers, the first case study has no exact solution, while the second case study has an exact solution.

The paper is organized as follows. Section 2 contains the main contribution of the present paper, with a description of the fractional order PI and PD optimized graphical tuning algorithms, while Section 3 presents the two case studies. The final section contains the concluding remarks.

2. Optimization Routine for Tuning Fractional Order Controllers

The transfer function of the fractional order PI (FO-PI) controller is given by:

$$H_{FO-PI}(s) = k_p\left(1 + \frac{k_i}{s^{\mu}}\right) \quad (1)$$

while the transfer function for the fractional order PD (FO-PD) controller is given by:

$$H_{FO-PD}(s) = k_p\left(1 + k_d s^{\lambda}\right) \quad (2)$$

where k_p, k_i and k_d are the proportional, integral and derivative gains and $\mu \in \Re$ and $\lambda \in \Re$ are the fractional order. If $\mu=1$, then the FO-PI controller in (1) is reduced to a traditional PI controller:

$$H_{FO-PI}(s)=k_p\left(1+\frac{k_i}{s}\right) \quad (3)$$

and the FO-PD is reduced to the classical PD controller by setting $\lambda=1$ in (2):

$$H_{FO-PD}(s)=k_p\left(1+k_d s\right) \quad (4)$$

A proper tuning of the FO-PI and FO-PD controllers in (1) and (3), as well as of the PI and PD controllers in (3) and (4), respectively, implies determining the values for the parameters, three in the case of the FO-PI and FO-PD controllers and two in the case of the traditional PI and PD controllers. For tuning FO-PI and FO-PD controllers, in order to uniquely determine the three parameters - μ, k_p and k_i in the case of the FO-PI and λ, k_p and k_d in the case of the FO-PD– three equations are used that describe the performance of the closed loop system. The general approach regarding the tuning of fractional order controllers is based on frequency domain performance specifications [8-10], which refer to imposing a gain crossover frequency, a phase margin and robustness to open loop gain variations.

For a general process transfer function $H_p(s)$, the open loop system when $s\to j\omega$ is written as:

$$H_{open-loop}(j\omega)=H_{FO-PI}(j\omega)H_P(j\omega) \quad (5)$$

where ω is the frequency.

In order for the open loop system to attain an imposed gain crossover frequency ω_{gc}, then the following relation must hold:

$$\left|H_{open-loop}(j\omega_{gc})\right|=1 \quad (6)$$

where$\left|\ \right|$denotes the modulus of the complex function.

The open loop phase margin, φ_m, is also computed at the gain crossover frequency as:

$$\angle H_{open-loop}(j\omega_{gc})=-\pi+\varphi_m \quad (7)$$

where $\angle$ denotes the phase of the complex function.

Finally, the last performance specification, robustness to gain variations, implies that the phase of the open loop system at the gain crossover frequency should be flat:

$$\frac{d\left(\angle H_{open-loop}(j\omega_{gc})\right)}{d\omega_{gc}}=0 \quad (8)$$

2.1. Optimization Routine for Tuning Fractional Order PI Controllers

The transfer function of the FO-PI controller, in the frequency domain, may be written as:

$$H_{FO-PI}(j\omega)=k_p\left[1+k_i\omega^{-\mu}\left(\cos\frac{\pi\mu}{2}-j\sin\frac{\pi\mu}{2}\right)\right] \quad (9)$$

in which

$$\frac{1}{s^{\mu}}\to\frac{1}{(j\omega)^{\mu}}=\omega^{-\mu}\left(\cos\frac{\pi\mu}{2}-j\sin\frac{\pi\mu}{2}\right) \quad (10)$$

Equations (6), (7) and (8) imply a certain behavior of the closed loop system, according to the specified values for the gain crossover frequency and the phase margin, and may further be used to determine all three values for the k_p, k_i and μ parameters of the FO-PI controller:

$$\left|k_p\left[1+k_i\omega_{gc}^{-\mu}\left(\cos\frac{\pi\mu}{2}-j\sin\frac{\pi\mu}{2}\right)\right]\right|=1/\left|H_p(j_{\omega_{gc}})\right| \quad (11)$$

$$\frac{k_i\sin\left(\frac{\pi\cdot\mu}{2}\right)}{\omega_{gc}^{\mu}+k_i\cos\left(\frac{\pi\cdot\mu}{2}\right)}=tg\left(\pi-\phi_m+\angle H_p(j\omega_{gc})\right) \quad (12)$$

$$\frac{\mu k_i\omega_{gc}^{-\mu-1}\sin\frac{\pi\mu}{2}}{1+2k_i\omega_{gc}^{-\mu}\cos\frac{\pi\mu}{2}+k_i^2\omega_{gc}^{-2\mu}}=-\frac{d\angle H_p(j\omega_{gc})}{d\omega_{gc}} \quad (13)$$

where $H_p(s)$ is the process transfer function.

Using solely equations (12) and (13), k_i and μ may be determined uniquely, while (11) may be then used to determine k_p. The simplest method for computing the FO-PI parameter values is based on a graphical approach [1, 5-7], which implies that k_i is computed and plotted as a function of μ using equations (12) and (13). The intersection point of the resulting two curves yields the final values for k_i and μ. Consequently, k_p is determined using (11) and the previously graphically selected values for k_i and μ. Such an example is given in **Figure 1**.

Although this method proves to be highly efficient and simple, the graphical approach is based upon the intersection of the curves resulting from (12) and (12). Such

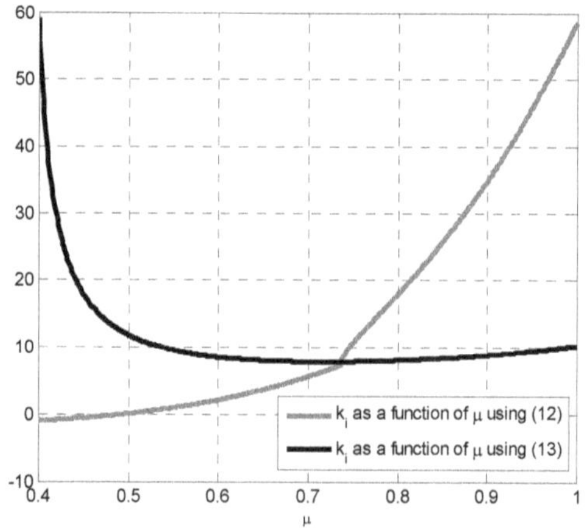

Figure 1. Selection of k_i and μ according to the intersection point of the curves.

an intersection point depends upon the imposed criteria for the gain crossover frequency and the phase margin. For a specified set of gain crossover frequency and phase margin, such an intersection point might not exist. Thus, the existing graphical methods cannot be used to compute the parameters and optimization algorithms need to be used instead.

In order to facilitate the use of the simplicity of the graphical methods in tuning the FO-PI controllers and to avoid complex optimization algorithms, a simple approach is proposed that combines the graphical methods with a very simple optimization routine. The idea behind the optimization routine consist in plotting the two curves for k_i as a function of μ and selecting the values that minimize the distance between the two plotted curves. The proposed tuning algorithm is given below:

for $\mu = 0:1$
  compute k_i using (12)
  store result in vector k_{i1}
  compute k_i using (13)
  store result in vector k_{i2}
end
  plot k_{i1} as a function of μ
  plot k_{i2} as a function of μ
  compute absolute value of *distance* = k_{i1}-k_{i2}
  determine *optim* = min(*distance*)
  return μ_{optim} corresponding to *optim*
  compute k_i using (13) and μ_{optim}
  compute k_p using (11)

The algorithm for computing PI controllers is based upon setting $\mu = 1$ and computing k_i using either (12) or (13) and k_p using (11). Since, the PI controller in (3) has only two design parameters, the tuning of the PI controller may be done using any combination of two performance criteria in (11), (12) or (13). Thus, imposing (11) and (12) means that (13) will not necessarily be ensured, which is the main drawback of traditional PI controllers as compared to FO-PI controllers.

2.2. Optimization Routine for Tuning Fractional Order PD Controllers

The tuning of the FO-PD controller is achieved in a similar manner to the FO-PI. The transfer function of the FO-PD controller, in the frequency domain, may be written as:

$$H_{FO-PI}(j\omega) = k_p\left[1 + k_d\omega^\lambda\left(\cos\frac{\pi\lambda}{2} + j\sin\frac{\pi\lambda}{2}\right)\right] \quad (14)$$

in which

$$s^\lambda \to (j\omega)^\lambda = \omega^\lambda\left(\cos\frac{\pi\lambda}{2} + j\sin\frac{\pi\lambda}{2}\right) \quad (15)$$

Similar to the FO-PI situation, equations (6), (7) and (8) may be used to determine the three parameters of the FO-PD controller, k_p, k_d and λ :

$$\left|k_p\left[1 + k_d\omega_{gc}^\lambda\left(\cos\frac{\pi\lambda}{2} + j\sin\frac{\pi\lambda}{2}\right)\right]\right| = 1/\left|H_p(j_{\omega_{gc}})\right| \quad (16)$$

$$\frac{k_d \sin\left(\frac{\pi\cdot\lambda}{2}\right)}{\omega_{gc}^{-\lambda} + k_d\cos\left(\frac{\pi\cdot\lambda}{2}\right)} = tg\left(\pi - \phi_m + \angle H_p(j\omega_{gc})\right) \quad (17)$$

$$\frac{\lambda k_d\omega_{gc}^{\lambda-1}\sin\frac{\pi\lambda}{2}}{1 + 2k_d\omega_{gc}^\lambda\cos\frac{\pi\lambda}{2} + k_d^2\omega_{gc}^{2\lambda}} = -\frac{d\angle H_p(j\omega_{gc})}{d\omega_{gc}} \quad (18)$$

Then, (17) and (18) may be employed to determine using the optimized graphical algorithm the parameters k_d and λ, and then, k_p may be computed directly using (16), as described below:

for $\lambda = 0:1$
  compute k_d using (17)
  store result in vector k_{d1}
  compute k_d using (18)
  store result in vector k_{d2}
end
  plot k_{d1} as a function of λ
  plot k_{d2} as a function of λ
  compute absolute value of *distance* = k_{d1}-k_{d2}
  determine *optim* = min(*distance*)
  return λ_{optim} corresponding to *optim*
  compute k_d using (18) and λ_{optim}
  compute k_p using (16)

The algorithm for computing PD controllers is based upon setting $\lambda = 1$ and computing k_d using either (17) or (18) and k_p using (16). Since, the PD controller in (4) has only two design parameters; the tuning of the PD controller may be done using any combination of two performance criteria in (16), (17) or (18). Thus, imposing (16) and (17) means that (18) will not necessarily be ensured, which is the main drawback of traditional PD controllers as compared to FO-PD controllers.

3. Case Studies

3.1. Tuning an FO-PI Controller for a First Order Process

The process transfer function is given by:

$$H_p(s) = \frac{27.5}{0.26s + 1} \quad (19)$$

For a gain crossover frequency of ω_{cg} =15 rad/s and a phase margin of φ_m=70°, the curves in **Figure 1** are obtained. Thus, the existing graphical methods may be used to determine the final values for the FO-PI controller parameters. Imposing slightly different performance criteria, such as ω_{cg}=30 rad/s, φ_m=70° and robustness to gain uncertainties, the two curves in **Figure 2** are obtained.

For these particular performance criteria, the two plots for k_i do not intersect. Nevertheless, using the algorithm proposed in Section 2, the minimum distance between the two curves is computed, yielding $\mu = 0.55$ and k_i= 5.69. Finally, using (11) the remaining FO-PI parameter is computed as k_p=0.1677.

The resulting (FO-PI) is:

$$H_{FO-PI}(s) = 0.1677\left(1 + \frac{5.69}{s^{0.55}}\right) \tag{20}$$

Figure 3 shows that the Bode plot of the open loop system with a FO-PI controller. It can be seen that the phase margin is slightly increased from 70° as imposed to74°. This is due to the optimization algorithm, in which the final value for k_i is chosen in order to meet the robustness criteria, rather than the phase margin criteria. However, thanks to the optimal choice for the fractional order μ, the phase margin criteria obtained does not vary significantly from the one imposed in the design phase. The Bode plot also indicates that the modulus crosses the zero axes at 30rad/s, as imposed in the design specifications. Most importantly, it can be seen, that changing the open loop gain will not reduce the phase margin, but rather increase it, which means that the overshoot of the closed loop system will not vary significantly from the nominal value. Hence, the closed loop system should behave robustly despite uncertainties in the gain variations.

The closed loop results considering ±50% gain uncertainty are given in **Figure 4**. It can be seen that the FO-PI controller maintains the overshoot below 5%, while the settling time varies slightly between 0.15-0.25 seconds.

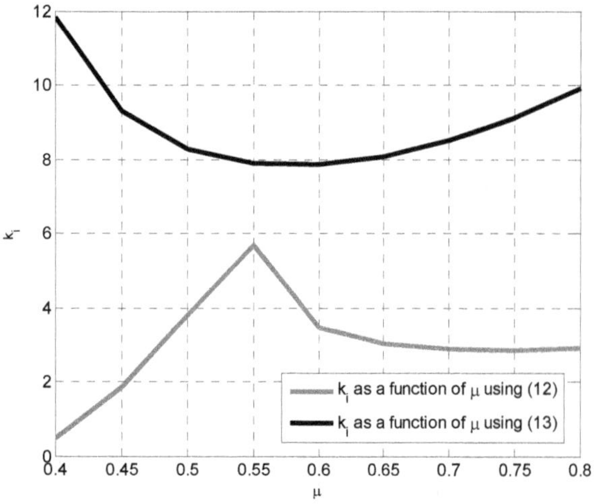

Figure 2. Plots of k_i as a function of μ using (12) and (13) for ω_{cg} = 25 rad/s and φ_m = 70°

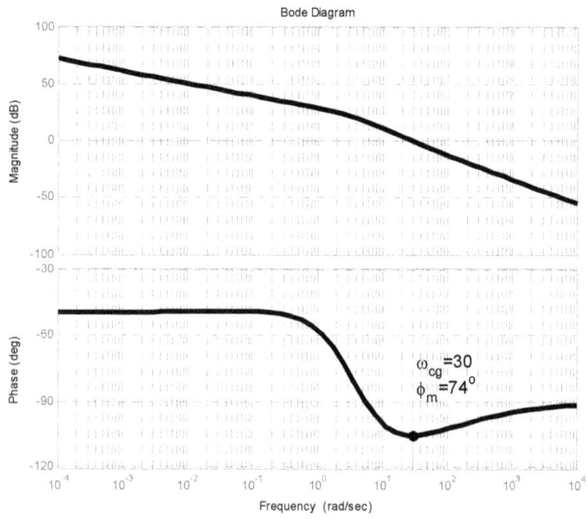

Figure 3. Open loop Bode diagram using FO-PI controller.

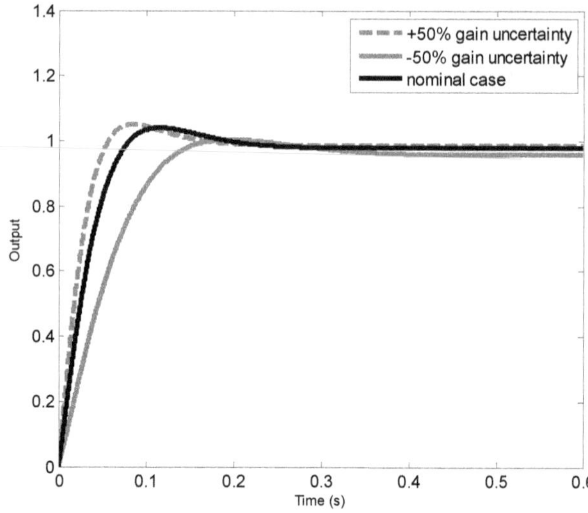

Figure 4. Closed loop results with FO-PI controller considering ±50% process gain variation.

3.2. Tuning an FO-PD Controller for a Second Order Process

The process transfer function is given as:

$$H_p(s) = \frac{1}{s(s+0.5)} \tag{21}$$

Taking ω_{cg}= 15 and φ_m = 50° and using the algorithm described in Section 2, the plots of k_d as a function of μ are derived as given in **Figure 5**. In this case, the algorithm presented in Section 2 yields the same result as any of the existing graphical methods, since the two curves intersect. **Figure 5** finally yields a fractional order λ= 0.573 and k_d =2.59.

Using (16), the following value is obtained for the k_p parameter, as a function of the previously determined μ and k_d values: k_p =17.5.

The resulting (FO-PD) is:

$$H_{FO-PD}(s) = 17.5\left(1 + 2.59s^{0.573}\right) \tag{22}$$

The Bode diagram of the open loop system using the previously determined FO-PD controller is given in **Figure 6**, while the closed loop system considering ±50% gain uncertainty is given in **Figure** 7.

The Bode diagram in **Figure 6** shows that variations of the open loop gain will not have a negative effect on the overshoot of the closed loop system, but only on the settling time, which demonstrates that the designed fractional order PD controller ensures the robustness of the closed loop system despite gain variations. As compared to the fractional order PI controller, the solution of the PD controller at the intersection of the two curves implies that all performance specifications are met: the gain crossover frequency is exactly 15 rad/sec, as specified, the phase margin is exactly 50° and the phase plot is flat around the gain crossover frequency.

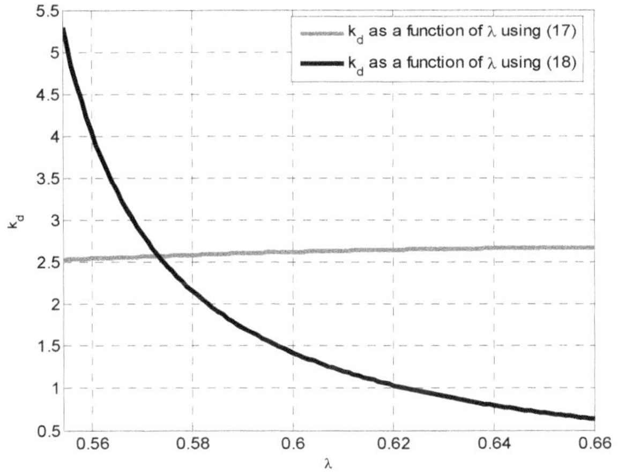

Figure 5. Selection of the fractional order μ and k_d parameter.

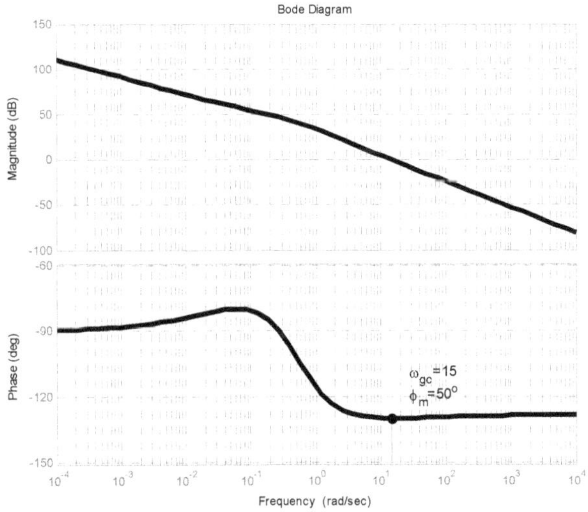

Figure 6. Bode diagram of the open loop system with FO-PD controller.

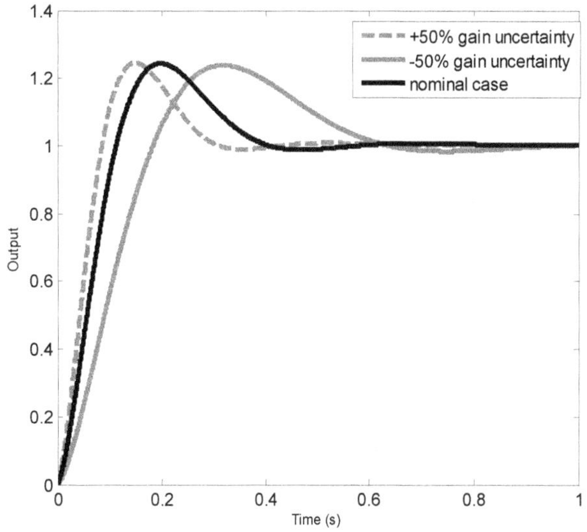

Figure 7. Closed loop responses considering ±50% gain uncertainty with a FO-PD controller.

As expected from the Bode plot, the overshoot is maintained in all three case scenarios at the value of 25%, while the settling time varies between 0.3-0.6 seconds.

4. Conclusions

The purpose of the present paper was to present a simple and efficient optimization algorithm for tuning fractional order PI and PD controllers. For specific performance criteria, the existing graphical methods may not yield an exact solution. Thus, optimization routines need to be used in order to tune the fractional order controllers. The paper shows that even in the case of no exact solution, the graphical methods may still be employed with a slight modification that implies computing and selecting the minimum distance between the possible solutions. It is shown through simulations that the fractional order controllers designed using the proposed method yield satisfactorily results in terms of closed loop performance and robustness.

5. Acknowledgements

This work was supported by a grant of the Romanian National Authority for Scientific Research, CNCS – UE-FISCDI, project number PN-II-RU-TE-2012-3-0307.

REFERENCES

[1] C.A. Monje, Y. Chen, B. M. Vinagre, D. Xue and V. Feliu, "Fractional-order Systems and Controls: Fundamentals and Applications," Springer, London, 2010.

[2] C. I. Pop (Muresan), C. M. Ionescu, R. De Keyser, E. H. Dulf, "Robustness Evaluation of Fractional Order Control for Varying Time Delay Processes," *Signal, Image and*

Video Processing, Vol. 6, 2012, pp. 453-461.

[3] A. Oustaloup," La Commande CRONE: Commande Robust d'ordre non entiere," Hermes, Paris, France, 1991

[4] C. A. Monje, B. Vinagre, Y. Chen and V. Feliu, "On Fractional PI^{λ}controllers: Some Tuning Rules for Robustness to Plant Uncertainties," *Nonlinear Dynam,* Vol. 38, 2004, pp. 369-381.

[5] C. I. Muresan, E. H. Dulf, R. Both, A. Palfi and M. Caprioru, "Microcontroller Implementation of a Multivariable Fractional Order PI Controller," *The 9th International Conference on Control Systems and Computer Science (CSCS19-2013)*, 29-31 May, Bucharest, Romania, Vol. 1, 2013, pp. 44-51.

[6] Y. Luo and Y. Chen, "Fractional Order Motion Controls," John Wiley & Sons, 2012.

[7] Y. Luo, Y. Chen, C.Y. Wang and Y. G. Pi, "Tuning Fractional Order Proportional Integral Controllers for Fractional Order Systems," *Journal of Process Control,* Vol. 20, 2010, pp. 823-831.

[8] C.A. Monje, B. Vinagre, Y. Chen and V. Feliu, "On Fractional PI^{λ}controllers: Some Tuning Rules for Robustness to Plant Uncertainties," *Nonlinear Dynam,* Vol. 38, 2004, pp. 369-381.

[9] Y. Q. Chen and K. L. Moore, "Discretization Schemes for Fractional-order Differentiators and Integrators," *IEEE T. Circuits.-I.*, Vol. 49, 2002, pp. 363-367.

[10] Y. Q. Chen, H. Dou, B. M. Vinagre and C. A. Monje, "A Robust Tuning Method for Fractional Order PI Controllers," *Proceedings of the 2nd IFAC Workshop on Fractional Differentiation and its Applications*, Portugal, 2006.

The Series of Reciprocals of Non-central Binomial Coefficients

Laiping Zhang, Wanhui Ji
Yinchan Energy College, Yinchuan, China

ABSTRACT

Utilizing Gamma-Beta function, we can build one series involving reciprocal of non-central binomial coefficients, then We can structure several new series of reciprocals of non-central binomial coefficients by item splitting, these new created denominator of series contain 1 to 4 odd factors of binomial coefficients. As the result of splitting items, some identities of series of numbers values of reciprocals of binomial coefficients are given. The method of splitting terms offered in this paper is a new combinatorial analysis way and elementary method to create new series.

Keywords: Binomial Coefficients; Split Terms; Reciprocals; Series; Non-central; Closed Form

1. Introduction

Binomial coefficient series plays an important role among Number theory, Graph Theory, mathematical statistics and Probability Theory, etc mathematics branches. Binomial coefficient series conversion also plays key role in research areas of combinatorial mathematics and mathematical analysis, it aroused great attention and can be referred to a large number of literature [1-7]. Paper [2-5] the author used is called Lehmer series identity

$$\sum_{n=1}^{\infty}\frac{(2x)^{2n}}{n\binom{2n}{n}}=\frac{2x\arcsin x}{\sqrt{1-x^2}},\quad |x|<1,$$

Some authors differential, integral, generating function, the white to Beta-gamma function, hyper geometric function series, recurrence and other mathematical tools to get important results on series of reciprocals of binomial coefficients. Paper [8] obtains alternating series involving reciprocals of central binomial coefficients. Paper [9] obtains series involving reciprocals of central binomial coefficients. We did the research the series of reciprocal of non-central binomial coefficient. Utilizing Gamma-beta function, we obtain one series involving reciprocal of non-central binomial coefficients. We can structure several new series of reciprocal of non-central binomial coefficients by splitting item; the new created denominators of series contain 1 to 4 odd factors of non-central binomial coefficients.

By continuously using the method of splitting items, we can get the series involving reciprocal of non-central.

Binomial coefficients of which denominator contain 5odd factor, 6 odd factor, … and p odd factor.

Therefore splitting term method in this paper is an elementary to build new series and also provide a new combinatorial analysis method. As the result of analysis some identities of series of number values of reciprocal of non-central binomial coefficients also provided in this paper.

2. Main Results and Proof of Theorem

Theorem 1 one series of reciprocals of non-central binomial coefficients

$$\sum_{k=0}^{\infty}\frac{x^k}{\binom{2k+1}{k}}=D$$

$$=\begin{cases}\dfrac{2}{4-x}+\dfrac{8}{(4-x)\sqrt{4x-x^2}}\arctan\dfrac{x}{\sqrt{4x-x^2}}, & 0<x<4\\[2ex] \dfrac{2}{4-x}+\dfrac{4}{(4-x)\sqrt{-4x+x^2}}\ln\dfrac{\sqrt{-4x+x^2}-x}{\sqrt{-4x+x^2}+x}, & -4<x<0\end{cases}\qquad(0)$$

Proof From power series theories easy to understand, the series

$$\sum_{k=0}^{\infty}\frac{x^k}{\binom{2k+1}{k}}=\sum_{k=0}^{\infty}\frac{x^k}{\binom{2k+1}{k+1}}$$

is convergence.

Over interval (-4, 4). Utilizing Gamma-Beta function relation and binomial coefficients,

$$\binom{n}{k}^{-1}=(n+1)\int_0^1 t^k(1-t)^{n-k}\,dt$$

we calculation sum function of series

$$\sum_{k=0}^{\infty}\frac{x^k}{\binom{2k+1}{k}}=\sum_{k=0}^{\infty}x^k(2k+2)\int_0^1 t^{k+1}(1-t)^k\,dt$$

$$=2\int_0^1\sum_{k=0}^{\infty}kx^kt^{k+1}(1-t)^k\,dt+2\int_0^1\sum_{k=0}^{\infty}x^kt^{k+1}(1-t)^k\,dt$$

$$=2\int_0^1\sum_{k=0}^{\infty}k\left[xt(1-t)\right]^k\,tdt+2\int_0^1\sum_{k=0}^{\infty}\left[xt(1-t)\right]^k\,tdt'$$

$$=2\int_0^1\frac{xt(1-t)}{[1-xt(1-t)]^2}\,dt+2\int_0^1\frac{tdt}{1-xt(1-t)}$$

$$=-2\int_0^1\frac{1-xt(1-t)+1}{[1-xt(1-t)]^2}\,tdt+2\int_0^1\frac{tdt}{1-xt(1-t)}$$

$$=2\int_0^1\frac{tdt}{[1-xt+xt^2]^2}$$

Using recursion formulas of integral [7]

$$=2\left[\frac{1}{4-x}+\frac{1}{4-x}\int_0^1\frac{1}{1-xt+xt^2}dt\right]$$

$$=\frac{2}{4-x}+\frac{2}{4-x}\int_0^1\frac{1}{1+xt-xt^2}dt$$

$$=\frac{2}{4-x}+\frac{8}{(4-x)\sqrt{4x-x^2}}\arctan\frac{x}{\sqrt{4x-x^2}},$$

for $0<x<4$

$$=\frac{2}{4-x}+\frac{4}{(4-x)\sqrt{-4x+x^2}}\ln\frac{\sqrt{-4x+x^2}-x}{\sqrt{-4x+x^2}+x},$$

for $-4<x<0$. This completes proof of Theorem 1.

Theorem 2 The series of reciprocals of non-central binomial coefficients

1) The denominator contains 1 odd factor of binomial coefficients in the series

$$\sum_{m=0}^{\infty}\frac{x^m}{\binom{2m+1}{m}(2m+3)}=\left(\frac{4}{x}-1\right)D-\frac{4}{x} \tag{1}$$

$$\sum_{m=0}^{\infty}\frac{x^m}{\binom{2m+1}{m}(2m+5)}=\left(\frac{32}{x^2}-\frac{12}{x}+1\right)D-\frac{32}{x^2}+\frac{4}{3x} \tag{2}$$

$$\sum_{m=0}^{\infty}\frac{x^m}{\binom{2m+1}{m}(2m+7)}=\left(\frac{512}{3x^3}-\frac{64}{x^2}+\frac{4}{x}+\frac{1}{3}\right)\times D-\frac{512}{3x^3}+\frac{64}{9x^2}+\frac{4}{15x} \tag{3}$$

$$\sum_{m=0}^{\infty}\frac{x^m}{\binom{2m+1}{m}(2m+9)}=\left(\frac{4096}{5x^4}-\frac{1536}{5x^3}+\frac{96}{5x^2}+\frac{4}{5x}+\frac{1}{5}\right)\times D-\frac{4096}{5x^4}+\frac{512}{15x^3}+\frac{32}{25x^2}+\frac{4}{35x} \tag{4}$$

2) The denominator contains 2 odd factors of binomial coefficients in the series

$$\sum_{m=0}^{\infty}\frac{x^m}{\binom{2m+1}{m}(2m+3)(2m+5)}=\left(-\frac{16}{x^2}+\frac{8}{x}-1\right)D+\frac{16}{x^2}-\frac{8}{3x} \tag{5}$$

$$\sum_{m=0}^{\infty}\frac{x^m}{\binom{2m+1}{m}(2m+3)(2m+7)}=\left(-\frac{128}{3x^3}+\frac{16}{x^2}-\frac{1}{3}\right)D+\frac{128}{3x^3}-\frac{16}{9x^2}-\frac{16}{15x} \tag{6}$$

$$\sum_{m=0}^{\infty}\frac{x^m}{\binom{2m+1}{m}(2m+5)(2m+7)}=\left(-\frac{256}{3x^3}+\frac{48}{x^2}-\frac{8}{x}+\frac{1}{3}\right)D+\frac{256}{3x^3}-\frac{176}{9x^2}+\frac{8}{15x} \tag{7}$$

$$\sum_{m=0}^{\infty}\frac{x^m}{\binom{2m+1}{m}(2m+3)(2m+9)}=\left(-\frac{2048}{15x^4}+\frac{256}{5x^3}-\frac{16}{5x^2}+\frac{8}{15x}-\frac{1}{5}\right)D+\frac{2048}{15x^4}-\frac{256}{45x^3}-\frac{16}{75x^2}-\frac{24}{35x} \tag{8}$$

$$\sum_{m=0}^{\infty}\frac{x^m}{\binom{2m+1}{m}(2m+5)(2m+9)}=\left(-\frac{1024}{5x^4}+\frac{384}{5x^3}+\frac{16}{5x^2}-\frac{16}{5x}+\frac{1}{5}\right)D+\frac{1024}{5x^4}-\frac{128}{15x^3}-\frac{208}{25x^2}+\frac{32}{105x} \tag{9}$$

$$\sum_{m=0}^{\infty}\frac{x^m}{\binom{2m+1}{m}(2m+7)(2m+9)}=\left(-\frac{2048}{5x^4}+\frac{3584}{15x^3}-\frac{208}{5x^2}+\frac{8}{5x}+\frac{1}{15}\right)D+\frac{2048}{5x^4}-\frac{512}{5x^3}+\frac{656}{225x^2}+\frac{8}{105x} \tag{10}$$

3) The denominator contains 3 odd factors of binomial coefficients in the series

$$\sum_{m=0}^{\infty}\frac{x^m}{\binom{2m+1}{m}(2m+3)(2m+5)(2m+7)}=\left(\frac{64}{3x^3}-\frac{16}{x^2}+\frac{4}{x}-\frac{1}{3}\right)D-\frac{64}{3x^3}+\frac{80}{9x^2}-\frac{4}{5x} \tag{11}$$

$$\sum_{m=0}^{\infty}\frac{x^m}{\binom{2m+1}{m}(2m+3)(2m+5)(2m+9)}=\left(\frac{512}{15x^4}-\frac{64}{5x^3}-\frac{16}{5x^2}+\frac{28}{15x}-\frac{1}{5}\right)D-\frac{512}{15x^4}+\frac{64}{45x^3}+\frac{304}{75x^2}-\frac{12}{105x} \tag{12}$$

$$\sum_{m=0}^{\infty}\frac{x^m}{\binom{2m+1}{m}(2m+3)(2m+7)(2m+9)}$$
$$=(\frac{1024}{15x^4}-\frac{704}{15x^3}+\frac{48}{5x^2}-\frac{4}{15x}-\frac{1}{15})D \quad (13)$$
$$-\frac{1024}{15x^4}+\frac{1088}{45x^3}-\frac{176}{225x^2}-\frac{4}{21x}$$

$$\sum_{m=0}^{\infty}\frac{x^m}{\binom{2m+1}{m}(2m+5)(2m+7)(2m+9)}$$
$$=(\frac{512}{5x^4}-\frac{1216}{15x^3}+\frac{112}{5x^2}-\frac{12}{5x}+\frac{1}{15})D \quad (14)$$
$$-\frac{512}{5x^4}+\frac{704}{15x^3}-\frac{1228}{225x^2}+\frac{4}{35x}$$

4) The denominator contains 4 odd factors of binomial coefficients in the series

$$\sum_{m=0}^{\infty}\frac{x^m}{\binom{2m+1}{m}(2m+3)(2m+5)(2m+7)(2m+9)}$$
$$=(-\frac{256}{15x^4}+\frac{256}{15x^3}-\frac{32}{5x^2}+\frac{16}{15x}-\frac{1}{15})D \quad (15)$$
$$+\frac{256}{15x^4}-\frac{512}{45x^3}+\frac{544}{225x^2}-\frac{16}{105x}$$

Proof of Theorem

1) Left sides of (0) splitting terms

$$\sum_{n=0}^{\infty}\frac{n!(n+1)!x^n}{(2n+1)!}=D$$

$$\sum_{n=0}^{\infty}\frac{n!(n+1)!x^n}{(2n+1)!}$$
$$=1+\sum_{n=0}^{\infty}\frac{(n-1)!n!(n)(n+1)x^n}{(2n-2+1)!(2n-1+1)(2n+1)}=D,$$

put $n-1=m$,

$$1+\sum_{m=0}^{\infty}\frac{m!(m+1)!(m+1)(m+2)x^{m+1}}{(2m+1)!(2m+2)(2m+3)}=D,$$

Multiply both sides by $\frac{4}{x}$, arrive to

$$\frac{4}{x}+\sum_{m=0}^{\infty}\frac{m!(m+1)!(2m+3+1)x^m}{(2m+1)!(2m+3)}=\frac{4}{x}D,$$
$$\frac{4}{x}+\sum_{m=0}^{\infty}\frac{m!(m+1)!x^m}{(2m+1)!}(1+\frac{1}{2m+3})=\frac{4}{x}D,$$

We obtain (1). Let right sides (1) be D_3 .(2). Left sides of (0) splitting terms

$$1+\frac{x}{3}$$
$$+\sum_{n=0}^{\infty}\frac{(n-2)!(n-1)!(n-1)n\bullet n(n+1)x^n}{(2n-4+1)!(2n-3+1)(2n-2+1)(2n-1+1)((2n+1)}$$
$$=D,$$

put $n-2=m$

$$1+\frac{x}{3}$$
$$+\sum_{m=0}^{\infty}\frac{m!(m+1)!(m+1)(m+2)(m+2)(m+3)x^{m+2}}{(2m+1)!(2m+2)(2m+3)(2m+4)(2m+5)}$$
$$=D,$$

Multiply both sides by $\frac{16}{x^2}$, arrive to

$$\frac{16}{x^2}+\frac{16}{3x}+\sum_{m=0}^{\infty}\frac{m!(m+1)!x^m}{(2m+1)!}$$
$$\times(1+\frac{1}{2m+3})(1+\frac{1}{2m+5})=\frac{16}{x^2}D$$
$$\frac{16}{x^2}+\frac{16}{3x}+\sum_{m=0}^{\infty}\frac{m!(m+1)!x^m}{(2m+1)!}$$
$$\times(1+\frac{1}{2m+3}+\frac{1}{2m+5}+\frac{1}{(2m+3)(2m+5)})=\frac{16}{x^2}D$$
$$\frac{16}{x^2}+\frac{16}{3x}+\sum_{m=0}^{\infty}\frac{m!(m+1)!x^m}{(2m+1)!}$$
$$\times(1+\frac{3/2}{2m+3}+\frac{1/2}{2m+5})=\frac{16}{x^2}D$$

Multiply both sides by 2, arrive to

$$\frac{32}{x^2}+\frac{32}{3x}+2D+3D_3$$
$$+\sum_{m=0}^{\infty}\frac{m!(m+1)!x^m}{(2m+1)!(2m+5)}=\frac{32}{x^2}D$$

We obtain (2). Let right sides (2) be D_5.

3) Left sides of (0) splitting terms

$$1+\frac{x}{3}+\frac{x^2}{10}$$
$$+\sum_{n=0}^{\infty}\frac{(n-3)!(n-2)!(n-2)(n-1)n\bullet(n-1)n(n+1)x^n}{(2n-6+1)!(2n-5+1)(2n-4+1)\cdots(2n+1)}$$
$$=D,$$

put $n-3=m$

$$1+\frac{x}{3}+\frac{x^2}{10}+\sum_{m=0}^{\infty}\frac{m!(m+1)!(m+1)(m+2)(m+3)(m+2)(m+3)(m+4)x^{m+3}}{(2m+1)!(2m+2)(2m+3)(2m+4)(2m+5)(2m+6)(2m+7)}=D,$$

$$1+\frac{x}{3}+\frac{x^2}{10}+\sum_{m=0}^{\infty}\frac{m!(m+1)![(m+2)(m+3)](m+4)x^m}{8(2m+1)!(2m+3)(2m+5)(2m+7)}=D,$$

Multiply both sides by $\frac{64}{x^3}$,

$$\frac{64}{x^3}+\frac{64}{3x^2}+\frac{16}{5x}+\sum_{m=0}^{\infty}\frac{m!(m+1)!x^m}{(2m+1)!}[1+\frac{3/2}{2m+3}+\frac{1/2}{2m+5}](1+\frac{1}{2m+7})=\frac{64}{x^3}D$$

$$\frac{64}{x^3}+\frac{64}{3x^2}+\frac{16}{5x}+\sum_{m=0}^{\infty}\frac{m!(m+1)!x^m}{(2m+1)!}[1+\frac{3/2}{2m+3}+\frac{1/2}{2m+5}+\frac{1}{2m+7}+\frac{3/2}{(2m+3)(2m+7)}+\frac{1/2}{(2m+5)(2m+7)}]=\frac{64}{x^3}D$$

$$\frac{64}{x^3}+\frac{64}{3x^2}+\frac{16}{5x}+\sum_{m=0}^{\infty}\frac{m!(m+1)!x^m}{(2m+1)!}(1+\frac{15/8}{2m+3}+\frac{3/4}{2m+5}+\frac{3/8}{2m+7})=\frac{64}{x^3}D,$$

Multiply both sides by $\frac{8}{3}$, arrive to

$$\frac{512}{3x^3}+\frac{512}{9x^2}+\frac{256}{15x}+\frac{8}{3}D+5D_3+2D_5+\sum_{m=0}^{\infty}\frac{m!(m+1)!x^m}{(2m+1)!(2m+7)}=\frac{512}{3x^3}D,$$

we obtain (3).

4) Let right sides of (3) be D_7, Left sides of (0) splitting terms

$$1+\frac{x}{3}+\frac{x^2}{10}+\frac{x^3}{35}+\sum_{n=0}^{\infty}\frac{(n-4)!(n-3)!(n-3)(n-2)(n-1)n(n-2)(n-1)n(n+1)x^n}{(2n-8+1)!(2n-7+1)(2n-6+1)\cdots(2n)((2n+1)}=D$$

Put $n-4=m$,

$$1+\frac{x}{3}+\frac{x^2}{10}+\frac{x^3}{35}+\sum_{m=0}^{\infty}\frac{m!(m+1)![(m+2)(m+3)(m+4)](m+5)x^{m+4}}{16(2m+1)!(2m+3)(2m+5)(2m+7)(2m+9)}=D.$$

Multiply both sides by $\frac{256}{x^4}$, arrive to

$$\frac{256}{x^4}+\frac{256}{3x^3}+\frac{128}{5x^2}+\frac{256}{35x}+\sum_{m=0}^{\infty}\frac{m!(m+1)!x^m}{(2m+1)!}(1+\frac{1}{2m+3})(1+\frac{1}{2m+5})(1+\frac{1}{2m+7})(1+\frac{1}{2m+9})=\frac{256}{x^4}D,$$

expand to

$$\frac{256}{x^4}+\frac{256}{3x^3}+\frac{128}{5x^2}+\frac{256}{35x}+\sum_{m=0}^{\infty}\frac{m!(m+1)!x^m}{(2m+1)!}[1+\frac{1}{2m+3}+\frac{1}{2m+5}+\frac{1}{2m+7}+\frac{1}{2m+9}+\frac{1}{(2m+3)(2m+5)}$$
$$+\frac{1}{(2m+3)(2m+7)}+\frac{1}{(2m+3)(2m+9)}+\frac{1}{(2m+5)(2m+7)}+\frac{1}{(2m+5)(2m+9)}+\frac{1}{(2m+7)(2m+9)}$$
$$+\frac{1}{(2m+3)(2m+5)(2m+7)}+\frac{1}{(2m+3)(2m+5)(2m+9)}+\frac{1}{(2m+5)(2m+7)(2m+9)}$$
$$+\frac{1}{(2m+3)(2m+5)(2m+7)(2m+9)}]=\frac{256}{x^4}D$$

In (0.1), there are 10 fractions containing 2 factors, 10 fractions containing 3 factors, 5 fractions containing 4 and one fraction containing 5 factors.

After the below calculation to (0.1), arrive one series of reciprocals of non-central binomial coefficients with the denominators 1, 2, 3, 4, 5 factors

1) All fractions of (0.1) divided into partial fractions, arrive to

$$\frac{256}{x^4}+\frac{256}{3x^3}+\frac{128}{5x^2}+\frac{256}{35x}+\sum_{m=0}^{\infty}\frac{m!(m+1)!x^m}{(2m+1)!}(1+\frac{15/8}{2m+3}+\frac{3/4}{2m+5}+\frac{3/8}{2m+7}+\frac{1}{2m+9}+\frac{15/8}{(2m+3)(2m+9)}$$
$$+\frac{3/4}{(2m+5)(2m+9)}+\frac{3/8}{(2m+7)(2m+9)}=\frac{256}{x^4}D,$$

arrive to

$$\frac{256}{x^4}+\frac{256}{3x^3}+\frac{128}{5x^2}+\frac{256}{35x}+\sum_{m=0}^{\infty}\frac{m!(m+1)!x^m}{(2m+1)!}[1+\frac{35/16}{2m+3}+\frac{15/16}{2m+5}+\frac{9/16}{2m+7}+\frac{5/16}{2m+9}]=\frac{256}{x^4}D$$

Multiply both sides by $\frac{16}{5}$, arrive to

$$\frac{4096}{5x^4}+\frac{4096}{15x^3}+\frac{2048}{25x^2}+\frac{4096}{175x}+\frac{16}{5}D+7D_3+3D_5+\frac{9}{5}D_7+\sum_{m=0}^{\infty}\frac{m!(m+1)!x^m}{(2m+1)!(2m+9)}=\frac{4096}{5x^4}D,$$

We obtain (4), Let right sides (4) be D_9

2) In (0.1), the factions of 2 factors retained, other fractions divided partial fractions. Then spit term the fraction containing 2 factions, reserve one during each split term, what left has been split into partial fraction, arrive to

A) $$\frac{256}{x^4}+\frac{256}{3x^3}+\frac{128}{5x^2}+\frac{256}{35x}+D_{3\cdot5}+D+\frac{27}{16}D_3+\frac{23}{16}D_5+\frac{9}{16}D_7+\frac{5}{16}D_9=\frac{256}{x^4}D$$

B) $$\frac{256}{x^4}+\frac{256}{3x^3}+\frac{128}{5x^2}+\frac{256}{35x}+D_{3\cdot7}+D+\frac{31}{16}D_3+\frac{15}{16}D_5+\frac{13}{16}D_7+\frac{5}{16}D_9=\frac{256}{x^4}D$$

C) $$\frac{256}{x^4}+\frac{256}{3x^3}+\frac{128}{5x^2}+\frac{256}{35x}+D_{3\cdot9}+D+\frac{97}{48}D_3+\frac{15}{16}D_5+\frac{9}{16}D_7+\frac{23}{48}D_9=\frac{256}{x^4}D$$

D) $$\frac{256}{x^4}+\frac{256}{3x^3}+\frac{128}{5x^2}+\frac{256}{35x}+D_{5\cdot7}+D+\frac{35}{16}D_3+\frac{7}{16}D_5+\frac{17}{16}D_7+\frac{5}{16}D_9=\frac{256}{x^4}D.$$

E) $$\frac{256}{x^4}+\frac{256}{3x^3}+\frac{128}{5x^2}+\frac{256}{35x}+D_{5\cdot9}+D+\frac{35}{16}D_3+\frac{11}{16}D_5+\frac{9}{16}D_7+\frac{9}{16}D_9=\frac{256}{x^4}D.$$

F) $$\frac{256}{x^4}+\frac{256}{3x^3}+\frac{128}{5x^2}+\frac{256}{35x}+D_{7\cdot9}+D+\frac{35}{16}D_3+\frac{15}{16}D_5+\frac{1}{16}D_7+\frac{13}{16}D_9=\frac{1024}{x^5}D.$$

Because D, D_3, D_5, D_7, D_9 is known, the calculation of (A) – (F), we get (5) – (10),

3) In (0.1), the factions of 3 factors retained, other fractions divided partial fractions.

Then spit term the fraction containing 3 factions, reserve one during each split term, what left has been split into partial fraction, arrive to

A) $$\frac{256}{x^4}+\frac{256}{3x^3}+\frac{128}{5x^2}+\frac{256}{35x}+D_{3\cdot5\cdot7}+D+\frac{33}{16}D_3+\frac{19}{16}D_5+\frac{7}{16}D_7+\frac{5}{16}D_9=\frac{256}{x^4}D.$$

B) $$\frac{256}{x^4}+\frac{256}{3x^3}+\frac{128}{5x^2}+\frac{256}{35x}+D_{3\cdot5\cdot9}+D+\frac{101}{48}D_3+\frac{17}{16}D_5+\frac{9}{16}D_7+\frac{13}{48}D_9=\frac{256}{x^4}D.$$

C) $$\frac{256}{x^4}+\frac{256}{3x^3}+\frac{128}{5x^2}+\frac{256}{35x}+D_{3\cdot7\cdot9}+D+\frac{103}{48}D_3+\frac{15}{16}D_5+\frac{11}{16}D_7+\frac{11}{48}D_9=\frac{256}{x^4}D.$$

D) $$\frac{256}{x^4}+\frac{256}{3x^3}+\frac{128}{5x^2}+\frac{256}{35x}+D_{5\cdot7\cdot9}+D+\frac{35}{16}D_3+\frac{13}{16}D_5+\frac{13}{16}D_7+\frac{3}{16}D_9=\frac{256}{x^4}D.$$

Because D, D_3, D_5, D_7, D_9 is known, the calculation of (A) – (D), we get (11) – (14),

4) In (0.1), the factions of 4 factors retained, other fractions divided partial fractions.

Then spit term the fraction containing 4 factions, reserve one during each split term, what left has been split into partial fraction, arrive to

$$\frac{256}{x^4}+\frac{256}{3x^3}+\frac{128}{5x^2}+\frac{256}{35x}+D_{3\cdot5\cdot7\cdot9}+D+\frac{13}{6}D_3+D_5+\frac{1}{2}D_7+\frac{1}{3}D_9=\frac{256}{x^4}D.$$

Because D, D_3, D_5, D_7, D_9, is known, the calculation of this expression , we get (15).

This completes proof of Theorem 2

3. Some Series of Number Values

In (1) - (15) of theorem 2, put $x=1$,

$$D=\sum_{m=0}^{\infty}\frac{1}{\binom{2m+1}{m}}=\frac{2}{3}+\frac{4\pi\sqrt{3}}{27}$$

Put $x=-1$,

$$D=\sum_{m=0}^{\infty}\frac{(-1)^m}{\binom{2m+1}{m}}=\frac{2}{5}+\frac{8\sqrt{5}}{25}\ln\varphi,\quad \varphi=\frac{\sqrt{5}+1}{2};$$

we have

Corollary 1 The series of number values of reciprocal of non-central binomial coefficients

1) $\sum_{m=0}^{\infty}\frac{1}{\binom{2m+1}{m}(2m+3)}=\frac{4\pi\sqrt{3}}{9}-2$;

2) $\sum_{m=0}^{\infty}\frac{1}{\binom{2m+1}{m}(2m+5)}=\frac{28\pi\sqrt{3}}{9}-\frac{50}{3}$;

3) $\sum_{m=0}^{\infty}\frac{1}{\binom{2m+1}{m}(2m+7)}=\frac{148\pi\sqrt{3}}{9}-\frac{41058}{45}$;

4) $\sum_{m=0}^{\infty}\frac{1}{\binom{2m+1}{m}(2m+9)}=\frac{2348\pi\sqrt{3}}{9}-\frac{225158}{525}$;

5) $\sum_{m=0}^{\infty}\frac{1}{\binom{2m+1}{m}(2m+3)(2m+5)}=-\frac{4\pi\sqrt{3}}{3}+\frac{22}{3}$;

6) $\sum_{m=0}^{\infty}\frac{1}{\binom{2m+1}{m}(2m+3)(2m+7)}=-4\pi\sqrt{3}+\frac{1078}{45}$;

7) $\sum_{m=0}^{\infty}\frac{1}{\binom{2m+1}{m}(2m+5)(2m+7)}=-\frac{20\pi\sqrt{3}}{3}+\frac{1634}{45}$;

8) $\sum_{m=0}^{\infty}\frac{1}{\binom{2m+1}{m}(2m+3)(2m+9)}=-\frac{196\pi\sqrt{3}}{15}+\frac{112054}{1575}$;

9) $\sum_{m=0}^{\infty}\frac{1}{\binom{2m+1}{m}(2m+5)(2m+9)}=-\frac{284\pi\sqrt{3}}{15}+\frac{18084}{175}$;

10) $\sum_{m=0}^{\infty}\frac{1}{\binom{2m+1}{m}(2m+7)(2m+9)}=-\frac{156\pi\sqrt{3}}{5}+\frac{267422}{1575}$;

11) $\sum_{m=0}^{\infty}\frac{1}{\binom{2m+1}{m}(2m+3)(2m+5)(2m+7)}=\frac{4\pi\sqrt{3}}{3}-\frac{326}{45}$;

12) $\sum_{m=0}^{\infty}\frac{1}{\binom{2m+1}{m}(2m+3)(2m+5)(2m+9)}=\frac{44\pi\sqrt{3}}{15}-\frac{45316}{1575}$;

13) $\sum_{m=0}^{\infty}\frac{1}{\binom{2m+1}{m}(2m+3)(2m+7)(2m+9)}=\frac{68\pi\sqrt{3}}{15}-\frac{38842}{1575}$;

14) $\sum_{m=0}^{\infty}\frac{1}{\binom{2m+1}{m}(2m+5)(2m+9)(2m+9)}=\frac{92\pi\sqrt{3}}{15}-\frac{52306}{1575}$;

15) $\sum_{m=0}^{\infty}\frac{1}{\binom{2m+1}{m}(2m+3)(2m+5)(2m+7)(2m+9)}=-\frac{4\pi\sqrt{3}}{5}+\frac{762}{175}$;

Corollary 2 Alternating series of number values of reciprocals of non-central binominal coefficients

1) $\sum_{m=0}^{\infty}\frac{(-1)^m}{\binom{2m+1}{m}(2m+3)}=-\frac{8\sqrt{5}}{5}\ln\varphi+2$;

2) $\sum_{m=0}^{\infty}\frac{(-1)^m}{\binom{2m+1}{m}(2m+5)}=\frac{72\sqrt{5}}{5}\ln\varphi-\frac{46}{3}$;

3) $\sum_{m=0}^{\infty}\frac{(-1)^m}{\binom{2m+1}{m}(2m+7)}=-\frac{1144\sqrt{5}}{15}\ln\varphi+\frac{3698}{45}$;

4) $\sum_{m=0}^{\infty}\frac{(-1)^m}{\binom{2m+1}{m}(2m+9)}=\frac{1832\sqrt{5}}{5}\ln\varphi-\frac{206938}{525}$;

5) $\sum_{m=0}^{\infty}\frac{(-1)^m}{\binom{2m+1}{m}(2m+3)(2m+5)}=-8\sqrt{5}\ln\varphi+\frac{26}{3}$;

6) $\sum_{m=0}^{\infty}\frac{(-1)^m}{\binom{2m+1}{m}(2m+3)(2m+7)}=\frac{56\sqrt{5}}{3}\ln\varphi-\frac{902}{45}$;

7) $\sum_{m=0}^{\infty}\frac{(-1)^m}{\binom{2m+1}{m}(2m+5)(2m+7)}=\frac{136\sqrt{5}}{3}\ln\varphi-\frac{2194}{45}$;

8) $\sum_{m=0}^{\infty}\frac{(-1)^m}{\binom{2m+1}{m}(2m+3)(2m+9)}=-\frac{184\sqrt{5}}{3}\ln\varphi+\frac{152294}{1575}$;

9) $\sum_{m=0}^{\infty}\frac{(-1)^m}{\binom{2m+1}{m}(2m+5)(2m+9)}=-88\sqrt{5}\ln\varphi+\frac{16574}{175}$;

10) $\sum_{m=0}^{\infty}\frac{(-1)^m}{\binom{2m+1}{m}(2m+7)(2m+9)}=-\frac{664\sqrt{5}}{3}\ln\varphi+\frac{375122}{1575}$;

REFERENCES

[1] B. Sury, T. N. Wang and F. Z. Zhao, "Some Identities Involving of Binomial Coefficients," *J. integer Sequences*, Vol. 7, 2004,Article 04.2.8

[2] J. H. Yang and F. Z. Zhao, "Sums Involving the Inverses of Binomial Coefficients, *Journal of Integer Sequences*, Vol. 9, 2006, Article 06.4.2

[3] S. Amghibech, "On Sum Involving Binomial Coefficient," *Journal of Integer Sequences,* Vol.10, 2007, Article 07.2.1

[4] T. Trif, "Combinatorial Sums and Series Involving Inverses of Binomial Coefficients," *Fibonacci Quarterly*, Vol. 38, No. 1, 2000, pp. 79-84.

[5] F.-Z Zhao and T. Wang, "Some Results for Sums of the Inverses of Binomial Coefficients," Integers: Electronic, *Journal of combinatorial Number Theory*, Vol. 5, No. 1,

2005, p. A22.

[6] R. Sprugnoli, "Sums of Reciprocals of the Central Binomial Coefficients," *Integers: Electronic Journal of Combinatorial Number Theory*, Vol. 6 ,2006, p. A27

[7] I. S. Gradshteyn and I. M. Zyzhik, "A Table of Integral, Series and Products," Academic Press is an Imprint of Elsevier, Seventh Edition, Vol. 56, No. 61.

[8] W. H. Ji and L. P. Zhang, "On Series Alternated with Positive and Negative Involving Reciprocals of Binomial Coefficients," *Pure Mathematics*, Vol. 2, No. 4, 2012, pp. 192-201.

[9] W. H. Ji and B. L. Hei, "The Series of Reciprocals of Binomial Coefficients Constructing by Splitting Terms," *Pure Mathematics*, 2013, Vol. 3, No. 1, p. 18.

Differential Games of Pursung in the Systems with Distributed Parameters and Geometrical Restrictions

M. Sh. Mamatov, E. B. Tashmanov, H. N. Alimov
Department "Geometry", National University of Uzbekistan Named After M. Ulugbek, Tashkent, Uzbekistan

ABSTRACT

A problem of pursuit in the controlled systems of elliptic type without mixed derivatives with variable coefficients was considered. The model of the considered system is described by partial differential equations. The players (opponents) control parameters occur on the right-hand side of the equation and are subjected to various constraints. The first player's goal is to bring the system from one state into another desired state; the second player's goal is to prevent this from happening. We represent new sufficient conditions for bringing the system from one state into another. The finite-difference method is used to solve this problem.

Keywords: Pursuit; Pursuer; Evader; Terminal Set; Pursuit Control; Evasion control

1. Introduction

Some problem formulations in the theory of differential games may be illustrated by motion of two controlled objects, pursuer and evader. Let in the course of motion the objects continuously observe each other and at each time instant correct their motions depending on the information about the adversary. Depending on the pursuer's aim, the problem of pursuit is then formulated as follows: using the information about the evader, at each time instant t select a control such that coincidence of the objects' spatial coordinates is reached as soon as possible.

The majority of studies consider the case where behavior of the lumped-parameter model described by a system of ordinary differential equations. This scheme encompasses many problems of differential games arising in diverse filed of the natural sciences. The mathematical issues of the differential games describing the lumped-parameter systems were developed in detail.

In many applications, however, the lumped-parameter models describe phenomena inadequately. It often turns out that a system which is optimal in the sense of a simplified model does not use the additional designed-in potentialities of control. The distributed-parameter models obeying the differential equations with partial derivatives offer a better, more adequate description. Use of these equations also gives rise to various game problems of which one is the subject matter of the present paper. It focuses only on the problem of pursuit. Therefore, we make an assumption about the nature of information for this problem.

2. Formulation of the Problem

The operated distributed system described by the elliptic equations (see, for example, [1,2]) is considered

$$a(x,y)\partial^2 z/\partial x^2 + b(x,y)\partial^2 z/\partial y^2 = f(u(x,y),\upsilon(x,y)), \quad (1)$$

$$\partial z/\partial\xi + \alpha(x,y)z = \varphi(x,y), \quad (x,y)\in\Gamma$$

where $z=z(x,y)$ – unknown function, $a(x,y)$, $b(x,y)$ – continuous functions in $\Omega=\{(x,y):0<x<1, 0<y<1\}$ with border Γ, $\varphi(x,y)$ – smooth function on Γ, ξ – external normal. It is supposed that there is a positive constant ν such that for any $(x,y)\in\Omega$ the inequality, $b(x,y)\geq\nu$, $u=u(x,y)$, $\upsilon=\upsilon(x,y)$ – operating functions is executed from a class $L_2(\Omega)$. The first (pursuing) player, (pursued or escaping) the player, $u\in\bar{P}$, $\upsilon\in\bar{Q}$, $\bar{P}$ and $\bar{Q}$ – nonempty compacts in R^1 disposes of function $\upsilon(x,y)$ second of function $u(x,y)$. The terminal set $\bar{M}_1\subset R^1$ s allocated.

Definition 1. In a task (1) it is possible ε – completion of $(\varepsilon>0)$ prosecutions from "boundary" situation $\phi(\cdot,\cdot)$, if exist function $u(\upsilon,x,y)\in\bar{P}$, $\upsilon\in\bar{Q}$, $(x,y)\in\Omega$, such that for any function $\upsilon_0(x,y)\in\bar{Q}$, $(x,y)\in\Omega$ the solution $z_0(x,t)$ of a task (1) where $u=u(\upsilon_0(x,y),x,y)$, $\upsilon=\upsilon_0(x,y)$, gets on a set $\varepsilon I+\bar{M}_1$, at some $(\tilde{x},\tilde{y})$, $(\tilde{x},\tilde{y})\in\Omega: z_0(\tilde{x},\tilde{y})\in\varepsilon I+\bar{M}_1$ where $I=(-1,1)$.

Decompose the Euclidean space R^2 of variables (x,y) by the planes $x_i=ih$, $h=1/r$, $i=0,1,\ldots$, and $y_j=jl$, $l=1/\theta$, $j=0,1,2,\ldots$, into parallelepipeds

$\Omega_{(i,j)} = \{(x,y): ih < x_i < (i+1)h, jl < y < (j+1)l\}$, r and θ being some natural numbers. The points (x_i, y_j) belonging to a set Ω are the nodes of the grid Ω_{hl}. Each node has its neighbors. If all these neighbor nodes also belong to the grid Ω_{hl}, then the node $(x_i, y_j)=(ih, jl)$ is referred to as "internal", otherwise, (x_i, y_j) is called the "boundary" node. The set of all boundary nodes is called as border of net area and is designated through Γ_h.

Replace the internal nodes of the derivatives (1) differential second-order accuracy of approximation ratios with formulas

$$\partial z/\partial x(x_i, y_j)=(z(x_{i+1}, y_j)-z(x_{i-1}, y_j))/2h+O(h)^2,$$
$$\partial z/\partial y(x_i, y_j)=(z(x_i, y_{j+1})-z(x_i, y_{j-1}))/2l+O(l)^2,$$
$$\partial^2 z/\partial x^2(x_i, y_j)=(z(x_{i+1}, y_j)-2z(x_i, y_j)+z(x_{i-1}, y_j))/h^2+O(h)^2,$$
$$\partial^2 z/\partial y^2(x_i, y_j)=(z(x_i, y_{j+1})-2z(x_i, y_j)+z(x_i, y_{j-1}))/l^2+O(l)^2,$$

Substituting these ratios in (1), having rejected an error of approximation of derivatives, we will receive the differential equations for unknown $z_{i,j}$

$$a_{i,j}(z_{i+1,j}-2z_{i,j}+z_{i-1,j})/h^2+b_{i,j}(z_{i,j+1}-2z_{i,j}+z_{i,j-1})/l^2=f_{ij}, \quad i=0,1,\dots,r-1;\ j=0,1,\dots,\theta-1. \tag{2}$$

where the following designations of values of coefficients and the right part in a hub (x_i, y_j):$a_{i,j}, b_{i,j}, c_{i,j}, d_{i,j}, g_{i,j}, f_{i,j}$, for example are entered

$$f_{i,j}=f(u(x_i, y_j), \upsilon(x_i, y_j)),\ (x_i, y_j)\in\Omega_{hl}$$

Ratios (2) contains except unknown $z_{i,j}$ in internal nodes also unknown $z_{i,j}$ on border of net area. For boundary nodes we will write down a ratio

$$\begin{aligned}
&(z_{1,j}-z_{0,j})/h+\alpha_{0,j}(z_{0,j}+z_{1,j})/2=\varphi_{0,j},\ (j=0,1,2,\dots,\theta-1)\\
&(z_{r,j}-z_{r-1,j})/h+\alpha_{r,j}(z_{r-1,j}+z_{r,j})/2=\varphi_{r,j},\\
&(z_{i,1}-z_{i,0})/l+\alpha_{i,0}(z_{i,1}+z_{i,0})/2=\varphi_{i,0},\ (i=0,1,2,\dots,r-1)\\
&(z_{i,\theta}-z_{i,\theta-1})/l+\alpha_{i,\theta}(z_{i,\theta}+z_{i,\theta-1})/2=\varphi_{i,\theta}
\end{aligned} \tag{3}$$

Thus, we will receive system of $r\theta+2(r+\theta)$ equations with the same number of unknown $z_{i,j}$.

Using boundary conditions (3), we will express $z_{i,1}$, $z_{i,\theta}$ through $z_{i,0}$, $z_{i,\theta-1}$. Let's have

$$\begin{aligned}
z_{i,1}&=(2-l\alpha_{i,0})/(2+l\alpha_{i,0})z_{i,0}+2l\varphi_{i,0}(2+l\alpha_{i,0})\\
z_{i,\theta}&=(2-l\alpha_{i,\theta})/(2+l\alpha_{i,\theta})z_{i,\theta-1}+2l\varphi_{i,\theta}(2+l\alpha_{i,\theta}).
\end{aligned} \tag{4}$$

Using these ratios, we will exclude in system (3) unknown $z_{i,1}$, $z_{i,\theta}$. If to enter designation $\gamma=h^2/l^2$, we will receive system

$$\left.\begin{aligned}
&z_{i+1,0}-(2+2\gamma-k_{i0}\gamma)z_{i,0}+\gamma\, z_{i,1}+z_{i-1,0}=F_{i,0},\\
&z_{i+1,j}+\gamma z_{i,j-1}-2(1+\gamma)z_{i,j}+\gamma z_{i,j+1}+z_{i-1,j}=F_{i,j}\\
&\qquad (j=1,2,\dots,\theta-2),\\
&z_{i+1,\theta-1}+\gamma\, z_{i,\theta-2}-(2+2\gamma-k_{i,\theta}\gamma)z_{i,\theta-1}+z_{i-1,\theta-1}=F_{i,\theta-1}\\
&\qquad (i=0,1,2,\dots,r-1),
\end{aligned}\right\} \tag{5}$$

where

$$\begin{aligned}
&F_{i,0}=h^2 f_{i0}-\gamma 2l\varphi_{i,0}/(2+l\alpha_{i,0});\ F_{ij}=h^2 f_{i,j},\\
&(j=1,2,\dots,\theta-2);\\
&F_{i,\theta-1}=h^2 f_{i,\theta-1}-\gamma 2l\varphi_{i,\theta}(2+l\alpha_{i,\theta}).
\end{aligned} \tag{6}$$

This system can shortly be written down in a look

$$z_{i+1}+A_i z_i+z_{i-1}=F_i,\ (i=0,1,2,\dots,r-1). \tag{7}$$

where

$$z_i=(z_{i,0}, z_{i,1},\dots, z_{i,\theta-1});\quad F_i=(F_{i,0}, F_{i,1},\dots, F_{i,\theta-1}),$$

$$A_i=\begin{pmatrix}
-2(1+\gamma)+k_{i,0}\gamma & \gamma & 0 & \dots 0 & 0\\
\gamma & -2(1+\gamma) & \gamma & \dots 0 & 0\\
0 & \gamma & -2(1+\gamma) & \dots 0 & 0\\
\dots & \dots & \dots & \dots\dots & \dots\\
0 & 0 & 0 & \dots \gamma & -2(1+\gamma)+k_{i,\theta}\gamma
\end{pmatrix} \tag{8}$$

Boundary conditions (3) and (5) can be copied in a look

$$\begin{aligned}
z_{1,j}&=(2-h\alpha_{0,j})/2+h\alpha_{0,j}z_{0,j}+(2h\varphi_{0,j})/(2+h\alpha_{0,j})\\
&=k_{0,j}z_{0,j}+y_{0,j},\ (j=0,1,2,\dots,\theta-1)\\
z_{r-1,j}&=(2+h\alpha_{r,j})/(2-h\alpha_{r,j})z_{r,j}-(2h\varphi_{r,j})/(2-h\alpha_{r,j})\\
&=k_{r,j}z_{r,j}+y_{r,j},\ (j=0,1,2,\dots,\theta-1)
\end{aligned} \tag{9}$$

where

$$\begin{aligned}
&k_{0,j}=(2-h\alpha_{0,j})/(2+h\alpha_{0,j});\ y_{0,j}=(2h\varphi_{0,j})/(2+h\alpha_{0,j});\\
&k_{r,j}=(2+h\alpha_{r,j})/(2-h\alpha_{r,j});\ y_{r,j}=-(2h\varphi_{r,j})/(2-h\alpha_{r,j}).
\end{aligned} \tag{10}$$

Having put

$$y_0=(y_{0,0}, y_{0,1},\dots, y_{0,\theta-1});\ y_r=(y_{r,0}, y_{r,1},\dots, y_{r,\theta-1}).$$

$$X_0=\begin{pmatrix}
k_{0,0} & 0 & 0 & \dots & 0 & 0\\
0 & k_{0,1} & 0 & \dots & 0 & 0\\
\dots & \dots & \dots & \dots & \dots & \dots\\
0 & 0 & 0 & \dots & 0 & k_{0,\theta-1}
\end{pmatrix};\quad
X_1=\begin{pmatrix}
k_{r,0} & 0 & 0 & \dots & 0 & 0\\
0 & k_{r,1} & 0 & \dots & 0 & 0\\
\dots & \dots & \dots & \dots & \dots & \dots\\
0 & 0 & 0 & \dots & 0 & k_{r,\theta-1}
\end{pmatrix}, \tag{11}$$

it is possible to write down systems (9) in such look:

$$\left.\begin{aligned}
z_1&=X_0 z_0+y_0,\\
z_{r-1}&=X_1 z_r+y_r.
\end{aligned}\right\} \tag{12}$$

Finally we have the following system of the equations:

$$\begin{aligned}
&z_1=X_0 z_0+y_0,\\
&z_{i+1}+A_i z_i-z_{i-1}=F_i\quad (i=1,2,\dots,r-1)\\
&z_{r-1}=X_1 z_r+y_r.
\end{aligned} \tag{13}$$

Instead of game (13) we will consider more the general game described by system of the equations

$$\begin{aligned}
&C_0 z_0-B_0 z_1=f_0\ ,\\
&-A_n z_{n-1}+C_n z_n-B_n z_{n+1}=f_n(u_n, \upsilon_n),\ 1\le n\le N-1,\\
&-A_N z_{N-1}+C_N z_N=f_N,
\end{aligned} \tag{14}$$

where $z_n\in R^m$, $n=\overline{0,N}$, A_n, C_n, B_n – $m\times m$ constant square matrixes, u_n, υ_n – operating parameters, u_n –

prosecution parameter, υ_n – beanie parameter, $u_n \in P_n \subset R^p$, $\upsilon_n \in Q_n \subset R^q$, P_n and Q_n – nonempty sets; f_n – the set function displaying $R^p \times R^q$ in R^m. Besides, in R^m the terminal set is M allocated.

Definition 2. We shall say that from "boundary" situation (f_0, f_N) it is possible to complete pursuit for N steps if from any sequence $\bar{\upsilon}_1, \bar{\upsilon}_2, \ldots, \bar{\upsilon}_{N-1}$ of the values of evasion controls it is possible to construct a sequence $\bar{u}_1, \bar{u}_2, \ldots, \bar{u}_{N-1}$ of values of the pursuit control values such that the solution $\{z_0, \bar{z}_1, \bar{z}_2, \ldots, \bar{z}_{N-1}, z_N\}$ of the equation

$$\begin{gathered} C_0 z_0 - B_0 z_1 = f_0, \\ -A_n z_{n-1} + C_n z_n - B_n z_{n+1} = f_n(\bar{u}_n, \bar{\upsilon}_n), \ 1 \le n \le N-1, \\ -A_N z_{N-1} + C_N z_N = f_N. \end{gathered} \tag{15}$$

Gets on $M: \bar{z}_i \in M$ for some i. Thus for finding of value $\bar{u}_n$ it is allowed to use values $\bar{\upsilon}_n$ and $\bar{z}_n$.

Note that the type of systems (14) is difference schemes for elliptic equations of second order with variable coefficients in any field of any number of dimensions [3-14].

Solution of problem (14) will be sought in the form

$$z_n = \alpha_{n+1} z_{n+1} + \beta_{n+1}, \ n = N-1, N-2, \ldots, 0, \tag{16}$$

where α_{n+1} – uncertain while a square matrix of the sizes $m \times m$, and β_{n+1} – a vector of dimension m. From a formula (16) and the equations of system (14) for $1 \le n \le N-1$ there are recurrent ratios for calculation of matrixes α_n and vectors β_n. Really from a formula (16) $z_{n-1} = \alpha_n z_n + \beta_n$ substituting it in (14) we will receive

$$\begin{gathered} -A_n(\alpha_n z_n + \beta_n) + C_n z_n - B_n z_{n+1} = f_n(u_n, \upsilon_n), \\ 1 \le n \le N-1; \\ (C_n - A_n \alpha_n) z_n = B_n z_{n+1} + f_n(u_n, \upsilon_n) + A_n \beta_n; \\ z_n = (C_n - A_n \alpha_n)^{-1} B_n z_{n+1} + (C_n - A_n \alpha_n)^{-1} [f_n(u_n, \upsilon_n) + A_n \beta_n]. \end{gathered}$$

Equating now the right parts of the last and (16) equalities we will receive

$$\begin{gathered} \alpha_{n+1} = (C_n - A_n \alpha_n)^{-1} B_n, \ n = 1, 2, \ldots, N-1; \\ \beta_{n+1} = (C_n - A_n \alpha_n)^{-1} [f_n(u_n, \upsilon_n) + A_n \beta_n], \\ n = 1, 2, \ldots, N. \end{gathered}$$

Further from (16) and the equations (14) for $n = 0, N$, there are the initial values α_1, β_1 and z_N, allowing beginning the account on recurrent ratios. From (14) and (16) for $n = 0$ we will have

$$z_0 = C_0^{-1} B_0 z_1 + C_0^{-1} f_0, \ z_0 = \alpha_1 z_1 + \beta_1,$$

And, therefore

$$\alpha_1 = C_0^{-1} B_0, \ \beta_1 = C_0^{-1} f_0.$$

In the same way for $n = N$ we have

$$-A_N(\alpha_N z_N + \beta_N) + C_N z_N = f_N$$

or

$$z_N = (C_N - A_N \alpha_N)^{-1} (f_N + A_N \beta_N).$$

Uniting, we will write out final formulas

$$\alpha_{n+1} = (C_n - A_n \alpha_n)^{-1} B_n, \ n = 1, 2, \ldots, N, \ \alpha_1 = C_0^{-1} B_0 \tag{17}$$

$$\begin{gathered} \beta_{n+1} = (C_n - A_n \alpha_n)^{-1} (f_n(u_n, \upsilon_n) + A_n \beta_n), \\ n = 1, 2, \ldots, N-1. \ \beta_1 = C_0^{-1} f_0, \end{gathered} \tag{18}$$

$$\begin{gathered} z_n = \alpha_{n+1} z_{n+1} + \beta_{n+1}(u_n, \upsilon_n), \\ n = N-1, N-2, \ldots, 0, \ z_N = \beta_{N+1} \end{gathered} \tag{19}$$

It is clear, that if in game (17), (18), (19) $z_{\bar{n}} \in M$ that in game (14) too game comes to the end. Therefore further instead of game (14) we will consider discrete game described by system of the equations (17), (18), (19).

Before giving determination of stability of algorithm (17), (18), (19), we will provide some data from linear algebra.

Let A – any square matrix $m \times m$ and $\|x\|_m$ be norm of a vector in R^m, then the norm A is defined by equality

$$\|A\| = \sup_{x \ne 0} \|Ax\|_m / \|x\|_m.$$

For a case of Euclidean norms in R^m we have $\|A\| = \sqrt{\rho}$, where ρ – maximum on the module own value of a matrix $A \cdot A$.

Without the proof we will give the following known lemma (see [15]).

Lemma 1. Let for some matrix norm the square matrix meet a condition $\|A\| \le q < 1$. Then there is a matrix $(E + A)^{-1}$ and $\|(E + A)^{-1}\| \le 1/(1-q)$.

Let's say that the algorithm is steady if the assessment $\|\alpha_j\| \le 1$ for $1 \le j \le N$ is carried out.

Lemma 2. If C_j for $0 \le j \le N$ – no degenerate matrixes and A_j and B_j – nonzero matrixes for $1 \le j \le N-1$ also are satisfied conditions

$$\begin{gathered} \|C_0^{-1} B_0\| \le 1, \ \|C_N^{-1} A_N\| \le 1, \\ \|C_j^{-1} A_j\| + \|C_j^{-1} B_j\| \le 1, \ 1 \le j \le N-1. \end{gathered}$$

And at least in one of inequalities the strict inequality takes place, there are return to the $C_j - A_j \alpha_j$. matrix and $\|\alpha_j\| \le 1$, here $\alpha_1 = C_0^{-1} B_0$,

$$\alpha_{j+1} = (C_j - A_j \alpha_j)^{-1} B_j, \ 1 \le j \le N-1.$$

Proof. $\|\alpha_1\| = \|C_0^{-1} B_0\| \le 1$, suppose, that $\|\alpha_j\| \le 1$ also we will show $\|\alpha_{j+1}\| \le 1$. After a course the proof of this fact we will receive existence of a matrix $(C_j - A_j \alpha_j)^{-1}$. Really from conditions of a lemma we will have

$$\|C_j^{-1} A_j \alpha_j\| \le \|C_j^{-1} A_j\| * \|\alpha_j\| \le \|C_j^{-1} A_j\| \le 1 - \|C_j^{-1} B_j\| < 1.$$

As $C_j^{-1} A_j \alpha_j$ square matrix that owing to a lemma 1 there are return to $E - C_j^{-1} A_j \alpha_j$ and $C_j - A_j \alpha_j$ matrixes and $\|(E - C_j^{-1} A_j \alpha_j)\| \le 1 / \|C_j^{-1} B_j\|$. From here and from (17) we will receive

$$\begin{gathered} \|\alpha_{j+1}\| \le \|(E - C_j^{-1} A_j \alpha_j)^{-1} C_j^{-1} B_j\| \\ \le \|(E - C_j^{-1} A_j \alpha_j)^{-1}\| \cdot \|C_j^{-1} B_j\| \le 1. \end{gathered}$$

The proof of the lemma is complete.

3. Main Results

Everywhere further it is supposed that $M=M_0+M_1$, where M_0 – linear subspace R^m, M_1 – a subset a subspace, L – orthogonal complement of M_0 in R^n. Denote Π we will designate a matrix of orthogonal design from R^m on L.

Let $W(0)=\{0\}$,

$$W(k)=\sum_{i=0}^{k-1} \bigcap_{\upsilon_{N-k+i}\in Q_{N-k+i}} \prod \alpha_{N-k}\ldots\alpha_{N-k+i}\beta_{N-k+i+1}(P_{N-k+i},\upsilon_{N-k+i})$$

$$W_1(k)=-M_1+W(k),\ 1\le k\le N. \tag{20}$$

Theorem 1. Let $\bar{N}$ be the smallest of the numbers *k*, such that

$$-\prod \alpha_{N-k}\alpha_{N-k+1}\ldots\alpha_N z_N\in W_1(k). \tag{21}$$

Then from "boundary" situation (f_0,f_N) it is possible to complete pursuit for $\bar{N}$ steps.

Let now $W_2(0)=-M_1$,

$$W_2(1)=\bigcap_{\upsilon_{N-k-1}\in Q_{N-k-1}}[W_2(0)+\prod\beta_{N-k}(P_{N-k-1},\upsilon_{N-k-1})],$$
$$W_2(k)=\bigcap_{\upsilon_{N-1}\in Q_{N-1}}[W_2(k-1)+\prod\alpha_{N-k}\alpha_{N-k+1}\ldots\alpha_{N-1}\beta_N(P_{N-1},\upsilon_{N-1})] \tag{22}$$

Theorem 2. If $\bar{N}$ be smallest of those numbers k, for each of which takes place inclusion

$$-\prod\alpha_{N-k}\alpha_{N-k+1}\ldots\alpha_N z_N\in W_2(k) \tag{23}$$

that of "boundary" situation (f_0,f_N) it is possible to complete pursuit for $\bar{N}$ steps.

Let

$$\gamma_k(\cdot)=\left\{\gamma_0,\gamma_1,\ldots,\gamma_{k-1}:\gamma_i\ge 0,\sum_{i=0}^{k-1}\gamma_i=1\right\}$$

and

$$W(\gamma_k(\cdot))$$
$$=\sum_{i=0}^{k-1}\bigcap_{\upsilon_{N-k+i}\in Q_{M-k+i}}[-\gamma_i M_1+\prod\alpha_{N-k}\ldots\alpha_{N-k+i}\beta_{N-k+i+1}(P_{N-k+i},\upsilon_{N-k+i})],$$
$$0\le k\le N,$$
$$W_3(0)=-M_1,\ W_3(k)=\bigcup_{\gamma_k(\cdot)}W(\gamma_k(\cdot)),\ 0\le k\le N. \tag{24}$$

Theorem 3. If M_1 – a convex set and $\bar{N}$ be smallest of those numbers k. For each of which inclusion takes place

$$-\prod\alpha_{N-k}\alpha_{N-k+1}\ldots\alpha_N z_N\in W_3(k). \tag{25}$$

That of "boundary" situation (f_0,f_N) it is possible to complete pursuit for $\bar{N}$ steps.

It is easy to be convinced [15] that the solution of $z_{i,j}$ differential task (2) meets to the solution z of an initial task (1), the following assessment of speed of convergence takes place

$$\|(z)_{hl}-z_{i,j}\|_{\Phi_{hl}}\le K_1h^2+K_2l^2, \tag{26}$$

where $(z)_{hl}$ – values of the exact decision a task (1) in grid functions, Φ_{hl} – spaces of net functions, $\|\cdot\|_{\Phi_{hl}}$ – is its norm and, K_1 and K_2 constants.

Theorem 4. Let in an inequality (26) $K_1h^2+K_2l^2<\varepsilon$, and in game (13) from a "boundary" situation $(f_0,f_N)=(-y_0,-y_N)$ completion of prosecution that is definitions 2 be possible. Then in game (1) from "boundary" situation $\partial z/\partial\xi+\alpha(x,y)z=\varphi(x,y)$, $(x,y)\in\Gamma$ it is possible to complete pursuit that are definitions 1.

4. Proof of Theorem

Proof of Theorem 1. Let $\bar{\upsilon}_1,\bar{\upsilon}_2,\ldots,\bar{\upsilon}_{N-1}$, $\bar{\upsilon}_i\in Q_i$, $1\le i\le N-1$ – any sequence. Instead of inclusion (21) we will consider other inclusion equivalent to it

$$-\prod\alpha_{N-k}\alpha_{N-k+1}\ldots\alpha_N z_N\in W_1(k-1)$$
$$+\bigcap_{\upsilon_{N-1}\in Q_{N-1}}\prod\alpha_{N-k}\alpha_{N-k+1}\ldots\alpha_{N-1}\beta_N(P_{N-1},\upsilon_{N-1})$$

Means, exists a_{N-1}

$$a_{N-1}\in\bigcap_{\upsilon_{N-1}\in Q_{N-1}}\prod\alpha_{N-k}\alpha_{N-k+1}\ldots\alpha_{N-1}\beta_N(P_{N-1},\upsilon_{N-1}).$$

Such that

$$-\prod\alpha_{N-k}\alpha_{N-k+1}\ldots\alpha_N z_N\in W_1(k-1)+a_{N-1}. \tag{27}$$

Now control of the pursuing player $\bar{u}_{N-1}$, the relevant control of the escaping player $\bar{\upsilon}_{N-1}$, we will construct as the solution of the following control

$$\prod\alpha_{N-k}\alpha_{N-k+1}\ldots\alpha_{N-1}\beta_N(u_{N-1},\bar{\upsilon}_{N-1})=a_{N-1}.$$

It is clear, that the equation has the decision. From here owing to (27) we have

$$-\prod\alpha_{N-k}\alpha_{N-k+1}\ldots\alpha_N z_N$$
$$\in W_1(k-1)+\alpha_{N-k}\alpha_{N-k+1}\ldots\alpha_{N-1}\beta_N(\bar{u}_{N-1},\bar{\upsilon}_{N-1})$$

We write down this inclusion in other look.

$$-\prod\alpha_{N-k}\alpha_{N-k+1}\ldots\alpha_{N-1}[\alpha_N z_N+\beta_N(\bar{u}_{N-1},\bar{\upsilon}_{N-1})]\in W_1(k-1). \tag{28}$$

As a result from equalities (18) and (28) we will receive

$$-\prod\alpha_{N-k}\alpha_{N-k+1}\ldots\alpha_{N-1}z_{N-1}\in W_1(k-1) \tag{29}$$

Done above a reasoning allow us to construct on the set control $\bar{\upsilon}_{N-1}$ providing inclusion (29). If now the control $\bar{\upsilon}_{N-2}$ becomes known that, we above can receive in the stated way control $\bar{u}_{N-1}$ providing inclusion

$$-\prod\alpha_{N-k}\alpha_{N-k+1}\ldots\alpha_{N-2}z_{N-2}\in W_1(k-2).$$

Repeating this process, further we can construct step by step control $\bar{u}_i$, proceeding from becoming known controls $\bar{\upsilon}_i$, therefore, that in any step inclusion takes place

$$-\prod z_{N-k-1}\in W_1(0)=-M_1.$$

It means that

$$\prod z_{N-k-1} \in M$$

As we set out to prove.

Proof of Theorem 2. Let $\bar{\upsilon}_1, \bar{\upsilon}_2, \ldots, \bar{\upsilon}_{N-1}$, $\bar{\upsilon}_i \in Q_i$, $1 \le i \le N-1$ – any sequence. For concrete $\bar{\upsilon}_{N-1}$ owing to (22) and (23) we will receive inclusion

$$\begin{aligned} &-\prod \alpha_{N-k}\alpha_{N-k+1}\ldots\alpha_N z_N \in W_2(k-1) \\ &+\prod \alpha_{N-k}\alpha_{N-k+1}\ldots\alpha_{N-1}\beta_N(P_{N-1}, \bar{\upsilon}_{N-1}) \end{aligned} \tag{30}$$

Now as $\bar{u}_{N-1}$ we take that element from P_{N-1} for which inclusion (30) remained. Then we will receive

$$\begin{aligned} &-\prod \alpha_{N-k}\alpha_{N-k+1}\ldots\alpha_N z_N \in W_2(k-1) \\ &+\prod \alpha_{N-k}\alpha_{N-k+1}\ldots\alpha_{N-1}\beta_N(\bar{u}_{N-1}, \bar{\upsilon}_{N-1}) \end{aligned}$$

From this it follows that

$$-\prod \alpha_{N-k}\alpha_{N-k+1}\ldots\alpha_{N-1}[\alpha_N z_N + \beta_N(\bar{u}_{N-1}, \bar{\upsilon}_{N-1})] \in W_2(k-1).$$

And therefore, owing to (19) we have

$$-\prod \alpha_{N-k}\alpha_{N-k+1}\ldots\alpha_{N-1} z_{N-1} \in W_2(k-1).$$

If now the control $\bar{\upsilon}_{N-1}$ becomes the stated way known that we above us can construct control $\bar{u}_{N-1}$ providing inclusion

$$-\prod \alpha_{N-k}\alpha_{N-k+1}\ldots\alpha_{N-2} z_{N-2} \in W_2(k-2).$$

Further arguing similarly in any step we will receive

$$-\prod z_{N-k-1} \in W_2(0) = -M_1,$$

that is

$$z_{N-k-1} \in M.$$

The theorem is proved completely.

Proof of Theorem 3. Instead of inclusion (25) meaning (24) we will consider inclusion equivalent to it

$$-\prod \alpha_{N-k}\alpha_{N-k+1}\ldots\alpha_N z_N \in W(\bar{\gamma}(\cdot)).$$

Existence $\bar{\gamma}(\cdot) = \left\{\bar{\gamma}_0, \bar{\gamma}_1, \ldots, \bar{\gamma}_{k-1}, \bar{\gamma}_i \ge 0, \sum_{i=0}^{k-1} \bar{\gamma}_i = 1\right\}$ follows from (24). From here follows

$$\begin{aligned} &-\prod \alpha_{N-k}\alpha_{N-k+1}\ldots\alpha_N z_N \in \sum_{i=0}^{k-2} \bigcap_{\upsilon_{N-k+i} \in Q_{N-k+i}} \\ &[-\bar{\gamma}_i M_1 + \prod \alpha_{N-k}\ldots\alpha_{N-k+i}\beta_{N-k+i+1}(P_{N-k+i}, \upsilon_{N-k+i})] \\ &+ \bigcap_{\upsilon_{N-1} \in Q_{N-1}} [-\bar{\gamma}_{k-1} M_1 + \prod \alpha_{N-k}\ldots\alpha_{N-1}\beta_N(P_{N-1}, \upsilon_{N-1})]. \end{aligned} \tag{31}$$

Let now $\bar{\upsilon}_1, \bar{\upsilon}_2, \ldots, \bar{\upsilon}_{N-1}$, $\bar{\upsilon}_i \in Q_i$, $1 \le i \le N-1$ – any sequence. Owing to (31) exists such a_{N-1} that

$$\upsilon_{N-1} \in \bigcap_{\upsilon_{N-1} \in Q_{N-1}} [-\bar{\gamma}_{k-1} M_1 + \prod \alpha_{N-k}\ldots\alpha_{N-1}\beta_N(P_{N-1}, \upsilon_{N-1})],$$

$$\begin{aligned} &-\prod \alpha_{N-k}\alpha_{N-k+1}\ldots\alpha_N z_N \in \sum_{i=0}^{k-2} \bigcap_{\upsilon_{N-k+i} \in Q_{N-k+i}} \\ &[-\bar{\gamma}_i M_1 + \prod \alpha_{N-k}\ldots\alpha_{N-k+i}\beta_{N-k+i+1}(P_{N-k+i}, \upsilon_{N-k+i})] + a_{N-1} \end{aligned} \tag{32}$$

Therefore, controls $\bar{u}_{N-1}$ we will construct as the solution of the following equation

$$-\bar{\gamma}_{k-1} m_1 + \prod \alpha_{N-k}\ldots\alpha_{N-1}\beta_N(u_{N-1}, \bar{\upsilon}_{N-1}) = a_{N-1}, \quad m_{1,k-1} \in M_1.$$

Further owing to (32) we have

$$\begin{aligned} &-\prod \alpha_{N-k}\alpha_{N-k+1}\ldots\alpha_N z_N \in \sum_{i=0}^{k-2} \bigcap_{\upsilon_{N-k+i} \in Q_{N-k+i}} \\ &[-\bar{\gamma}_i M_1 + \prod \alpha_{N-k}\ldots\alpha_{N-k+i}\beta_{N-k+i+1}(P_{N-k+i}, \upsilon_{N-k+i})] \\ &+\prod \alpha_{N-k}\ldots\alpha_{N-1}\beta_N(\bar{u}_{N-1}, \bar{\upsilon}_{N-1}) - \bar{\gamma}_{k-1} m_{1,k-1}. \end{aligned}$$

It is equivalent to the following

$$\begin{aligned} &-\prod \alpha_{N-k}\alpha_{N-k+1}\ldots\alpha_{N-1}[\alpha_N z_N + \beta_N(\bar{u}_{N-1}, \bar{\upsilon}_{N-1})] \in \bar{\gamma}_{k-1} m_{1,k-1} \\ &+\sum_{i=0}^{k-2} \bigcap_{\upsilon_{N-k+i} \in Q_{N-k+i}} [-\bar{\gamma}_i M_1 + \prod \alpha_{N-k}\ldots\alpha_{N-k+i}\beta_{N-k+i+1}(P_{N-k+i}, \upsilon_{N-k+i}) \end{aligned}$$

Therefore owing to (32) we have

$$\begin{aligned} &-\prod \alpha_{N-k}\alpha_{N-k+1}\ldots\alpha_{N-1} z_{N-1} \in \bar{\gamma}_{k-1} m_{1,k-1} \\ &+\sum_{i=0}^{k-2} \bigcap_{\upsilon_{N-k+i} \in Q_{N-k+i}} [-\bar{\gamma}_i M_1 + \prod \alpha_{N-k}\ldots\alpha_{N-k+i}\beta_{N-k+i+1}(P_{N-k+i}, \upsilon_{N-k+i})]. \end{aligned}$$

In the same way, if the control $\bar{\upsilon}_{N-2}$ becomes the stated way known that we above us can construct controls $\bar{u}_{N-2}$ providing inclusion

$$\begin{aligned} &-\prod \alpha_{N-k}\alpha_{N-k+1}\ldots\alpha_{N-2} z_{N-2} \in \bar{\gamma}_{k-1} m_{k-1} + \bar{\gamma}_{k-2} m_{k-2} \sum_{i=0}^{k-3} \bigcap_{\upsilon_{N-k+i} \in Q_{N-k+i}} \\ &[-\bar{\gamma}_i M_1 + \prod \alpha_{N-k}\ldots\alpha_{N-k+i}\beta_{N-k+i+1}(P_{N-k+i}, \upsilon_{N-k+i})] \end{aligned}$$

etc. Thus, we will receive

$$\begin{aligned} &-\prod \alpha_{N-k}\alpha_{N-k+1}\ldots\alpha_{N-2} z_{N-2} \in \bar{\gamma}_{k-1} m_{k-1} + \bar{\gamma}_{k-2} m_{k-2} \sum_{i=0}^{k-3} \bigcap_{\upsilon_{N-k+i} \in Q_{N-k+i}} \\ &[-\bar{\gamma}_i M_1 + \prod \alpha_{N-k}\ldots\alpha_{N-k+i}\beta_{N-k+i+1}(P_{N-k+i}, \upsilon_{N-k+i})] \end{aligned}$$

from here we receive

$$z_{N-k-1} \in M.$$

The theorem is proved completely.

Proof of Theorem 4. Let in game (13) one be able to complete the pursuit from "boundary" situation $(f_0, f_N) = (-y_0, -y_N)$ in N steps. Then, it follows from Definition 2 that from any sequence $\bar{\upsilon}_0, \bar{\upsilon}_1, \ldots, \bar{\upsilon}_{N-1}, \bar{\upsilon}_k \in Q$, $0 \le k \le N-1$, of the evasion control it is possible to construct a sequence $\bar{u}_0, \bar{u}_1, \ldots, \bar{u}_{N-1}, \bar{u}_k \in P$, $0 \le k \le N-1$, of pursuit control such that the solution $(z_0, \bar{z}_1, \ldots, z_{N-1}, z_N)$ of the equation $z_1 = X_0 z_0 + y_0$, $z_{n+1} + A_n z_n - z_{n-1} = F_n$, $1 \le n \le N-1$, $z_{N-1} = X_1 z_N + y_N$, for some $d \le N$ hits $M : \bar{z}_d \in M$. Let now in game (2) $\upsilon = \bar{\upsilon}(x, y) \in \bar{Q}$, $(x, y) \in \Omega$, be an arbitrary control of an evader from the class $L_2(\Omega)$. With the knowledge of the evader control $\upsilon = \bar{\upsilon}(x, y)$, it is possible to determine $\bar{\upsilon}_{i,k}$ as the values of this function at the node points of the grid Ω_{hl}, that is,

$$\upsilon_k = \bar{\upsilon}_k = (\bar{\upsilon}_{1,k}, \bar{\upsilon}_{2,k}, \ldots, \bar{\upsilon}_{r-1,k}).$$

Whence it follows that in virtue of Theorem 4 we can construct the pursuer control in game (13) providing completion of pursuit

$$u_k = \bar{u}_k = (\bar{u}_{1,k}, \bar{u}_{2,k}, \ldots, \bar{u}_{r-1,k}).$$

Now in game (2) we construct the pursuer control $u = \bar{u}(x, y)$ as follows: $\bar{u}(x, y) = \{\bar{u}_{i,k} = \bar{u}_{i,k} : ih \le x_i << (i+1)h,$ $i = 0, 1, \ldots, r-1,\ kl \le y < (k+1)l,\ k = 0, 1, \ldots, \theta - 1\}$.

Obviously, $u \in P$ and $\bar{u}(x, y) \in L_2(\Omega)$. By substituting $\upsilon = \bar{\upsilon}(x, y)$ and $u = \bar{u}(x, y)$ in (2), we obtain a differential equation. Similarly, by substituting $\bar{\upsilon}_{i,k}$ and $\bar{u}_{i,k}$ in (3), we obtain a grid equation approximating equation (2).

Let $(\bar{z})_{hl}$ be the value of the exact solution corresponding to the controls $\upsilon=\bar{\upsilon}(x,y)$ and $u=\bar{u}(x,y)$ of problem (2) at the nodes of the grid Ω_{hl}, $\bar{z}_{i,k}$ be the solution corresponding to the controls $\bar{\upsilon}_{i,k}$ and $\bar{u}_{i,k}$ of the difference problem (3). Then, we obtain from (13) and the condition of Theorem 4 that

$$\left\|(z)_{hl}-\bar{z}_{i,k}\right\|_{\Phi_{hl}}\le K_1 l+K_2 h^2<\varepsilon.$$

From this fact and $\bar{z}_{i,k}\in\bar{M}_1$, we obtain $(z)_{hl}-\bar{z}_{i,k}\in\varepsilon S$, $(z)_{hl}\in\varepsilon S+\bar{z}_{i,k}$, $(z)_{hl}\in\varepsilon S+\bar{M}_1$, which proves the theorem.

5. Conclusions

Thus, to solve the game problem of pursuit in the form (1) we pass to the discrete game (13) or (14), and Theorems 1-3 establish the sufficient condition for such problems. Theorem 4 establishes the sufficient conditions for solving the problem of pursuit (1). Here, the difference $(z)_{hl}-z_{i,j}$ (see Section 3) plays the main part in the solution of problem and implies that the solutions of the grid equation (2) are stable.

The problem of stability of the grid equation (2) lies in determining the conditions under which the numerical error $p_{i,j}\equiv(z)_{hl}-z_{i,j}$ tends to zero with growing j uniformly in all i, $0\le i\le n$, or at least remains bounded.

Equation (2) is called stable if the round off errors generated in the course of calculations have tendency to decrease or at least not to increase. Otherwise, the accumulated errors may reach a value such that the numerical solution $(z)_{hl}$ has nothing in common with the exact solution of the grid problem (2). It goes without saying that such unstable grid equations cannot be used for numerical solution of the differential games.

Theorems 1-4 are easily generalized to a wider class of differential games, for example, when

$$\sum_{\alpha=1}^{n} a_\alpha(x_1,x_2,\dots,x_n)\frac{\partial^2 z}{\partial x_\alpha^2}=f(u(x_1,x_2,\dots,x_n),\ \upsilon(x_1,x_2,\dots,x_n))$$

with discontinuous coefficients.

REFERENCES

[1] O. A. Ladyzhenskaya, "Kraevye Zadachi Matematicheskoi Fiziki," (Boundary Problems of Mathematical Physics), Moscow, Nauka, 1973.

[2] O. A. Ladyzhenskaya, V. A. Solonnikov and N. N. Ural'tseva, "Lineinye I Kvazilineinye Uravneniya Parabolicheskogo Tipa," (Linear and Quasi linear Functions of Parabolic Type), Moscow, Nauka, 1967.

[3] V. A. Il'in, "Boundary Control of String Oscillations at One End with Other End Fixed, Provided that Finite Energy Exists," *Dokl. Ross. Akad. Nauk*, Vol. 378, No. 6, 2001, pp. 743-747.

[4] V. A. Il'in and V. V. Tikhomirov, "Wave Equation with Boundary Control at Two Ends and Problem of Complate Oscillation Damping ," *Diff. Uravn.*, Vol. 35, No. 5, 1999, pp. 692-704.

[5] Yu. S. Osipov and S. P. Okhezin, "On the Theory of Differential Games in Parabolic Systems," *Dokl. Akad. Nauk SSSR*, Vol. 226, No. 6, 1976, pp. 1267-1270.

[6] F. L. Chernous'ko, "Bounded Controls in Distributed-parameter Systems," *Prikl. Mat. Mekh.*, Vol. 56, No. 5, 1992, pp. 810-826.

[7] N. Satimov and M. Sh. Mamatov, "On a Class of Linear Differential and Discrete Games between Groups of Pursuers and Evaders," *Diff. Uravn.*, Vol. 26, No. 9, 1990, pp. 1541-1551.

[8] N. Satimov and M. Tukhtasinov, "On some Game Problems in the Distributed Controlled Systems," *Prikl. Mat. Mekh.*, Vol. 69, No. 6, 2005, pp. 997-1003.

[9] N. Satimov and M. Tukhtasinov, "On some Game Problems in Controlled First-order Evolutionary Equations," *Diff. Uravn.*, Vol. 41, No. 8, 2005, pp. 1114 -1121.

[10] M. Sh. Mamatov, "On the Theory of Differential Pursuit Games in Distributed Parameter Systems," *Automatic Control and Computer Sciences*, Vol. 43, No. 1, 2009, pp. 1-8.

[11] M. Sh. Mamatov, "About Application of a Method of Final Differences to the Decision a Prosecution Problem in Systems with the Distributed Parameters," *Automation and Remote Control*, Vol. 70, No. 8, 2009, pp. 1376-1384.

[12] M. Tukhtasinov and M. Sh. Mamatov, "On Pursuit Problems in Controlled Distributed Systems," *Mathematical notes*, Vol. 84, No. 2, 2008, pp. 273-280.

[13] M. Tukhtasinov and M. Sh. Mamatov, "About Transition Problems in Operated Systems," *Diff. Uravn.*, Vol. 45, No.3, 2009, pp. 1-6.

[14] M. Sh. Mamatov and M. Tukhtasinov, "Pursuit Problem in Distributed Control Systems," *Cybernetics and Systems Analysis*, Vol. 45, No. 2, 2009, pp. 297-302.

[15] G. I. Marchuk, "Metody Vychislitel'noi MateMatiki," (Methods of computational Mathematics), Moscow, Nauka, 1989.

8

A New Recombination Tree Algorithm for Mean-Reverting Interest-Rate Dynamics

Peter C. L. Lin
Department of Mathematical Sciences & Financial Engineering Program,
Stevens Institute of Technology, Hoboken, USA

ABSTRACT

In light of the fact that no existing tree algorithms can guarantee the recombination property for general Ornstein-Uhlenbeck processes with time-dependent parameters, a new trinomial recombination-tree algorithm is designed in this research. The proposed algorithm enhances the existing mechanisms in interest-rate modelings with the comparisons to [1,2] methodologies, and the proposed framework provides a more efficient way in discrete-time mean-reverting simulations.

Keywords: Natural Asset; Financial Value; Neural Network

1. Introduction

A general Ornstein-Uhlenbeck process is defined such that

$$dG(t) = k(t)(a(t) - G(t))dt + b(t)d\tilde{W}(t)$$

where $W(t)$ is a standard Brownian motion, and $k(t)$, $a(t)$, $b(t)$ are time-dependent deterministic parameters. The parameter $a(t)$ is the mean-reversion term indicating the long-term mean-reverting level with a rate $k(t)$ at time *t*. The source of randomness is described by the Brownian motion $W(t)$ multiplied by a volatility term $b(t)$.

A tree is an acyclic structure where each node has zero to multiple descendant nodes and one parent node. A recombination tree is a special tree structure of which the size grows linearly. Therefore the investing decisions, if computed recursively, have time complexity[1] at most $O(n^2)$, which is much more efficient than a general simulation method which may cost exponential amount of time. For example, a recombination tree can help us to efficiently determine the price and the buy/sell timing for an American style option by comparing the derivatives value at each tree node with its children nodes (see [4] for more details). However, designing a recombination tree algorithm for modeling interest rates is far from trivial. Here are two examples:

In [1], Hull and White provided a heuristic two-stage method for constructing an interest rate tree based on the extended-Vasicek short-rate model.[2] In the first stage the algorithm builds the framework of the tree, and in the second stage the algorithm calibrates the tree to the current interest-rate term structure. The algorithm is designed for a short-rate model; hence the tree cannot be adjusted or updated according to the markets. Also, their method cannot deal with stochastic mean-reverting parameters, and there is no guarantee that the tree is a recombination tree especially when the volatility term of the short-rate process is a decreasing function. Therefore, Hull-White's algorithm is not a good candidate.

In [2], Black, Derman, and Toy (BDT hereafter) also provided a recombination tree algorithm for short rates. Their tree is constructed recursively and calibrated to zero-coupon bond volatilities and current interest rate structure. Though the BDT tree guarantees a recombination structure, the tree is not designed for a general Ornstein-Uhlenbeck process. Therefore, the BDT tree is not a good candidate either.

In light of the fact that no existing tree algorithms can guarantee the recombination property for general Orn-

[1]For the terminology of computational complexity please see [3].

[2]For more definitions of short-rate models, see [5]. Yet, for this article we should focus only on the mathematical modeling for general Ornstein-Uhlenbeck Processes.

stein-Uhlenbeck processes, we propose a new recombination trinominal tree algorithm. The idea is to modify a standard trinominal tree (**Exhibit 1(a)**) by adding extra branches at each node. First, we denote the tree structure in black color the *center path*. Then, given a node V directly above (below) the center path, define $\langle V\rangle$ the set of nodes containing V and all the nodes down to (up to) the center path. Then we modify the tree according to the following rules: 1) the center path remains unchanged; 2) given a node V above the center path, we connect the node to all the descendant (children) nodes stemming from $\langle V\rangle$; 3) given a node V below the center path, we connect the node to all the descendant nodes stemming from all the nodes from $\langle V\rangle$. The modified tree structure is shown in **Exhibit 1(b)**. We will use the names, *spanning nodes and spanning branches*, to identify those nodes not on the center path, and branches not emanating from a center node.

The crucial key of modifying a standard trinominal tree is that we can further simplify and still keep the tree structure by adding *sibling branches*. Sibling branches are one-way streets through which we can only move up or down at a given time epoch, but not in both directions. A spanning node above the center path can reach the center path and all the nodes in between only by moving down through the sibling branches; a spanning node below the center path can reach the center path and all the nodes in between only by moving up through the sibling branches. As a result, by adding the sibling branches, each node can reach all but one descendent nodes via its sibling branches. So, in the simplified tree structure, each spanning node will have only one time transition descendant branch. The final tree structure is shown in **Exhibit 1(c)**. The algorithm is given below. The proof of the correctness of the algorithm for simulation is given in Section 3.

2. Algorithm

Now we give a full description of the algorithm. Let $g_{j,j}$ denote the node on the center path at time t_j, and let $g(t_j, j)$ denote the value of node $g_{j,j}$. Therefore, if $g:\mathbb{R}\times\mathbb{Z}\mapsto\mathbb{R}$ is represented as a function, then it indicates the value of the node. Let $g_{j,j+k}$ and $g(t_j, j+k)$ denote the k-th node above center node $g_{j,j}$ and the value of $g_{j,j+k}$ respectively. Similarly, let $g_{j,j-k}$ and $g(t_j, j-k)$ denote the k-th node below center node $g_{j,j}$ and the value of $g_{j,j-k}$ respectively. Moreover, if we use capital letter $G(t_j,\omega)$, then it represents a random variable of the tree value at time t_j. To shorthand the notation, the expectation value $G(t_{j+1})$ conditional on the position of $G(t_j)$ is written as

$$\mathrm{E}\left[G(t_{j+1})\middle|G(t_j)\right]. \quad (1)$$

Define the conditional expectation at node $g_{j,j}$ to be

$$M(t_j, j) = E\left[G(t_{j+1})\middle|G(t_j) = g_{j,j}\right] \quad (2)$$

Since the volatility term in stochastic-splines model is assumed to be a deterministic function, the conditional variance is the same for all nodes at a given time, *i.e.* at time t_j, the conditional variance

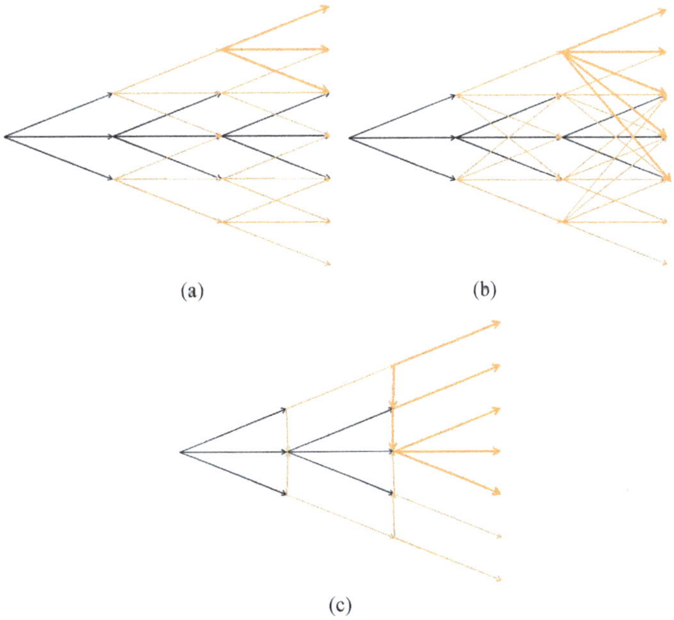

(a) (b) (c)

Exhibit 1. (a) original recombination trinominal tree; (b) modified tree; (c) simplified Tree.

Algorithm 1 Recombination Tree

Require: MAXLEVEL (Tree Size)

{Stage One – Center Path}

For $j = 1$ to **MAXLEVEL** do

$$\Delta x_j \leftarrow \sqrt{V_j}$$

$$h \leftarrow round\left(\frac{M(t_j, j)}{\Delta x_j}\right)$$

$$\text{Set } \begin{cases} g(t_j, j) \leftarrow h \times \Delta x_j \\ g(t_j, j+1) \leftarrow (h+1) \times \Delta x_j \\ g(t_j, j-1) \leftarrow (h-1) \times \Delta x_j \end{cases}.$$

$$\text{Set } \begin{cases} p_u \leftarrow \frac{1}{6} + \frac{\left(M(t_j, j) - g(t_j, j)\right)^2}{6V_{t_j}^2} + \frac{M(t_j, j) - g(t_j, j)}{2\sqrt{3}V_{t_j}} \\ p_n \leftarrow \frac{2}{3} - \frac{\left(M(t_j, j) - g(t_j, j)\right)^2}{3V_{t_j}^2} \\ p_d \leftarrow \frac{1}{6} + \frac{\left(M(t_j, j) - g(t_j, j)\right)^2}{6V_{t_j}^2} - \frac{M(t_j, j) - g(t_j, j)}{2\sqrt{3}V_{t_j}} \end{cases}.$$

end for

{Stage Two – Spanning Branches}

for $i = 1$ to **MAXLEVEL** do

for $j = 1$ to $2 \times j - 1$ do

{Move to the j-th vertex below the center path}

Find $x \in \left[M_{i,j-1} - \sqrt{V_i}, M_{i,j-1} + \sqrt{V_i}\right]$ where

$$\frac{x - M_{i,j-1}}{x - M_{i,j}} = \frac{V_i - \left(x - M_{i,j-1}\right)^2}{V_j + \left(M_{i,j} - M_{i,j-1}\right)^2 - \left(x - M_{i,j-1}\right)^2}$$

$$g(t_{i+1}, j-2) \leftarrow x$$

$$p\left(g_{i,j-1}, g_{i,j}\right) \leftarrow \frac{x - M_{i,j-1}}{x - M_{i,j}}$$

$$p\left(g_{i,j-1}, g_{i+1,j-2}\right) \leftarrow 1 - \frac{x - M_{i,j-1}}{x - M_{i,j}}$$

{Move to the j-th vertex above the center path}

Find $x \in \left[M_{i,j+1} - \sqrt{V_i}, M_{i,j+1} + \sqrt{V_i}\right]$ where

$$\frac{x - M_{i,j+1}}{x - M_{i,j}} = \frac{V_i - \left(x - M_{i,j+1}\right)^2}{V_j + \left(M_{i,j} - M_{i,j+1}\right)^2 - \left(x - M_{i,j+1}\right)^2}$$

Continued

$$g(t_{i+1}, j+2) \leftarrow x$$

$$p(g_{i,j+1}, g_{i,j}) \leftarrow \frac{x - M_{i,j+1}}{x - M_{i,j}}$$

$$p(g_{i,j+1}, g_{i+1,j+2}) \leftarrow 1 - \frac{x - M_{i,j+1}}{x - M_{i,j}}$$

end for

end for

$$V_j = \mathrm{Var}\left(G(t_{j+1}) \middle| G(t_j)\right). \tag{3}$$

The idea of the recombination algorithm is to construct the center path first including the node values and branch probabilities, then determine the values of the spanning nodes, the probabilities on the spanning branches, and the probabilities on the sibling branches. The details are given in Algorithm 1. However, the algorithmshows that each tree is designed for simulating one coefficient process; if we have *N* coefficient, we will need to build *N* trees altogether if coefficient processes are correlated. After constructing the coefficient recombination "forest", we can simulate the interest rate curve efficiently.

The justification of the algorithm is given in the next Section.

3. Verification

First we look at the *first part* of the algorithm and some notations. The first stage of the algorithm follows the standard Hull-White methodology (see [1]) and provides the backbone of the tree. Let $g_{j,j}$ denote the node on the center path at time t_j, and let $g(t_j, j)$ denote the value of node $g_{j,j}$. Therefore, if g is represented as a function, then it indicate the value of the node. Let $g_{j,j+k}$ and $g(t_j, j+k)$ denote the *k*-th node above center node $g_{j,j}$ and the value of $g_{j,j+k}$ respectively. Similarly, let $g_{j,j-k}$ and $g(t_j, j-k)$ denote the *k*-th node below center node $g_{j,j}$ and the value of $g_{j,j-k}$ respectively. Moreover, if we use capital letter $G(t_j)$, then it represents a random variable of the tree value at time t_j. To shorthand the notation, the expected value $G(t_{j+1})$ conditional on the position of $G(t_j)$ is written as

$$\mathrm{E}\left[G(t_{j+1}) \middle| G(t_j)\right]. \tag{4}$$

Now we move to the *second part* of the algorithm. The tree branches besides the central path are called spanning branches and spanning nodes. The second stage of the algorithm adopts the ideas of the law of total expectations and the law of total variances to assign the values and probabilities of spanning nodes and branches. The procedure is done recursively. Therefore we just need to look at the cases when $k = -1$ and $k = 1$. Given a node $g_{j+1,j+2}$ spanning from node $g_{j,j+1}$ at time t_j, we denote the conditional expectation and conditional variance at node $g_{j,j+1}$ to be $M_{j,j+1}$ and $V_{j,j+1}$ respectively. Denote the probability p to be the probability moving down from $g_{j,j+1}$ to $g_{j,j}$ and $(1-p)$ to be the probability moving through the spanning branch from $g_{j,j+1}$ to $g_{j+1,j+2}$.

We can recall the law of total expectation which states

$$\mathrm{E}[X] = \sum p(Y = y)\mathrm{E}[X|Y = y]. \tag{5}$$

If we let

$$X = \tilde{\mathrm{E}}\left[G(t_{j+1}) \middle| G(t_j) = g_{j,j+1}\right], \tag{6}$$

and *Y* denotes the random variable such that

$$Y = \begin{cases} 0, & \text{if moving to the spanning brance} \\ 1, & \text{otherwise} \end{cases}, \tag{7}$$

then

$$\begin{aligned} M_{j,j+1} &= \tilde{\mathrm{E}}\left[G(t_{j+1}) \middle| G(t_j) = g_{j,j+1}\right] \\ &= p\tilde{\mathrm{E}}\left[G(t_j) \middle| G(t_j) = g_{j,j}\right] \\ &\quad + (1-p)\tilde{\mathrm{E}}\left[G(t_{j+1}, j+2) \middle| G(t_j) = g_{j,j+1}\right] \\ &= pM_{j,j} + (1-p)g(t_{j+1}, j+2), \end{aligned} \tag{8}$$

which shows

$$p = \frac{g(t_{j+1}, j+2) - M_{j,j+1}}{g(t_{j+1}, j+2) - M_{j,j}}. \tag{9}$$

The task of deriving the relationship between $g(j+1, j+2)$ and p from the law of total variance is more complicated. We will show the the result first, then break into each part. Recall the law of total variance which states

$$\mathrm{Var}(X) = \mathrm{E}\left[\mathrm{Var}(X|Y)\right] + \mathrm{Var}\left(\mathrm{E}\left[X|Y\right]\right). \quad (10)$$

Similarly we let

$$X = \tilde{\mathrm{E}}\left[G(t_{j+1})\middle|G(t_j)\right], \quad (11)$$

and Y denotes the random variable such that

$$Y = \begin{cases} 0, & \text{if moving to spanning brance} \\ 1, & \text{otherwise} \end{cases}. \quad (12)$$

If the following statement is true:

$$\begin{aligned}
&\widetilde{\mathrm{Var}}\left(\tilde{\mathrm{E}}\left[G(t_{j+1})\middle|G(t_j) = g_{j,j+1}\right]\right) \\
&= \tilde{\mathrm{E}}\left[\widetilde{\mathrm{Var}}\left(\tilde{\mathrm{E}}\left[G(t_{j+1})\middle|G(t_j) = g_{j,j+1}\right]\middle|Y\right)\right] \\
&\quad + \widetilde{\mathrm{Var}}\left(\tilde{\mathrm{E}}\left[\tilde{\mathrm{E}}\left[G(t_{j+1})\middle|G(t_j) = g_{j,j+1}\right]\middle|Y\right]\right) \quad (13) \\
&= pV_j + p\left(M_{j,j} - M_{j,j+1}\right)^2 \\
&\quad + (1-p)\left(g(j, j+1) - M_{j,j+1}\right)^2,
\end{aligned}$$

then we have

$$p = \frac{V_j - \left(g(j, j+1) - M_{j,j+1}\right)^2}{V_j + \left(M_{j,j} - M_{j,j+1}\right)^2 - \left(g(j, j+1) - M_{j,j+1}\right)^2}. \quad (14)$$

Examining the first term, pV_j, on the right-hand-side of Equation (13),

$$\begin{aligned}
&\widetilde{\mathrm{Var}}\left(\tilde{\mathrm{E}}\left[G(t_{j+1})\middle|G(t_j) = g_{j,j+1}\right]\middle|Y = 0\right) \\
&= \widetilde{\mathrm{Var}}\left(g(j+1, j+2)\right) = 0
\end{aligned} \quad (15)$$

since there is only one choice moving from $g_{j,j+1}$ to $g_{j+1,j+2}$. On the other hand,

$$\widetilde{\mathrm{Var}}\left(\tilde{\mathrm{E}}\left[G(t_{j+1})\middle|G(t_j) = g_{j,j+1}\right]\middle|Y = 1\right) = V_j, \quad (16)$$

and we know this value recursively. So

$$\begin{aligned}
&\tilde{\mathrm{E}}\left[\widetilde{\mathrm{Var}}\left(\tilde{\mathrm{E}}\left[G(t_{j+1})\middle|G(t_j) = g_{j,j+1}\right]\middle|Y\right)\right] \\
&= (1-p)\widetilde{\mathrm{Var}}\left(\tilde{\mathrm{E}}\left[G(t_{j+1})\middle|G(t_j) = g_{j,j+1}\right]\middle|Y = 0\right) \\
&\quad + p\widetilde{\mathrm{Var}}\left(\tilde{\mathrm{E}}\left[G(t_{j+1})\middle|G(t_j) = g_{j,j+1}\right]\middle|Y = 1\right) \\
&= p\widetilde{\mathrm{Var}}\left(\tilde{\mathrm{E}}\left[G(t_{j+1})\middle|G(t_j) = g_{j,j+1}\right]\middle|Y = 1\right) = pV_j.
\end{aligned} \quad (17)$$

Next, the second and third terms. Since

$$\begin{aligned}
&\tilde{\mathrm{E}}\left[\tilde{\mathrm{E}}\left[\tilde{\mathrm{E}}\left[G(t_{j+1})\middle|G(t_j) = G(t_j, j+1)\right]\middle|Y\right]\right] \\
&= \tilde{\mathrm{E}}\left[G(t_{j+1})\middle|G(t_j) = G(t_j, j+1)\right] = M_{j,j+1},
\end{aligned} \quad (18)$$

we have

$$\begin{aligned}
&\widetilde{\mathrm{Var}}\left(\tilde{\mathrm{E}}\left[\tilde{\mathrm{E}}\left[G(t_{j+1})\middle|G(t_j) = g_{j,j+1}\right]\middle|Y\right]\right) \\
&= p\left(\tilde{\mathrm{E}}\left[\tilde{\mathrm{E}}\left[G(t_{j+1})\middle|G(t_j) = g_{j,j+1}\right]\middle|Y = 1\right] - M_{j,j+1}\right)^2 \\
&\quad + (1-p)\left(\tilde{\mathrm{E}}\left[\tilde{\mathrm{E}}\left[G(t_{j+1})\middle|G(t_j) = g_{j,j+1}\right]\middle|Y = 0\right] - M_{j,j+1}\right)^2 \\
&= p\left(\tilde{\mathrm{E}}\left[G(t_{j+1})\middle|G(t_j) = g_{j,j+1}\right] - M_{j,j+1}\right)^2 \\
&\quad + (1-p)\left(g(j+1, j+2) - M_{j,j+1}\right)^2 \\
&= p\left(M_{j,j} - M_{j,j+1}\right)^2 + (1-p)\left(g(j+1, j+2) - M_{j,j+1}\right)^2.
\end{aligned} \quad (19)$$

Now we have two equations and two unknown p and $g(t_{j+1}, j+2)$ in the following

$$\begin{cases} p = \dfrac{g(t_{j+1}, j+2) - M_{j,j+1}}{g(t_{j+1}, j+2) - M_{j,j}} \\ q = \dfrac{V_j - \left(g(j+1, j+2) - M_{j,j+1}\right)^2}{V_j + \left(M_{j,j} - M_{j,j+1}\right)^2 - \left(g(j+1, j+2) - M_{j,j+1}\right)^2} \end{cases}. \quad (20)$$

However, p must be a number between 0 and 1. And we now show that the equations indeed yield a solution such that $p \in [0,1]$. First, the case where $M_{j,j+1} \geq M_{j,j}$ and write

$$f_1(x) = \frac{x - M_{j,j+1}}{x - M_{j,j}} \quad (21)$$

and

$$f_2(x) = \frac{V_j - \left(x - M_{j,j+1}\right)^2}{V_j + \left(M_{j,j} - M_{j,j+1}\right)^2 - \left(x - M_{j,j+1}\right)^2}. \quad (22)$$

Since $M_{j,j+1} \geq M_{j,j}$, for any $x \geq M_{j,j+1}$

$$0 \leq \frac{x - M_{j,j+1}}{x - M_{j,j}} \leq 1 \quad (23)$$

and $f_1(x)$ is continuous and monotonically increasing. On the other hand, for any $x \in \left[M_{j,j+1}, M_{j,j+1} + \sqrt{V_j}\right]$,

$$0 \leq \frac{V_j - \left(x - M_{j,j+1}\right)^2}{V_j + \left(M_{j,j} - M_{j,j+1}\right)^2 - \left(x - M_{j,j+1}\right)^2} \leq 1 \quad (24)$$

and $f_2(x)$ is a continuous and monotonically decreasing function. Since

$$f_1\left(M_{j,j+1}\right) = 0, \qquad f_1\left(M_{j,j+1} + \sqrt{V_j}\right) > 0 \quad (25)$$

and

$$f_2\left(M_{j,j+1}\right) > 0, \qquad f_x\left(M_{j,j+1} + \sqrt{V_j}\right) = 0, \quad (26)$$

we know that there must exist a unique $\hat{x} \in \left[M_{j,j+1}, M_{j,j+1} + \sqrt{V_j}\right]$ such that

$$f_1(\hat{x}) = f_x(\hat{x}) = p \in [0,1]. \quad (27)$$

Alternatively, the proof is similar for the case when $M_{j,j+1} < M_{j,j}$ except the solution exists in $\left[M_{j,j+1} - \sqrt{V_j}, M_{j,j+1}\right]$. The uniqueness and existence of the solution p and $g\left(t_{j+1}, j+2\right)$ help us solve the equations fast.

4. Conclusion

This research proposes a new trinomial recombination-tree algorithm for simulating general Ornstein-Uhlenbeck processes with time-dependent parameters. We show that there is an equivalent recombination-tree structure to simulate the mean-reverting interest-rate dynamics. Detailed algorithm and justification are given.

REFERENCES

[1] J. Hull and A. White, "Pricing Interest-Rate-Derivative Securities," *Review of Financial Studies*, Vol. 3, No. 4, 1990, pp. 573-592.

[2] F. Black, E. Derman and W. Toy, "A One-Factor Model of Interest Rates and Its Application to Treasury Bond Options," *Financial Analysts Journal*, Vol. 46, No. 1, 1990, pp. 33-39.

[3] T. H. Cormen, C. E. Leiserson, R. L. Rivest and C. Stein, "Introduction to Algorithms," 3rd Edition, The MIT Press, Cambridge, 2009.

[4] J. Hull, "Options, Futures, and Other Derivatives," 7th Edition, Prentice Hall, Upper Saddle River, 2008.

[5] D. Brigo and F. Mercurio, "Interest Rate Models-Theory and Practice," 2nd Edition, Springer, New York, 2006.

Characterization of Periodic Eigenfunctions of the Fourier Transform Operator

Comlan de Souza[1], David W. Kammler[2]
[1]Department of Mathematics, California State University at Fresno, Fresno, USA
[2]Department of Mathematics, Southern Illinois University at Carbondale, Carbondale, USA

ABSTRACT

Let the generalized function (tempered distribution) f on $\mathbb{R}$ be a p-periodic eigenfunction of the Fourier transform operator $\mathcal{F}$, *i.e.*, $f(x+p)=f(x), \mathcal{F}f=\lambda f$, for some $\lambda \in \mathbb{C}$. We show that $\lambda = 1,-i,-1,$ or $+i$, that $p=\sqrt{N}$ for some $N=1,2,\cdots,$ and that f has the representation $f(x)=\sum_{m=-\infty}^{\infty}\sum_{n=0}^{N-1}\gamma[n]\delta\left(x-\frac{n}{p}-mp\right)$ where δ is the Dirac functional and γ is an eigenfunction of the discrete Fourier transform operator $\mathcal{F}_N$ with $(\mathcal{F}_N\gamma)[k]=\frac{1}{N}\sum_{n=0}^{N-1}\gamma[n]\mathrm{e}^{-2\pi ikn/N}=\frac{\lambda}{\sqrt{N}}\gamma[k]$, $k=0,1,\cdots,N-1$. We generalize this result to p_1,p_2-periodic eigenfunctions of $\mathcal{F}$ on $\mathbb{R}^2$ and to p_1,p_2,p_3-periodic eigenfunctions of $\mathcal{F}$ on $\mathbb{R}^3$.

Keywords: Eigenfunction; Fourier Transform Operator

1. Introduction

In this paper, we will study certain generalizations of the Dirac comb (or III functional, see [1])

$$\mathrm{III}(x):=\sum_{n=-\infty}^{\infty}\delta(x-n) \tag{1}$$

where δ is the Dirac functional. We work within the context of the Schwartz theory of distributions [2] as developed in [1,3-7]. For purposes of manipulation we use "function" notation for δ, III and related functionals. Various useful proprieties of δ and III are developed in [1,3-5].

The III functional is used in the study of sampling, periodization, etc., see [1,4,5]. We will illustrate this process using a notation that can be generalized to an n-dimensional setting. Let $a_1 \in \mathbb{R}$ with $a_1 \neq 0$, and let $A_1 := \frac{1}{a_1}$. We define the lattice

$$\mathcal{L}_{a_1} := \{na_1 : n \in \mathbb{Z}\}$$

and the corresponding a_1-periodic Dirac comb

$$\mathrm{grid}_{a_1}(x) := \sum_{a\in\mathcal{L}_{a_1}}\delta(x-a). \tag{2}$$

The Fourier transform of the a_1-periodic Dirac comb is

$$\mathrm{grid}_{a_1}^{\wedge}(s)=|A_1|\mathrm{grid}_{A_1}(s). \tag{3}$$

Let g be any univariate distribution with compact support. We can periodize g by writing

$$f(x):=\mathrm{grid}_{a_1}(x)*g(x), \tag{4}$$

where $*$ represents the convolution product, to obtain the weakly convergent Fourier series

$$f(x)=\sum_{k=-\infty}^{\infty}|A_1|g^{\wedge}(kA_1)\mathrm{e}^{2\pi ikA_1x}. \tag{5}$$

We observe that grid_{a_1} has support at the points $na_1, n = 0, \pm 1, \pm 2, \cdots$ of the lattice $\mathcal{L}_{a_1}$, while the Fourier transform $|A_1|\text{grid}_{A_1}$ has support at the points $\frac{n}{a_1}, n = 0, \pm 1, \pm 2, \cdots$ of the lattice $\mathcal{L}_{A_1}$. It follows that

$$\text{grid}_{a_1}^{\wedge} = \text{grid}_{a_1}$$

if and only if

$$a_1 = \pm 1,$$

i.e., if and only if

$$\text{grid}_{a_1} = \text{III}. \tag{6}$$

Let $\mathcal{F}$ be the Fourier transform operator on the space of tempered distributions. It is well known [1,4,5], that $\mathcal{F}$ is linear and that

$$\mathcal{F}^4 = \mathcal{I}, \tag{7}$$

where $\mathcal{I}$ denotes the identity operator on the space of tempered distributions. We are interested in tempered distributions f such that

$$\mathcal{F}f = \lambda f, \tag{8}$$

where λ is a scalar. Any distribution f that satisfies (8), and that we will call eigenfunction of $\mathcal{F}$, must also satisfy the following equation

$$\mathcal{F}^n f = \lambda^n f, \quad n \in \mathbb{N} \tag{9}$$

due to the linearity of the operator $\mathcal{F}$. When $n = 4$, then $\mathcal{F}^4 f = \lambda^4 f$. Thus the eigenvalues of the operator $\mathcal{F}$ are $1, -1, i, -i$.

Eigenvectors of $\mathcal{F}_N$

We first consider the eigenvectors of the discrete Fourier transform operator $\mathcal{F}_N$ since, as we will see later, they can be used to construct all periodic eigenfunctions of the Fourier transform operator $\mathcal{F}$ [8,9].

Definition 1. *Let* $N = 1, 2, \cdots$. *The matrix*

$$\mathcal{F}_N := \frac{1}{N}\begin{bmatrix} 1 & 1 & 1 & \cdots & 1 \\ 1 & \omega & \omega^2 & \cdots & \omega^{N-1} \\ 1 & \omega^2 & \omega^4 & \cdots & \omega^{2N-2} \\ \vdots & \vdots & \vdots & \ddots & \vdots \\ 1 & \omega^{N-1} & \omega^{2N-2} & \cdots & \omega^{(N-1)(N-1)} \end{bmatrix},$$

$\omega := e^{-2\pi i/N}$, is said to be the discrete Fourier transform operator.

It is easy to verify the operator identity

$$\mathcal{F}_N^2 = \frac{1}{N}\mathcal{R}_N$$

where

$$\mathcal{R}_N := \begin{bmatrix} 1 & 0 & 0 & \cdots & 0 & 0 \\ 0 & 0 & 0 & \cdots & 0 & 1 \\ 0 & 0 & 0 & \cdots & 1 & 0 \\ \vdots & \vdots & \vdots & \ddots & \vdots & \vdots \\ 0 & 0 & 1 & \cdots & 0 & 0 \\ 0 & 1 & 0 & \cdots & 0 & 0 \end{bmatrix}$$

is the reflection operator. It is easy to verify

$$\mathcal{F}_N^4 = \left[\frac{1}{N}\mathcal{R}_N\right]^2 = \frac{1}{N^2}\mathcal{R}_N^2 = \frac{1}{N^2}I_N$$

where I_N is the $N \times N$ identity matrix. In this way we see that if

$$\mathcal{F}_N f = \lambda f, f \neq 0,$$

then

$$\lambda^4 - \frac{1}{N^2} = 0,$$

so λ must take one of the values $\pm 1/\sqrt{N}, \pm i/\sqrt{N}$.

Let $M_r(N)$ be the multiplicity of the eigenvalue

$$\lambda = \frac{(-i)^r}{\sqrt{N}}$$

of $\mathcal{F}_N$, $r = 0, 1, 2, 3$, and let

$$f_{N,r,\mu}[n], \mu = 1, 2, \cdots, M_r(N) \tag{10}$$

be orthonormal eigenvectors of $\mathcal{F}_N$ corresponding to the eigenvalue

$$\lambda = \frac{(-i)^r}{\sqrt{N}}, r = 0, 1, 2, 3.$$

Example 1. $N = 2$

The matrix

$$\mathcal{F}_2 = \frac{1}{2}\begin{bmatrix} 1 & 1 \\ 1 & -1 \end{bmatrix}$$

has the eigenvalues $\lambda_1 = 1/\sqrt{2}$, $\lambda_2 = -1/\sqrt{2}$ *with corresponding eigenvectors*

$$\begin{bmatrix} 1 \\ -1+\sqrt{2} \end{bmatrix}, \begin{bmatrix} 1 \\ -1-\sqrt{2} \end{bmatrix}.$$

We normalize these vectors to obtain

$$f_{2,0,1}[0] = \frac{1}{\sqrt{4-2\sqrt{2}}}, f_{2,0,1}[1] = \frac{-1+\sqrt{2}}{\sqrt{4-2\sqrt{2}}},$$

$$f_{2,2,1}[0] = \frac{1}{\sqrt{4+2\sqrt{2}}}, f_{2,2,1}[1] = -\frac{1+\sqrt{2}}{\sqrt{4+2\sqrt{2}}}.$$

2. The Main Results

A generalized function $f, f \neq 0$, is said to be an eigenfunction of the Fourier transform operator $\mathcal{F}$ if

$$\mathcal{F}f = \lambda f$$

For $\lambda = \pm 1, \pm i$. We would like to characterize all periodic eigenfunctions f of the Fourier transform operator $\mathcal{F}$, *i.e.*,

$$\mathcal{F}f = \lambda f, f \neq 0,$$

within the context of 1,2,3 dimensions.

2.1. Periodic Eigenfunctions of $\mathcal{F}$ or $\mathbb{R}$

Let f be a *p*-periodic generalized function on $\mathbb{R}$, $p > 0$, and assume that

$$F := \mathcal{F}f = \lambda f$$

where $\lambda = \pm 1, \pm i$ and $f \neq 0$. The 2-periodic function

$$f(x) = \frac{1}{2}\left\{ \text{III}\left(\frac{x}{2}\right) + \text{III}\left(\frac{x-1/2}{2}\right) - \text{III}\left(\frac{x-2/2}{2}\right) + \text{III}\left(\frac{x-3/2}{2}\right) \right\},$$

is such an eigenfunction, constructed from the eigenvector $f_{4,0,2}$ of $\mathcal{F}_4$. We will now characterize all such periodic eigenfunctions.

Since f is *p*-periodic, f is represented by its weakly convergent Fourier series

$$f(x) = \sum_{k=-\infty}^{\infty} \Gamma[k] e^{2\pi i k x/p} \tag{11}$$

We Fourier transform term by term to obtain the weakly convergent series

$$F(s) = \sum_{k=-\infty}^{\infty} \Gamma[k] \delta\left(s - \frac{k}{p}\right) \tag{12}$$

for the Fourier transform of f. Now since $F = \lambda f$ and $\lambda \neq 0$, F must also be *p*- periodic with

$$F(s) = \left\{ \sum_{0 \le k < p^2} \Gamma[k] \delta\left(s - \frac{k}{p}\right) \right\} * \sum_{m=-\infty}^{\infty} \delta(s - mp) = \left\{ \sum_{0 \le k < p^2} \Gamma[k] \delta\left(s - \frac{k}{p}\right) \right\} * \frac{1}{p} \text{III}\left(\frac{s}{p}\right).$$

We recognize this as the Fourier transform of

$$f(x) = \left\{ \sum_{0 \le k < p^2} \Gamma[k] e^{2\pi i k x/p} \right\} (px) = \left\{ \sum_{0 \le k < p^2} \Gamma[k] e^{2\pi i k x/p} \right\} \frac{1}{p} \sum_{n=-\infty}^{\infty} \delta\left(x - \frac{n}{p}\right)$$

$$= \frac{1}{p} \sum_{n=-\infty}^{\infty} \sum_{0 \le k < p^2} \Gamma[k] e^{2\pi i k x/p} \delta\left(x - \frac{n}{p}\right) = \frac{1}{p} \sum_{n=-\infty}^{\infty} \left\{ \sum_{0 \le k < p^2} \Gamma[k] e^{2\pi i k n/p^2} \right\} \delta\left(x - \frac{n}{p}\right).$$

We define

$$\gamma[n] := \sum_{0 \le k < p^2} \Gamma[k] e^{2\pi i k n/p^2}$$

and write

$$f(x) = \frac{1}{p} \sum_{n=-\infty}^{\infty} \gamma[n] \delta\left(x - \frac{n}{p}\right). \tag{13}$$

Now if the term

$$\gamma[n] \delta\left(x - \frac{n}{p}\right), \gamma[n] \neq 0$$

appears in the sum (13) then (since f is *p*-periodic)

$$\gamma(n) \delta\left(x - p - \frac{n}{p}\right)$$

must also appear. Thus

$$\gamma[n'] \delta\left(x - \frac{n'}{p}\right) = \gamma[n] \delta\left(x - \frac{p^2 + n}{p}\right)$$

for some integer n'. It follows that

$$\frac{n'}{p} = \frac{p^2 + n}{p},$$

i.e,,

$$p^2 = n' - n,$$

and

$$\gamma[n] = \gamma[n'].$$

thus

$$p^2 = N$$

for some $N = 1, 2, \cdots$, and since $\gamma[n]$ is *N*-periodic, we can use (13) to write

$$\begin{aligned} f(x) &= \frac{1}{\sqrt{N}} \sum_{n=-\infty}^{\infty} \gamma[n] \delta\left(x - \frac{n}{\sqrt{N}}\right) \\ &= \frac{1}{\sqrt{N}} \sum_{m=-\infty}^{\infty} \sum_{n=0}^{N-1} \gamma[n] \delta\left(x - \frac{n}{\sqrt{N}} - m\sqrt{N}\right) \end{aligned} \tag{14}$$

where

$$\gamma[n] = \sum_{k=0}^{N-1}\Gamma[k]\mathrm{e}^{2\pi ikn/N}$$

is the inverse Fourier transform of the N-periodic sequence of Fourier coefficients Γ. Since $F = \lambda f$ we can use (12), (14) to see that

$$\Gamma[k] = (\mathcal{F}_N\gamma)[k] = \frac{\lambda}{\sqrt{N}}\gamma[k],\ k = 0,1,\cdots,N-1,$$

i.e., that γ is an eigenvector of the discrete Fourier transform operator $\mathcal{F}_N$ associated with the eigenvalue $\frac{\lambda}{\sqrt{N}}$. In this way we prove the following

Theorem 1. *Let the generalized function f on $\mathbb{R}$ be a p-periodic eigenfunction of the Fourier transform operator $\mathcal{F}$ with eigenvalue $\lambda = 1,-i,-1$, or $+i$. Then $p = \sqrt{N}$ for some integer $N = 1,2,\cdots$ and f has the representation*

$$f(x) = \sum_{m=-\infty}^{\infty}\sum_{n=0}^{N-1}\gamma[n]\delta\left(x - \frac{n}{p} - mp\right) \quad (15)$$

where γ is an eigenvector of the discrete Fourier transform operator $\mathcal{F}_N$ with

$$\begin{aligned}(\mathcal{F}_N\gamma)[k] &= \frac{1}{N}\sum_{n=0}^{N-1}\gamma[n]\mathrm{e}^{-2\pi ikn/N} \\ &= \frac{\lambda}{\sqrt{N}}\gamma[k],\ k = 0,1,\cdots,N-1.\end{aligned}$$

Example 2. *When $N = 1$ we obtain the corresponding 1-periodic*

$$f(x) = \sum_{n=-\infty}^{\infty}\delta(x-n) = \text{III}(x),$$

with

$$\mathcal{F}\text{III} = \text{III}.$$

Of course, this particular result is well known, see [1]. Our argument shows that a periodic eigenfunction of the Fourier transform operator that has one singular point per unit cell must be a scalar multiple of the Dirac comb III.

Example 3. *When $N = 2$, we obtain the $\sqrt{2}$-periodic eigenfunctions*

$$\begin{aligned}f_1(x) &= \frac{1}{\sqrt{4-2\sqrt{2}}}\sum_{n=-\infty}^{\infty}\delta\left(x-n\sqrt{2}\right) + \frac{-1+\sqrt{2}}{\sqrt{4-2\sqrt{2}}}\sum_{n=-\infty}^{\infty}\delta\left(x-\frac{1}{\sqrt{2}}-n\sqrt{2}\right) \\ &= \frac{1}{\sqrt{2\left(4-2\sqrt{2}\right)}}\left(\frac{x}{\sqrt{2}}\right) + \frac{-1+\sqrt{2}}{\sqrt{2\left(4-2\sqrt{2}\right)}}\left(\frac{x}{\sqrt{2}}-\frac{1}{2}\right),\end{aligned}$$

and

$$\begin{aligned}f_2(x) &= \frac{1}{\sqrt{4+2\sqrt{2}}}\sum_{n=-\infty}^{\infty}\delta\left(x-n\sqrt{2}\right) - \frac{1+\sqrt{2}}{\sqrt{4+2\sqrt{2}}}\sum_{n=-\infty}^{\infty}\delta\left(x-\frac{1}{\sqrt{2}}-n\sqrt{2}\right) \\ &= \frac{1}{\sqrt{2\left(4+2\sqrt{2}\right)}}\left(\frac{x}{\sqrt{2}}\right) - \frac{1+\sqrt{2}}{\sqrt{2\left(4+2\sqrt{2}\right)}}\left(\frac{x}{\sqrt{2}}-\frac{1}{2}\right)\end{aligned}$$

from the eigenvectors $f_{2,0,1}$ and $f_{2,2,1}$ for $\mathcal{F}_2$. It is easy to verify that

$$(\mathcal{F}f_1)(s) = f_1(s),\ (\mathcal{F}f_2)(s) = -f_2(s).$$

Characterization of periodic eigenfunctions of $\mathcal{F}$ on $\mathbb{R}^2$

Let f be a bivariate generalized function and assume that f is an eigenfunction of $\mathcal{F}$, *i.e.*,

$$F := \mathcal{F}f = \lambda f$$

with $\lambda = 1,-i,-1$, or $+i$, (and $f \neq 0$). Assume further that f is a_1,a_2-periodic, *i.e.*,

$$f(x+a_1) = f(x),\ f(x+a_2) = f(x).$$

Here a_1,a_2 are linearly independent vectors in $\mathbb{R}^2$.

We simplify the analysis by rotating the coordinate system as necessary so as to place a shortest vector from the lattice $\mathcal{L}_{a_1,a_2}$ along the positive x axis. We can and do further assume with no loss of generality that a_1,a_2 have the form

$$a_1 = (\alpha_1,0)^{\mathrm{T}},\ a_2 = (\beta_1,\beta_2)^{\mathrm{T}}$$

where

$$\alpha_1 > 0 \quad (16)$$

$$\alpha_1^2 \leq \beta_1^2 + \beta_2^2 \quad (17)$$

$$\beta_2 > 0 \quad (18)$$

$$0 \leq \beta_1 < \alpha_1. \quad (19)$$

The dual vectors then have the representation

$$A_1 = \frac{1}{\alpha_1\beta_2}(\beta_2, -\beta_1)^{\mathrm{T}},\ A_2 = \frac{1}{\alpha_1\beta_2}(0, \alpha_1)^{\mathrm{T}},$$

and

$$\mathrm{grid}_{a_1,a_2}(x) = \sum_{n_1=-\infty}^{\infty}\sum_{n_2=-\infty}^{\infty}\delta(x - n_1a_1 - n_2a_2)$$

has the Fourier transform

$$\mathrm{grid}^{\wedge}_{a_1,a_2}(s) = \Delta\sum_{k_1=-\infty}^{\infty}\sum_{k_2=-\infty}^{\infty}\delta(s - k_1A_1 - k_2A_2)$$

where $\Delta = |\det(A_1, A_2)|$. Now since f is a_1, a_2-periodic, f can be represented by the weakly convergent Fourier series

$$f(x) = \sum_{k_1=-\infty}^{\infty}\sum_{k_2=-\infty}^{\infty}\Gamma[k_1,k_2]\mathrm{e}^{2\pi i x\cdot(k_1A_1+k_2A_2)}. \tag{20}$$

We Fourier transform the series (20) to obtain the weakly convergent series

$$F(s) = \sum_{k_1=-\infty}^{\infty}\sum_{k_2=-\infty}^{\infty}\Gamma[k_1,k_2]\delta(s - k_1A_1 - k_2A_2). \tag{21}$$

From (21), we see that the support of F lies on the lattice $\mathcal{L}_{A_1,A_2}$ and since $F = \lambda f$, F must also be a_1, a_2-periodic so we can write

$$F(s) = \left\{\sum_{k_1A_1+k_2A_2\in\mathcal{U}}\Gamma[k_1,k_2]\delta(s - k_1A_1 - k_2A_2)\right\} * \mathrm{grid}_{a_1,a_2}(s) \tag{22}$$

where

$$\mathcal{U} := \{x_1'a_1 + x_2'a_2 : 0 \le x_1' < 1, 0 \le x_2' < 1\}$$

is a primitive unit cell associated with the lattice $\mathcal{L}_{a_1,a_2}$, where x_1', x_2' are affine coordinates, and $*$ is the bivariate convolution product. Using the bivariate inverse Fourier transform, we see that

$$\begin{aligned} f(x) &= F(x)\cdot\mathrm{grid}^{\wedge}_{a_1,a_2}(x) \\ &= \Delta\sum_{n_1=-\infty}^{\infty}\sum_{n_2=-\infty}^{\infty}\sum_{k_1A_1+k_2A_2\in\mathcal{U}}\left\{\Gamma[k_1,k_2]\cdot\mathrm{e}^{2\pi i(k_1A_1+k_2A_2)\cdot(n_1A_1+n_2A_2)}\cdot\delta(x - n_1A_1 - n_2A_2)\right\} \\ &= \frac{1}{\alpha_1\beta_2}\sum_{n_1=-\infty}^{\infty}\sum_{n_2=-\infty}^{\infty}\left\{\left\{\sum_{k_1A_1+k_2A_2\in\mathcal{U}}\Gamma[k_1,k_2]\cdot\mathrm{e}^{2\pi i\left\{(\beta_1^2+\beta_2^2)n_1k_1 - \alpha_1\beta_1(n_1k_2+n_2k_1)+\alpha_1^2n_2k_2\right\}/(\alpha_1^2\beta_2^2)}\right\}\right. \\ &\quad\left.\cdot\delta\left(x_1 - \frac{n_1\beta_2}{\alpha_1\beta_2}, x_2 + \frac{n_1\beta_1 - n_2\alpha_1}{\alpha_1\beta_2}\right)\right\}. \end{aligned}$$

We define

$$\gamma[n_1,n_2] := \sum_{k_1A_1+k_2A_2\in\mathcal{U}}\left\{\Gamma[k_1,k_2]\cdot\mathrm{e}^{2\pi i\left\{(\beta_1^2+\beta_2^2)n_1k_1 - \alpha_1\beta_1(n_1k_2+n_2k_1)+\alpha_1^2n_2k_2\right\}/(\alpha_1^2\beta_2^2)}\right\} \tag{23}$$

and write

$$f(x_1,x_2) = \frac{1}{\alpha_1\beta_2}\sum_{n_1=-\infty}^{\infty}\sum_{n_2=-\infty}^{\infty}\left\{\gamma[n_1,n_2]\cdot\delta\left(x_1 - \frac{n_1\beta_2}{\alpha_1\beta_2}, x_2 + \frac{n_1\beta_1 - n_2\alpha_1}{\alpha_1\beta_2}\right)\right\} \tag{24}$$

Now f is a_1, a_2-periodic, so if $\gamma[n_1,n_2] \ne 0$ for some integers n_1, n_2, then the term

$$\gamma[n_1,n_2]\delta\left(x_1 - \alpha_1 - \frac{n_1\beta_2}{\alpha_1\beta_2}, x_2 + \frac{n_1\beta_1 - n_2\alpha_1}{\alpha_1\beta_2}\right)$$

equals the term

$$\gamma[n_1',n_2']\delta\left(x_1 - \frac{n_1'\beta_2}{\alpha_1\beta_2}, x_2 + \frac{n_1'\beta_1 - n_2'\alpha_1}{\alpha_1\beta_2}\right)$$

and the term

$$\gamma[n_1,n_2]\delta\left(x_1 - \beta_1 - \frac{n_1\beta_2}{\alpha_1\beta_2}, x_2 - \beta_2 + \frac{n_1\beta_1 - n_2\alpha_1}{\alpha_1\beta_2}\right)$$

equals the term

$$\gamma[n_1'',n_2'']\delta\left(x_1 - \frac{n_1''\beta_2}{\alpha_1\beta_2}, x_2 + \frac{n_1''\beta_1 - n_2''\alpha_1}{\alpha_1\beta_2}\right)$$

for some integers n_1', n_2', n_1'', n_2''. From the supports of these δ-functions we see that

$$\alpha_1 + \frac{n_1\beta_2}{\alpha_1\beta_2} = \frac{n_1'\beta_2}{\alpha_1\beta_2},$$

i.e.,

$$\alpha_1^2 = n_1' - n_1$$
$$\alpha_1^2 = N_1$$

for some $N_1 = 1, 2, \cdots$. Likewise, we see in turn that

$$n_1\beta_1 - n_2\alpha_1 = n_1'\beta_1 - n_2'\alpha_1,$$
$$(n_1' - n_1)\beta_1 = (n_2' - n_2)\alpha_1,$$
$$\alpha_1^2\beta_1 = (n_2' - n_2)\alpha_1,$$
$$\alpha_1\beta_1 = n_2' - n_2 = M$$

for some $M = 0, \pm 1, \pm 2, \cdots$, and analogously

$$\beta_1 + \frac{n_1\beta_2}{\alpha_1\beta_2} = \frac{n_1''\beta_2}{\alpha_1\beta_2},$$
$$\alpha_1\beta_1 = n_1'' - n_1 = M.$$

Finally,

$$\beta_2 - \frac{n_1\beta_1 - n_2\alpha_1}{\alpha_1\beta_2} = -\frac{n_1''\beta_1 - n_2''\alpha_1}{\alpha_1\beta_2},$$
$$\beta_2^2 + (n_1'' - n_1)\frac{\beta_1}{\alpha_1} = n_2'' - n_2,$$
$$\beta_2^2 + \alpha_1\beta_1\frac{\beta_1}{\alpha_1} = n_2'' - n_2,$$
$$\beta_2^2 + \beta_1^2 = N_2$$

for some $N_2 = 1, 2, \cdots$. Using these expressions we can now write

$$\alpha_1 = \sqrt{N_1},\ \beta_1 = \frac{M}{\sqrt{N_1}},$$
$$\beta_2 = \frac{\sqrt{N_1N_2 - M^2}}{\sqrt{N_1}}$$
$$a_1 = \frac{1}{\sqrt{N_1}}(N_1, 0)^{\mathrm{T}}$$
$$a_2 = \frac{1}{\sqrt{N_1}}\left(M, \sqrt{N_1N_2 - M^2}\right)^{\mathrm{T}}$$
$$A_1 = \frac{1}{\sqrt{N_1(N_1N_2 - M^2)}}\left(\sqrt{N_1N_2 - M^2}, -M\right)^{\mathrm{T}}$$
$$A_2 = \frac{1}{\sqrt{N_1(N_1N_2 - M^2)}}(0, N_1)^{\mathrm{T}}$$

where, in view of (16)-(19)

$$N_1 \le N_2,\quad 0 \le M < N_1$$

and

$$\|a_1\| = \sqrt{N_1},\quad \|a_2\| = \sqrt{N_2}.$$

From (21), (23) we also have

$$\gamma[n_1, n_2] = \sum_{k_1A_1 + k_2A_2 \in \mathcal{U}} \left\{ \Gamma[k_1, k_2] \cdot \mathrm{e}^{2\pi i\left\{N_2n_1k_1 - M(n_1k_2 + n_2k_1) + N_1n_2k_2 / (N_2N_1 - M^2)\right\}} \right\} \tag{25}$$

$$F(s) = \sum_{k_1=-\infty}^{\infty}\sum_{k_2=-\infty}^{\infty} \Gamma[k_1, k_2]\delta(s - k_1A_1 - k_2A_2)$$
$$= \sum_{k_1=-\infty}^{\infty}\sum_{k_2=-\infty}^{\infty} \left\{ \Gamma[k_1, k_2] \cdot \delta\left(s_1 - \frac{k_1}{\sqrt{N_1}}, s_2 + \frac{k_1M - k_2N_1}{\sqrt{N_1(N_1N_2 - M^2)}} \right) \right\}. \tag{26}$$

We will now consider separately the cases $M = 0, M > 0$.

Case $M = 0$

When $M = 0$ the vectors a_1, a_2 are orthogonal and f has the corresponding periods

$$\alpha_1 = \sqrt{N_1},\ \beta_2 = \sqrt{N_2},$$

along the x-axis and y-axis, respectively. The function γ is represented by the synthesis equation

$$\gamma[n_1, n_2] = \sum_{k_1=0}^{N_1-1}\sum_{k_2=0}^{N_2-1} \left\{ \Gamma[k_1, k_2] \cdot \mathrm{e}^{2\pi i(n_1k_1/N_1 + n_2k_2/N_2)} \right\}, \tag{27}$$

and by using (24) and (26), in turn we write

$$F(s_1, s_2)$$
$$= \sum_{k_1=-\infty}^{\infty}\sum_{k_2=-\infty}^{\infty} \left\{ \Gamma[k_1, k_2] \cdot \delta\left(s_1 - \frac{k_1}{\sqrt{N_1}}, s_2 - \frac{k_2}{\sqrt{N_2}} \right) \right\}$$
$$= \lambda f(s_1, s_2) = \frac{\lambda}{\sqrt{N_1N_2}} \sum_{n_1=-\infty}^{\infty}\sum_{n_2=-\infty}^{\infty} \left\{ \gamma[n_1, n_2] \cdot \delta\left(s_1 - \frac{k_1}{\sqrt{N_1}}, s_2 - \frac{k_2}{\sqrt{N_2}} \right) \right\}.$$

In this way we conclude that

$$\Gamma[k_1, k_2] = \frac{\lambda}{\sqrt{N_1N_2}} \gamma[k_1, k_2]. \tag{28}$$

Thus γ must be an eigenvector of the bivariate discrete Fourier transform $\mathcal{F}_{N_1,N_2}$ associated with the eigenvalue $\frac{\lambda}{\sqrt{N_1N_2}}$, ($\lambda = 1, -i, -1$, or $+i$). Since γ is an N_1, N_2-periodic sequence of complex numbers, we can write

$$f(x) = \sum_{m_1=-\infty}^{\infty}\sum_{m_2=-\infty}^{\infty}\sum_{n_1=0}^{N_1-1}\sum_{n_2=0}^{N_2-1} \left\{ \gamma[n_1, n_2] \cdot \delta\left(x_1 - \frac{n_1}{\sqrt{N_1}} - m_1\sqrt{N_1}, x_2 - \frac{n_2}{\sqrt{N_2}} - m_2\sqrt{N_2} \right) \right\}.$$

Case $M \neq 0$

We observe that

$$a_1 = \frac{1}{\sqrt{N_1}}(N_1, 0)^{\mathrm{T}},$$

$$N_1 a_2 - M a_1 = \sqrt{N_1}\left(M, \sqrt{N_1 N_2 - M^2}\right)^{\mathrm{T}} - M\left(\sqrt{N_1}, 0\right)^{\mathrm{T}}$$
$$= \frac{1}{\sqrt{N_1}}\left(0, N_1\sqrt{N_1 N_2 - M^2}\right)^{\mathrm{T}}.$$

Since f is a_1, a_2-periodic, then f is also $a_1, N_1 a_2 - M a_1$-periodic. Thus f has the periods

$$\alpha_1 = \sqrt{N_1}, \text{ and } \beta_2' = \sqrt{N_1\left(N_1 N_2 - M^2\right)}$$

along the x-axis and the y-axis, respectively, a situation covered by the analysis from the $M = 0$ case. In this way we prove

Theorem 2. *Let the generalized function f on $\mathbb{R}^2$ be an a_1, a_2-periodic eigenfunction of the Fourier transform operator $\mathcal{F}$ with eigenvalue $\lambda = 1, -i, -1$, or $+i$. Assume that the linearly independent periods a_1, a_2 from $\mathbb{R}^2$ have been chosen as small as possible subject to the constraint that $0 < \|a_1\| \le \|a_2\|$. Then there are positive integers $N_1 \le N_2$ such that*

$$\|a_1\| = \sqrt{N_1}, \|a_2\| = \sqrt{N_2}$$

and there is a nonnegative integer $M < N_1$ such that a_1 is orthogonal to

$$a_2' := N_1 a_2 - M a_1$$

with

$$\|a_2'\| = \sqrt{N_2'}, N_2' := N_1\left(N_1 N_2 - M^2\right).$$

The generalized function f is a_1, a_2'-periodic and there is an orthogonal transformation Q such that

$$f_Q(x) := f(Qx)$$

is $\left(\sqrt{N_1}, 0\right)^{\mathrm{T}}, \left(0, \sqrt{N_2'}\right)^{\mathrm{T}}$-periodic with the representation

$$f_Q(x) = \sum_{m_1=-\infty}^{\infty}\sum_{m_2=-\infty}^{\infty}\sum_{n_1=0}^{N_1-1}\sum_{n_2=0}^{N_2'-1}\gamma[n_1, n_2]$$
$$\cdot\delta\left(x_1 - \frac{n_1}{\sqrt{N_1}} - m_1\sqrt{N_1}, x_2 - \frac{n_2}{\sqrt{N_2'}} - m_2\sqrt{N_2'}\right).$$

Here γ is an eigenfunction of $\mathcal{F}_{N_1, N_2'}$ with

$$\left(\mathcal{F}_{N_1, N_2'}\gamma\right)[k_1, k_2]$$
$$= \frac{1}{N_1 N_2'}\sum_{n_1=0}^{N_1-1}\sum_{n_2=0}^{N_2'-1}\gamma[n_1, n_2]\cdot e^{-2\pi i\left(k_1 n_1/N_1 + k_2 n_2/N_2'\right)}$$
$$= \frac{\lambda}{\sqrt{N_1 N_2'}}\gamma[k_1, k_2]$$

for $0 \le k_1 \le N_1 - 1, 0 \le k_2 \le N_2' - 1$.

Note that the $N_1 N_2$ normalized eigenfunctions γ denoted by

$$f_{N_1, r_1, \mu_1; N_2, r_2, \mu_2}[n_1, n_2] := f_{N_1, r_1, \mu_1}[n_1]\cdot f_{N_2, r_2, \mu_2}[n_2], \quad (29)$$

with $\mu_k = 1, \cdots, M_{r_k}(N_k), k = 1, 2$ of $\mathcal{F}_{N_1, N_2}$ serve as an orthonormal basis for the $N_1 N_2$ dimensional space $\mathbb{P}_{N_1, N_2}$ of N_1, N_2-periodic discrete real valued functions. Here (29) has the corresponding eigenvalue

$$\lambda = \frac{(-i)^{r_1}}{\sqrt{N_1}}\frac{(-i)^{r_2}}{\sqrt{N_2}}, r_1, r_2 = 0, 1, 2, 3.$$

Theorem 3. *Let the generalized function f on $\mathbb{R}^3$ be an a_1, a_2, a_3-periodic eigenfunction of the Fourier transform operator $\mathcal{F}$ with eigenvalue $\lambda = 1, -i, -1$, or $+i$. Assume that the linearly independent periods a_1, a_2, a_3 from $\mathbb{R}^3$ have been chosen as small as possible subject to the constraint that $0 < \|a_1\| \le \|a_2\| \le \|a_3\|$. Then there are positive integers $N_1 \le N_2 \le N_3$ such that*

$$\|a_1\| = \sqrt{N_1}, \|a_2\| = \sqrt{N_2}, \|a_3\| = \sqrt{N_3}$$

and there are nonnegative integers

$$0 \le M_1 < N_1, 0 \le M_2 < N_1, 0 \le M_3 < N_1 + N_2$$

such that a_1,

$$a_2' := N_1 a_2 - M_1 a_1,$$

and

$$a_3' := N_1\left[\left(M_1 M_3 - N_2 M_2\right)a_1 - \left(N_1 M_3 - M_1 M_2\right)a_2 + \left(N_1 N_2 - M_1^2\right)a_3\right]$$

are pairwisely orthogonal with

$$\|a_2'\| = \sqrt{N_2'}, \|a_3'\| = \sqrt{N_3'}$$

where

$$N_2' := N_1\left(N_1 N_2 - M_1^2\right),$$

$$N_3' := N_1^2\left(N_1 N_2 - M_1^2\right)\left[N_1 N_2 N_3 + 2M_1 M_2 M_3 - \left(N_1 M_3^2 + N_2 M_2^2 + N_3 M_1^2\right)\right]$$

The generalized function f is a_1, a_2', a_3'-periodic, and there is an orthogonal transformation Q such that

$$f_Q(x) := f(Qx)$$

is

$$\left(\sqrt{N_1}, 0, 0\right)^{\mathrm{T}}, \left(0, \sqrt{N_2'}, 0\right)^{\mathrm{T}}, \left(0, 0, \sqrt{N_3'}\right)^{\mathrm{T}}$$

-periodic with the representation

$$f_Q(x) = \sum_{m_1,m_2,m_3=-\infty}^{\infty} \sum_{n_1=0}^{N_1-1} \sum_{n_2=0}^{N_2'-1} \sum_{n_3=0}^{N_3'-1} \left\{ \gamma[n_1,n_2,n_3] \cdot \delta\left(x_1 - \frac{n_1}{\sqrt{N_1}} - m_1\sqrt{N_1}, x_2 - \frac{n_2}{\sqrt{N_2'}} - m_2\sqrt{N_2'}, x_3 - \frac{n_3}{\sqrt{N_3'}} - m_3\sqrt{N_3'} \right) \right\}. \quad (30)$$

Here

$$\left(\mathcal{F}_{N_1,N_2',N_3'}\gamma\right)[k_1,k_2,k_3] = \sum_{n_1=0}^{N_1-1} \sum_{n_2=0}^{N_2'-1} \sum_{n_3=0}^{N_3'-1} \left\{ \sigma \cdot \gamma[n_1,n_2,n_3] \cdot \mathrm{e}^{-2\pi i (k_1 n_1/N_1 + k_2 n_2/N_2' + k_3 n_3/N_3')} \right\} = \frac{\lambda}{\sqrt{N_1 N_2' N_3'}} \gamma[k_1,k_2,k_3]$$

where

$$\sigma = \frac{1}{N_1 N_2' N_3'}$$

for

$$0 \le k_1 \le N_1 - 1, 0 \le k_2 \le N_2' - 1,$$

and

$$0 \le k_3 \le N_3' - 1.$$

2.2. Some Quasiperiodic Eigenfunctions of the Fourier Transform Operator on $\mathbb{R}^2$

In this section we will construct some quasiperiodic eigenfunctions of the Fourier transform operator. A generalized function f is said to be quasiperiodic if the Fourier transform $f^\wedge$ is a weighted sum of Dirac δ functionals with isolated support [10].

Lemma 1 *Let* a_1, a_2 *be linearly independent vectors in* $\mathbb{R}^2$. *If*

$$\left|\det\left[a_1 \quad a_2\right]\right| = 1,$$

and grid_{a_1,a_2} is distinct from $\mathrm{grid}^\wedge_{a_1,a_2}$, then

$$f_+(x) := \mathrm{grid}_{a_1,a_2}(x) + \mathrm{grid}^\wedge_{a_1,a_2}(x) \quad (31)$$

$$f_-(x) := \mathrm{grid}_{a_1,a_2}(x) - \mathrm{grid}^\wedge_{a_1,a_2}(x) \quad (32)$$

are eigenfunctions of the Fourier transform operator $\mathcal{F}$ associated with $\lambda = 1, \lambda = -1$, respectively.

Quasiperiodic eigenfunctions of $\mathcal{F}$ on $\mathbb{R}^2$ with *m*-fold rotational symmetry.

Let

$$\alpha = 1\Big/\sqrt{\sin\left(\frac{2\pi}{n}\right)} \quad (33)$$

for some $n = 3, 4, \cdots,$ and let

$$a_k = \alpha\left(\cos(2\pi k/n), \sin(2\pi k/n)\right)^{\mathrm{T}} \quad (34)$$

where $0 \le k \le n-1$, be the vertices of a regular $n-gon$ with center at the origin. The parameter α has been chosen so that

$$\det\left[a_k \quad a_{k+1}\right] = 1$$

for each $k = 1, 2, \cdots, n-1$. Thus

$$\mathrm{grid}^\wedge_{a_k,a_{k+1}} = \mathrm{grid}_{Qa_k,Qa_{k+1}}, k = 0, 1, \cdots, n-1$$

(with $a_n := a_0$) where

$$Q = \begin{bmatrix} 0 & -1 \\ 1 & 0 \end{bmatrix}$$

is a quarter turn rotation. We will use this fact to generate quasiperiodic eigenfunctions of $\mathcal{F}$ on $\mathbb{R}^2$ with rotational symmetry.

We will now construct a family of quasiperiodic eigenfunctions of $\mathcal{F}$ that have rotational symmetry. Let $n = 3, 4, \cdots$, and $a_k, k = 0, 1, 2, \cdots, n-1$ be given by (34), let α be given by (33), and let

$$f_{n+}(x) := \sum_{k=0}^{n-1} \mathrm{grid}_{a_k,a_{k+1}}(x) + \mathrm{grid}^\wedge_{a_k,a_{k+1}}(x), \quad (35)$$

and

$$f_{n-}(x) := \sum_{k=0}^{n-1} \mathrm{grid}_{a_k,a_{k+1}}(x) - \mathrm{grid}^\wedge_{a_k,a_{k+1}}(x), \quad (36)$$

(with $a_n := a_0$). **Figures 1** and **2** show representations of such eigenfunctions with $n = 5$ and $n = 7$ respectively. Filled circles correspond to negatively scaled Dirac δ's, and unfilled circles correspond to positively scaled Dirac δ's. The radius of each circle is proportional to the square root of the modulus of the scale factor for the corresponding δ. By construction,

$$f^\wedge_{n+} = f_{n+} \text{ and } f^\wedge_{n-} = -f_{n-}.$$

3. Representation of Some Quasiperiodic Eigenfunctions

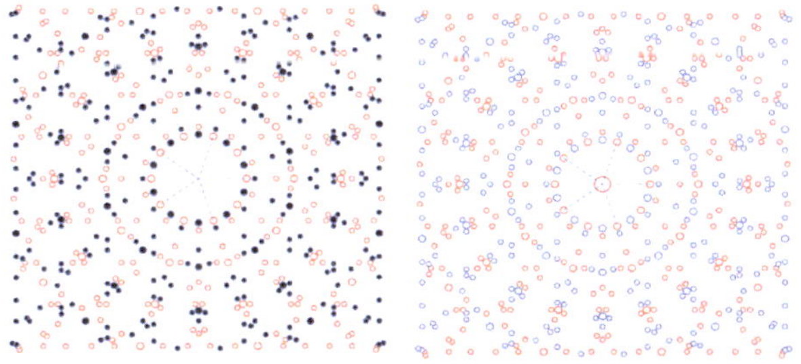

(a) (b)

Figure 1. (a) f_{5-}; (b) f_{5+}; The quasiperiodic eigenfunctions f_{5-} with 10-fold rotational symmetry, and f_{5+} with 20-fold rotational symmetry for the Fourier transform operator $\mathcal{F}$ with respectively $\lambda = -1$, and $\lambda = 1$.

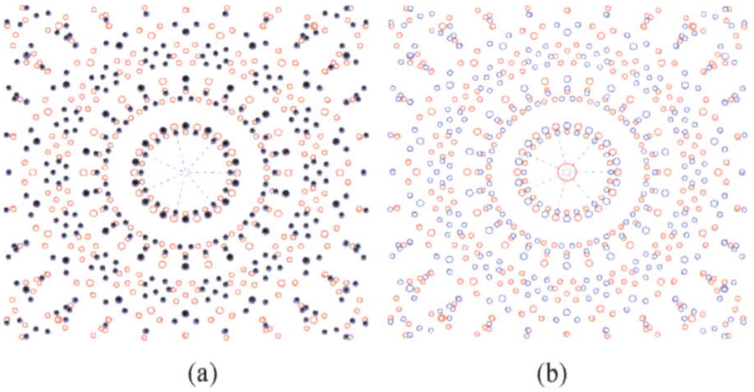

Figure 2. (a) f_{7-} ; (b) f_{7+} ; The quasiperiodic eigenfunctions f_{7-} with 14-fold rotational symmetry, and f_{7+} with 28-fold rotational symmetry for the Fourier transform operator $\mathcal{F}$ with respectively $\lambda=-1$, and $\lambda=1$.

REFERENCES

[1] R. N. Bracewell, "Fourier Analysis and Imaging," Kluwer Academic, New York, 2003.

[2] L. Schwartz, "Theorie des Distributions," Hermann, Paris, 1950.

[3] R. Strichartz, "A Guide to Distribution Theory and Fourier Transforms," CRC Press, Inc., Boca Raton, 1994.

[4] B. Osgood, "The Fourier Transform and Its Applications," Lecture Notes, Stanford University, 2005.

[5] D. W. Kammler, "A First Course in Fourier Analysis," Prentice Hall, New Jersey, 2000.

[6] J. I. Richards and H. K. Youn, "Theory of Distributions: A Non-technical Introduction," Cambridge University Press, Cambridge, 1990.

[7] M. J. Lighthill, "An Introduction to Fourier Analysis and Generalized Functions," Cambridge University Press, New York, 1958.

[8] L. Auslander and R. Tolimieri, "Is Computing with the Finite Fourier Transform Pure or Applied Mathematics?" *Bulletin of the American Mathematical Society*, Vol. 1, 1979, pp. 847-897.

[9] J. H. McClellan and T. W. Parks, "Eigenvalue and Eigenvector Decomposition of the Discrete Fourier Transform," *IEEE Transactions on Audio and Electroacoustics*, Vol. 20, No. 1, 1972, pp. 66-74.

[10] M. Senechal, "Quasicrystals and Geometry," Cambridge University Press, New York, 1995.

The Correlation and Linear Regression Analysis between Annual GDP Growth Rate and Money Laundering in Albania during the Period 2007-2011*

Llambrini Sota[1#], Fejzi Kolaneci[2]
[1]University "Pavaresia" in Vlora, Vlora, Albania
[2]University of New York in Tirana, Tirana, Albania

ABSTRACT

This study is the first attempt to investigate the relationship between the annual GDP growth rate and money laundering in the Republic of Albania during the period 2007-2011. The main result of the study: there is a negative correlation between money laundering process and economic growth rate in Albania during the specified period; there is a negative correlation between money laundering and import, but there is a positive correlation between money laundering and the government expenditure, as well a positive correlation between money laundering and export. More concretely: 1) The coefficient of correlation between cases referred in the prosecutor's office for money laundering (X) and the annual GDP growth rate (Y) is $r = -0.74$. The equation of the linear regression is $y = 7.2827 - 0.0585x$. 2) The coefficient of correlation between cases reported in the police for money laundering (X) and the annual GDP growth rate (Y) is $r = -0.81$, while the equation of the linear regression is $y = 7.3223 - 0.0249x$. 3) The coefficient of correlation between reported CAA (Cases for Alleged Activity) for money laundering (X) and the annual GDP growth rate (Y) is $r = -0.49$, while the equation of the linear regression is $y = 7.1411 - 0.0107x$. 4) The coefficient of correlation between government expenditures and the cases referred in the prosecutor's office for money laundering is $r = 0.72$. 5) The coefficient of correlation between government expenditures and the cases reported in the police for money laundering is $r = 0.79$. 6) The coefficient of correlation between net export and the cases referred in the prosecutor's office for money laundering is $r = 0.89$. 7) The coefficient of correlation between the cases referred in the prosecutor's office for money laundering and unemployment rate is $r = 0.64$. 8) The coefficient of correlation between the cases reported in the police for money laundering and unemployment rate is $r = 0.60$. 9) The coefficient of correlation between the cases referred in the prosecutor's office for money laundering and inflation rate is $r = 0.22$. 10) The coefficient of correlation between the cases reported in the police for money laundering and inflation rate is $r = 0.136$. 11) The coefficient of correlation between the cases referred in the prosecutor's office for money laundering and the investments in Albania is $r = -0.979$. 12) The coefficient of correlation between the cases reported in the police for money laundering and inflation rate is $r = -0.9736$. 13) The coefficient of correlation between the cases referred in the prosecutor's office for money laundering and the foreign direct investments in Albania is $r = 0.95$. 14) The coefficient of correlation between the cases reported in the police for money laundering and the foreign direct investments in Albania is $r = 0.8941$.

Keywords: Illicit Money; Money Laundering; GDP Growth Rate; Linear Regression; Pearson's Coefficient of Correlation

1. Introduction

Nowadays one of the major issues in the world is anti-money laundering. According to the IMF (International Monetary Funds), money laundering has become one of the most serious activities faced by the international financial community [1]. It became harmful because of its necessary coexistence with crime. The value of global

*There is an indecisive relationship between money laundering and GDP growth rate.

#Corresponding author.

money laundering amounted to 2.5 trillion USD (see [2]). However, in 2011, it is estimated to be over 3.0 trillion USD. But before going further, it is necessary to clarify some of the concepts that will be used in this paper.

Definition 1. Illicit money refers to the money that is originated from illicit activities, especially from the criminal ones [2].

Definition 2. Money laundering is the process of obscuring the source, ownership or use of funds, usually cash, that are profits of illicit activity [3].

In other words, money laundering is the process of creating the appearance that large amounts of money obtained from serious crimes, such as drug trafficking, human trafficking, weapons trafficking, corruption, counterfeiting of currency, or terrorist activity, originated from legitimate source (see [1,2]).

Definition 3. Guilty of money laundering is that a person hides or disguises the true origin, the source, movement or alienation of money, for which he (she) has knowledge that directly or indirectly derives from illegal activities, especially criminal (see [2,3]).

Remark. This definition is according to the EU's legislation, as well as UN's legislation. Furthermore, it is also used by the World Bank and Interpol.

Definition 4. GDP (Gross Domestic Product) is the market value of all finished goods and services within a country during a given period of time. The GDP is given by the formula: Y = C + I + G + NX, where C denotes private consumption, I denotes gross investments, G denotes government spending and NX denotes Net Exports = Exports − Imports (see [1,4]).

According to some recent estimation, money laundering constitutes about 6% - 8% of GDP in USA, 7% - 9% of GDP in UK, Germany, France, Italy, etc. However, some scientists claim that the amount of official data is underestimated compared to the reality of money laundering process (see [2,3]). Actually, to fight money laundering in nations like America, where the relationship between the amount of money laundering and the annual GDP growth rate is positive, can be a double-edged sword. This happens because if illicit money is not turned into legal capital, they cannot be used into the economy, but only as a capital for illegal activities. While, in other countries of Europe such as France and Germany, the relationship between money laundering and GDP is negative (see [1,2,5]). Regarding the Republic of Albania there is not, yet, any study which analyses with a mathematical statistic method the relationship between money laundering and the annual GDP growth. Hence, our study is a first oriented toward this topic.

2. Mathematical Model

The linear regression analysis is used to estimate the impact that the process of money laundering has on the economic growth, in the case of Albania over the period January 2007-December 2011. The random variable X denotes the annual number of cases of money laundering, while the random variable Y denotes the corresponding rate of annual real GDP growth. X represents the explanatory variable (input). Y represents the dependent variable (output). The sources of the data are INSTAT (Albanian Institute of Statistics), Bank of Albania (BoA), see [6-8] and General Directory for the Prevention of Money Laundering, Albanian Financial Intelligence Unit, AFIU 2011, see [9]. The **Table 1** contains the data sets for the random variables X and Y.

3. Main Results

Coefficient of correlation $r = -0.74$

Coefficient of determination d = 0.55

Linear regression equation $\boldsymbol{y = 7.2827 - 0.0585x}$

In **Figure 1** using the linear correlation and regression analysis, as well as the given data set, we obtain the following results:

The coefficient of correlation between cases referred in the prosecutor's office for money laundering and the annual GDP growth rate is $r = -0.74$, which (according to Gelfand's classification) indicates a moderate negative correlation between the two random variables. The coefficient of determination is $d = r^2 = 55\%$, which implies that 55% of the total variation in annual GDP growth rate can be explained by the variation in the number of cases of money laundering, while 45% of the total variation in GDP growth rate must be explained by the impact of other factors. The linear regression equation is $\boldsymbol{y = 7.2827 - 0.0585x}$, where x denotes the number of cases referred in prosecutor's office for money laundering and y denotes the annual GDP growth rate for Albania during the period of January 2007-December 2011.

Coefficient of correlation $r = -0.81$

Coefficient of determination d = 0.66

Linear regression equation $\boldsymbol{y = 7.3223 - 0.0249x}$

In **Figure 2** we have used the correlation analysis to measure the strength of the relationship between the cases referred in police for money laundry and annual GDP

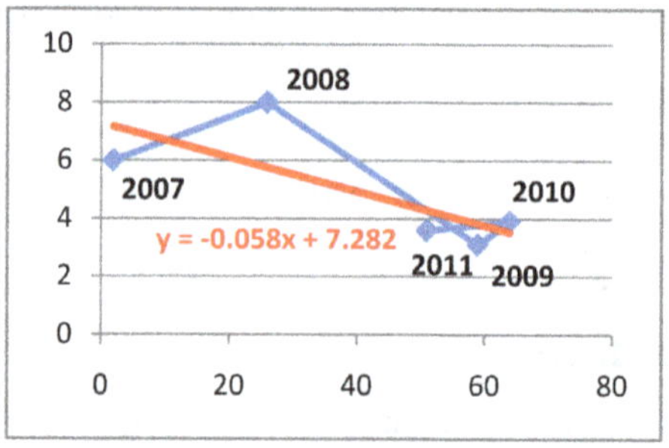

Figure 1. Relationship between the cases referred in the prosecutor's office for money laundering and the annual GDP growth rate during 2007-2011.

Table 1. Annual number of cases of money laundering and rate of annual real GDP growth or its components.

Year	Cases Referred in the Prosecutor's Office	Cases Referred in Police	reported CAA	Referred CAA	Rate of Annual Real DGP Growth (%)	Government Expenditure (million euro)	Export (million euro)	Import (million euro)	Net Export (million euro)	Annual Unemployment rate (%)	Annual Inflation rate (%)	Investments in Albania	Direct foreign Investment in Albania
2007	2	5	-	-	6.0	-	-	-	-	13.2	2.7	4454	481
2008	26	46	152	29	8.0	351492	917.5	3570.2	−2652.7	13	3.4	3505	665
2009	59	135	186	21	3.1	379863	750.6	3248.8	−2498.2	13.8	2.2	124	717
2010	64	137	211	71	3.9	362752	1171.5	3474.1	−2302.6	13.4	3.3	68	793
2011	51	160	383	115	3.6	-	1405.5	3905.3	−2499.8	13.3	3.5	480	742

The pearson's coefficient of correlation is $r = -0.81$, which (according to Gelfand's classification) indicates a strong negative correlation between the two random variables. The coefficient of determination is $d = r^2 = 66\%$. This result can be interpreted in this way: 66% of the total variation in GDP growth rate can be explained by the variation in the number of cases referred in police for money laundering. But, 34% of the total variation in GDP growth rate must be explained by the impact of other factors.

The linear regression equation is $\mathbf{y = 7.27 - 0.024x}$, where x denotes the number of cases referred in police for money laundry, and y denotes the annual GDP growth rate for Albania during the period January 2007-December 2011.

Coefficient of correlation $r = -0.49$

Coefficient of determination d = 0.24

Linear regression equation $\mathbf{y = 7.1411 - 0.0107x}$

In **Figure 3** the coefficient of correlation $r = -0.49$ indicates a moderate negative correlation between the two random variables. While, the coefficient of determination $d = r^2 = 24\%$. This means that 24% of the total variation in GDP growth rate is explained by the variation of reported CAA (Cases for Alleged Activity), whereas the other 66% is explained by the other factors.

The equation for the linear regression analysis is $\mathbf{y = 7.1411 - 0.0107x}$, where x denotes the reported CAA and y denotes the annual GDP growth rate for Albania

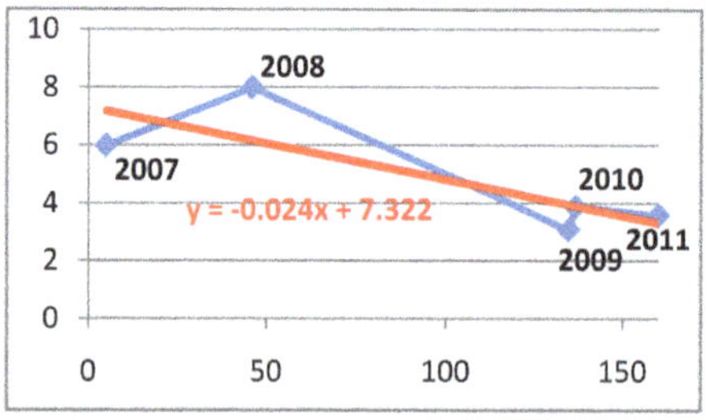

Figure 2. Relationship between the cases reported in the police for money laundering and the annual GDP growth rate during 2007-2011.

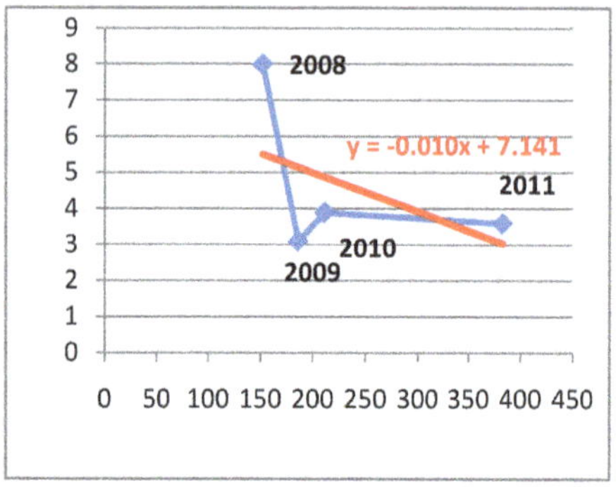

Figure 3. Relationship between reported CAA and the annual GDP growth during 2008-2011.

growth rate.

during the period January 2008- December 2011.

Coefficient of correlation $r = -0.36$

Coefficient of determination d = 0.13%

Linear regression equation $\mathbf{y = 5.7977 - 0.0195x}$

In **Figure 4** using the linear correlation and the regression analysis we have obtained that there is a relationship between the CAA (Cases for Alleged Activity) referred for money laundry and the annual GDP growth rate.

The coefficient of correlation is $r = -0.36$, which indicates a moderate negative correlation between the two random variables. While the coefficient of determination is $d = r^2 = 13\%$.

We can interpret this result in this way: 13% of the total variation in GDP growth rate can be explained by the variation in the referred CAA for money laundering. But 87% of the total variation in GDP growth rate must be explained by the impact of other factors.

The linear regression equation is $\mathbf{y = 5.7977 - 0.0195x}$, where x denotes the number of referred CAA, and y denotes the annual GDP growth rate for Albania during the period January 2008-December 2011.

Coefficient of correlation $r = 0.72$

Coefficient of determination d = 0.52

Equation of linear regression $\mathbf{y = 339877 + 499.83x}$

Coefficient of correlation $r = 0.79$

Coefficient of determination d = 0.62

Equation of linear regression $\mathbf{y = 341708 + 216.92x}$

Using the regression and correlation analysis we have concluded that there is a relationship between government spending and cases referred prosecutor's office and reported in the police for money laundering.

In **Figure 5** the coefficient of correlation is correspondingly $\mathbf{r = 0.72}$, which indicate a strong positive correlation between the variables. While, the coefficient of determination is respectively $\mathbf{d = r^2 = 52\%}$ (or $\mathbf{d = 62\%}$). This result is interpreted in this way: **52%** (or **62%**) of the total government spending can be explain by the cases referred in the prosecutor's office (or the ones reported

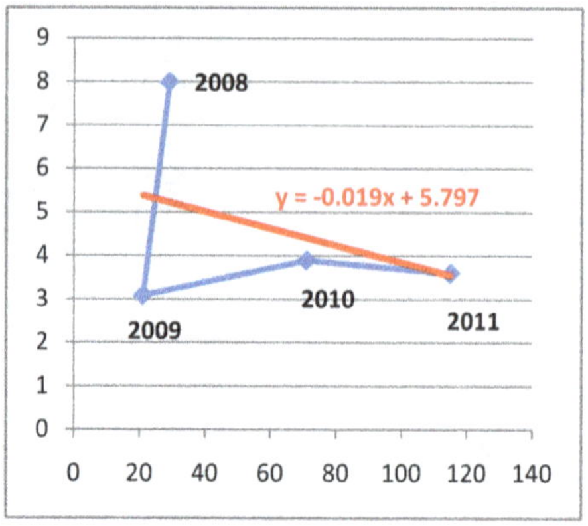

Figure 4. Relation between referred CAA and the annual GDP growth rate during 2008-2011.

in the police) for money laundering. But, 48% (or 38%) of the total government spending is explain by other factors.

The equation of the linear regression for the cases referred in the prosecutor`s office is y = **339877 + 499.83**x, where x denotes the cases referred in prosecutor's office, and y denotes the government spending in the Republic of Albania during the period of time January 2008-December 2010.

In **Figure 6** the equation of the linear regression for the cases reported in the police is y = **341708 + 216.92**x, where x denotes the cases reported in the police, and y denotes the government spending in the Republic of Albania during the period of time January 2008-December 2010.

Coefficient of correlation $r = 0.17$

Coefficient of determination d = 0.03

Equation of linear regression $y = 912.45 + 2.9765x$

Coefficient of correlation $r = 0.50$

Coefficient of determination d = 0.25

Equation of linear regression $y = 720.8 + 2.8491x$

Using the linear regression and correlation analysis we have concluded that there is a relationship between export and cases referred in the prosecutor's office and the ones reported in the police for money laundering.

The coefficient of correlation between cases referred in the prosecutor's office and the export is $r = 0.17$, while between cases reported in the police and export is $r = 0.50$, which indicates, respectively, a weak and a moderate positive correlation between the variables. The coefficient of determination is respectively **d** = r^2 = **3%** or **d = 25%.** In **Figure 7** the equation of the linear regression for the cases referred in the prosecutor's office is y = **912.45 + 2.9765**x**,** where x denotes the cases referred in the prosecutor's office for money laundering, and y denotes the export in the Republic of Albania during the period of January 2008-December 2011. In **Figure 8** the equation of the linear regression for the cases reported in the police for money laundering is y = **720.8 + 2.8491**x, where x indicates the cases reported in the police, and y indicates the export during the period of January 2008-December 2011.

Coefficient of correlation $r = -0.28$

Coefficient of determination d = 0.08

Equation of linear regression $y = 3778.1 - 4.5701x$

Coefficient of correlation $r = 0.17$

Coefficient of determination d = 0.03

Equation of linear regression $y = 3440.8 + 0.9103x$

Using the linear regression and correlation analysis we have concluded that there is a relationship between import and cases referred in the prosecutor's office, as well as the ones reported in the police for the money laundering.

The coefficient of correlation is respectively $r = -0.28$ **(or** $r = 0.17$**)**, which indicates, correspondingly, a weak negative and positive correlation between the given values. The coefficient of determination is respectively **d** = r^2 = **8%** or **d = 3%.**

In **Figure 9** the equation of the linear regression for the cases referred in the prosecutor's office is y = **3778.1 − 4.5701**x, where x denotes the cases referred in the

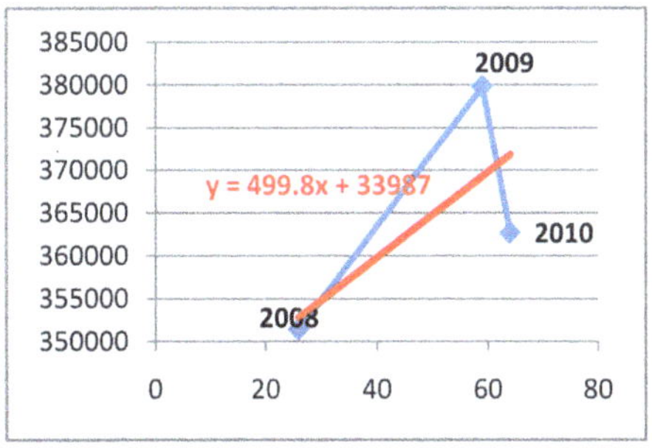

Figure 5. Relation between government spending and cases referred in prosecutor's office during 2008-2010.

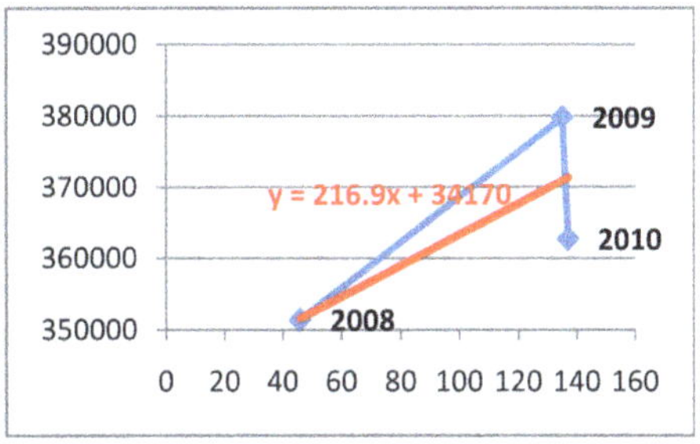

Figure 6. Relationship between government spending and cases referred in the police during 2008-2010.

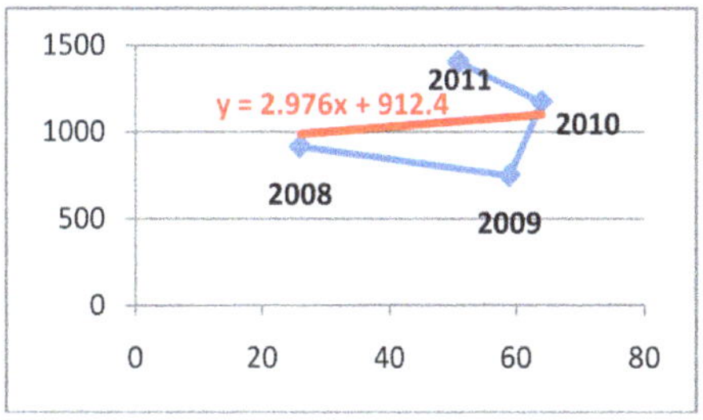

Figure 7. Relation between export and cases referred in prosecutor's office for money laundering.

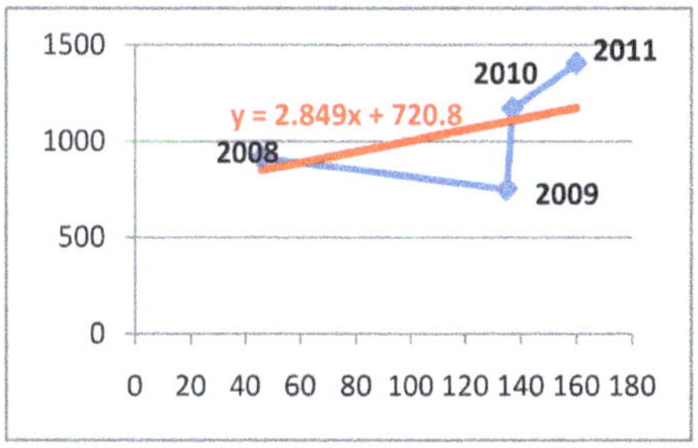

Figure 8. Relationship between export and cases reported in the police for money laundering during 2008-2011.

prosecutor's office for money laundering, and y denotes the import in the Republic of Albania during the period of January 2008-December 2011. In **Figure 10**, the equation of the linear regression for the cases reported in the police for money laundering is y = **3440.8 + 0.9103**x, where x indicates the cases reported in the police, and y indicates the import during the period of January 2008-December 2011.

Coefficient of correlation $r = 0.89$

Coefficient of determination d = 0.79

Equation of linear regression $y = -2865.7 + 7.5466x$

Coefficient of correlation $r = 0.68$

Coefficient of determination d = 0.46

Equation of linear regression $y = -2720 + 1.9389x$

Using the regression and correlation analysis we have concluded that there is a relationship between net export and cases referred in the prosecutor's office and reported in the police for money laundering.

In **Figure 11**, the coefficient of correlation is correspondingly $r = 0.89$, which indicate a strong and moderate positive correlation between the variables. While, the coefficient of determination is respectively $d = r^2 = 79\%$ (or **d = 46%**). This result is interpreted in this way: **79% (or 46%)** of the total net export can be explained by the cases referred in prosecutor's office (or the ones reported in the police) for money laundering. But, 21% (or 54%) of the total net export is explain by other factors.

The equation of the linear regression for the cases referred in the prosecutor's office is $y = -2865.7 + 7.5446x$, where x denotes the cases referred in prosecutor's office, and y denotes the net export in the Republic of Albania during the period of time January 2008-December 2011.

In **Figure 12** the equation of the linear regression for the cases reported in the police is $y = -2720 + 1.9389x$, where x denotes the cases reported in the police, and y denotes the net export in the Republic of Albania during the period of time January 2008-December 2011.

Coefficient of correlation $r = 0.22$

Coefficient of determination d = 0.05

Equation of linear regression $y = 0.0012x + 2.9712$

Coefficient of correlation $r = 0.136$

Coefficient of determination d = 0.02

Equation of linear regression $y = 0.0011x + 2.9113$

Using the regression and correlation analysis we have concluded that there is a relationship between annual inflation rate and the cases referred in the prosecutor's office and reported in the police for money laundering.

The coefficient of correlation is correspondingly $r = 0.22$ (or $r = 0.136$), which indicate a weak positive correlation between the variables. While, the coefficient of determination is respectively $d = r^2 = 5\%$ (or **d = 2%**). This result is interpreted in this way: **5% (or 2%)** of the total net export can be explained by the cases referred in prosecutor's office (or the ones reported in the police) for money laundering. But, 95% (or 98%) of the total net export is explain by other factors.

In **Figure 13** the equation of the linear regression for the cases referred in the prosecutor's office is $y = 0.0012x + 2.9712$, where x denotes the cases referred in prosecutor's office, and y denotes the net export in the Republic of Albania during the period of time

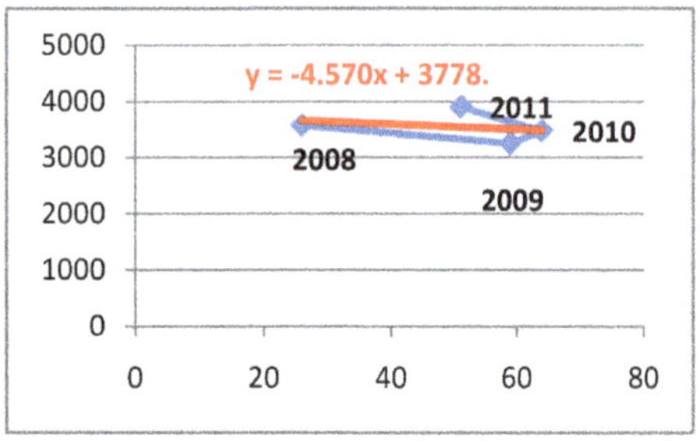

Figure 9. Relationship between import and cases referred in prosecutor's office for money laundering.

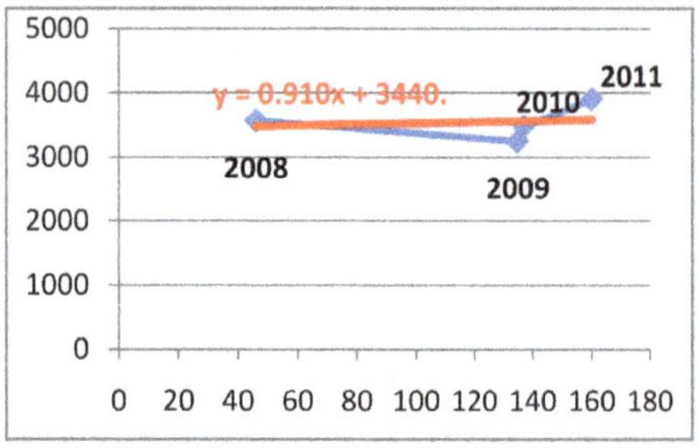

Figure 10. Relationship between import and cases reported in the police for money laundering.

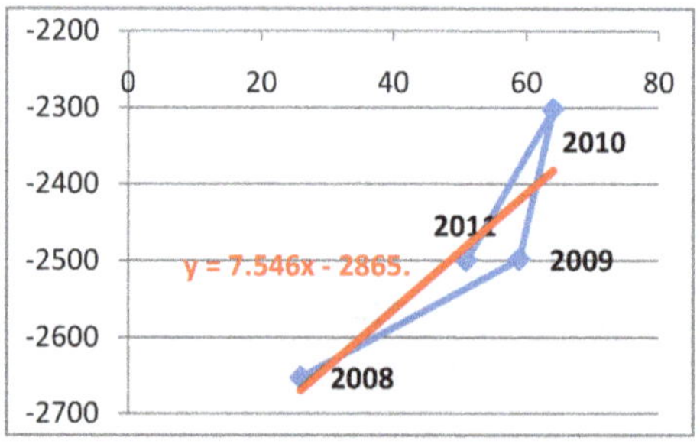

Figure 11. Relationship between net export and cases referred in prosecutor's office for money laundering.

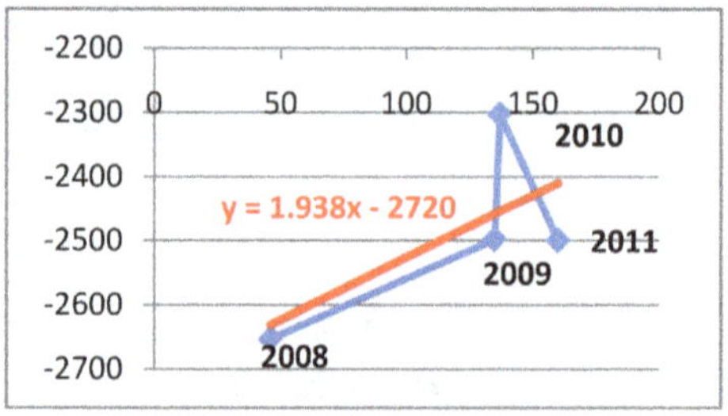

Figure 12. Relationship between net export and cases reported in the police for money laundering during 2008-2011.

January 2008- December 2011.

In **Figure 14** the equation of the linear regression for the cases reported in the police is $\mathbf{y = 0.0011x + 2.9113}$ where x denotes the cases reported in the police, and y denotes the net export in the Republic of Albania during the period of time January 2008-December 2011.

Coefficient of correlation $r = 0.64$

Coefficient of determination d = 0.41

Equation of linear regression $\mathbf{y = -710.73 + 56.307x}$

Coefficient of correlation $r = 0.60$

Coefficient of determination d = 0.36

Equation of linear regression $\mathbf{y = 13.086 + 0.0026x}$

Using the regression and correlation analysis we have concluded that there is a relationship between unemployment and cases referred in the prosecutor's office and reported in the police for money laundering.

The coefficient of correlation is correspondingly $\mathbf{r = 0.64}$ (or $\mathbf{r = 0.60}$), which indicate a moderate positive correlation between the variables. While, the coefficient of determination is respectively $\mathbf{d = r^2 = 41\%}$ (or $\mathbf{d = 36\%}$). This result is interpreted in this way: **41% (or 36%)** of the total unemployment can be explained by the cases referred in prosecutor's office (or the ones reported in the police) for money laundering. But, 59% (or 64%) of the total unemployment is explain by other factors, which were not taken in consideration during this study.

In **Figure 15** the equation of the linear regression for the cases referred in the prosecutor's office is $\mathbf{y = -710.73 + 56.307x}$, where x denotes the cases referred in prosecutor's office, and y denotes the unemployment in the Republic of Albania during the period of time January 2008-December 2011.

In **Figure 16** the equation of the linear regression for the cases reported in the police is $\mathbf{y = 13.086 + 0.0026x}$, where x denotes the cases reported in the police, and y denotes the unemployment in the Republic of Albania during the period of time January 2008- December 2011.

Coefficient of correlation $\mathbf{r = -0.979}$

Coefficient of determination $\mathbf{d = 0.9584}$

Equation of linear regression $\mathbf{y = 4909.2 - 78.788x}$

Coefficient of correlation $r = -0.9736$

Coefficient of determination d = 0.9479

Equation of linear regression $\mathbf{y = 4651.4 - 30.281x}$

Using the regression and correlation analysis we have concluded that there is a relationship between investments and cases referred in the prosecutor's office and reported in the police for money laundering. The coefficient of correlation between the cases reported in the prosecutor's office and investments is $\mathbf{r = -0.979}$, while the relationship between the cases reported in the police and the investments is $\mathbf{r = -0.9736}$, which indicate a strong negative e correlation between the variables. While, the coefficient of determination is respectively $\mathbf{d = r^2 = 0.9584}$ or $\mathbf{d = 0.9479}$.

In **Figure 17** the equation of the linear regression for the cases referred in the prosecutor's office is $\mathbf{y = 4909.2 - 78.788x}$, where x denotes the cases referred in prosecutor's office, and y denotes the unemployment in the Republic of Albania during the period of time January 2008-December 2011.

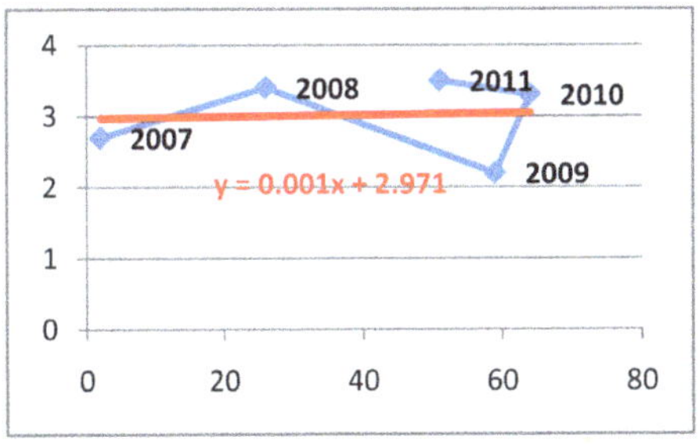

Figure 13. The relationship between the cases referred in the prosecutor's office and the annual inflation rate in Albania.

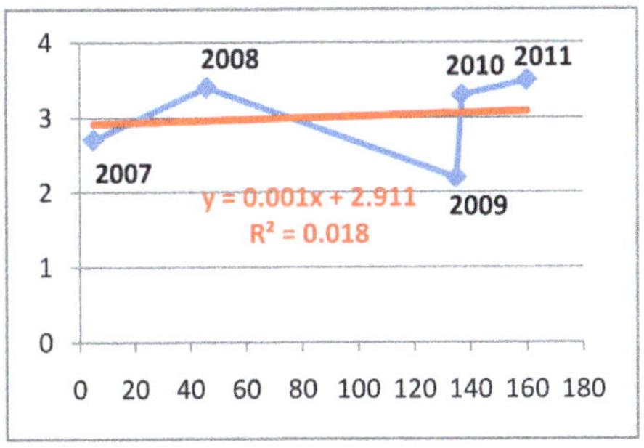

Figure 14. Relation between cases reported in the police for money laundering and the annual inflation rate in Albania.

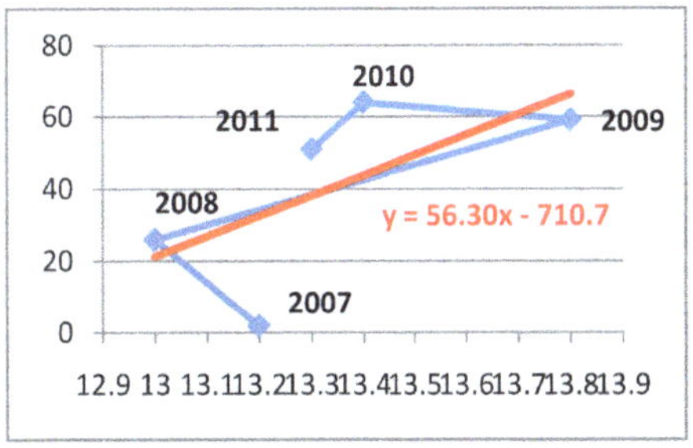

Figure 15. Relationship between cases referred in prosecutor's office and the annual unemployment rate in Albania.

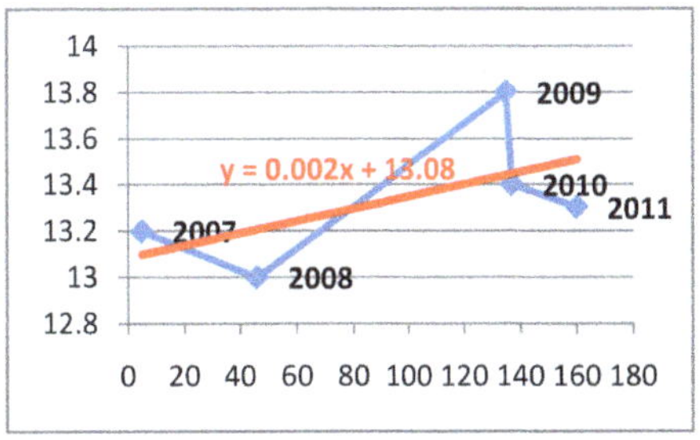

Figure 16. Relationship between cases reported in the police and the annual unemployment rate in Albania.

In **Figure 18** the equation of the linear regression for the cases reported in the police is $y = 4651.4 - 30.281x$, where x denotes the cases reported in the police, and y denotes the unemployment in the Republic of Albania during the period of time January 2008- December 2011.

Coefficient of correlation $r = 0.95$

Coefficient of determination d = 0.9036

Equation of linear regression $y = 501.81 + 4.4008x$

Coefficient of correlation $r = 0.8941$

Coefficient of determination d = 0.7995

Equation of linear $y = 525.14 + 1.5989x$

Using the regression and correlation analysis we have concluded that there is a relationship between the foreign direct investments and cases referred in the prosecutor's office and reported in the police for money laundering.

The coefficient of correlation between the cases reported in the prosecutor's office and the foreign direct investments is $r = 0.9505$, while the relationship between the cases reported in the police and the foreign direct investments is $r = 0.894$, which indicate a strong positive correlation between the variables. While, the coefficient of determination is respectively $d = r^2 = 0.9036$ or **d = 7995**.

In **Figure 19** the equation of the linear regression or the cases referred in the prosecutor's office is $y = 501.81 + 4.4008x$, where x denotes the cases referred in prosecutor's office, and y denotes the unemployment in the Republic of Albania during the period of time January 2008-December 2011.

In **Figure 20** the equation of the linear regression for the cases reported in the police is $y = 525.14 + 1.5989x$, where x denotes the cases reported in the police, and y denotes the unemployment in the Republic of Albania during the period of time January 2008-December 2011.

The main difficulty (and restriction) of the study is missing of the official data regarding to the money laundering before 2007 in Albania.

Let us reconsider the relationship between the annual number of cases referred in prosecutor's office for money laundering (denoted by x) and the rate of annual real GDP growth (denoted by y) during the period 1 January 2007-31 December 2011 in Albania.

Given the data set containing n = 5 observations:
$x_1 = 2$, $x_2 = 26$, $x_3 = 59$, $x_4 = 64$, $x_5 = 51$,
$y_1 = 6$, $y_2 = 8$, $y_3 = 3.1$, $y_4 = 3.9$, $y_5 = 3.6$,
see **Table 1**.

Compute the following statistics:

$$S_{xx} = \sum_{k=1}^{5}(x_k - \bar{x})^2 = 2697.2$$

$$S_{xy} = \sum_{k=1}^{5}(x_k - \bar{x})(y_k - \bar{y}) = -157.74\,.$$

$$S_{yy} = \sum_{k=1}^{5}(y_k - \bar{y})^2 = 16.748\,.$$

$$SSE = S_{yy} - S_{xy} = 174.488$$

$$S^2 = \frac{SSE}{3} = 58.163\,.$$

$$S = \sqrt{\frac{SSE}{3}} = 7.626$$

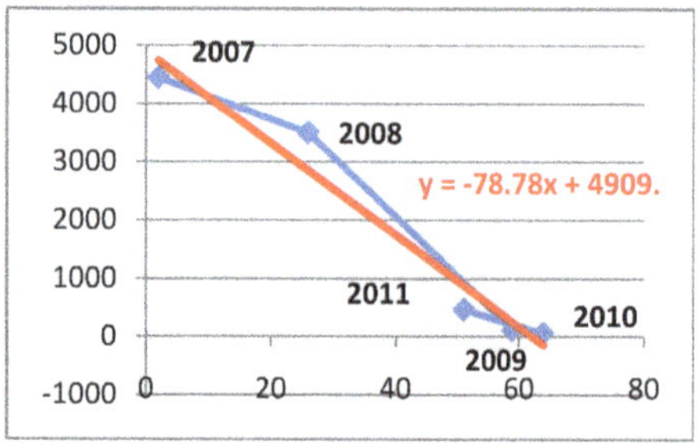

Figure 17. Relation between the cases referred in the prosecutor's office and the investments in Albania.

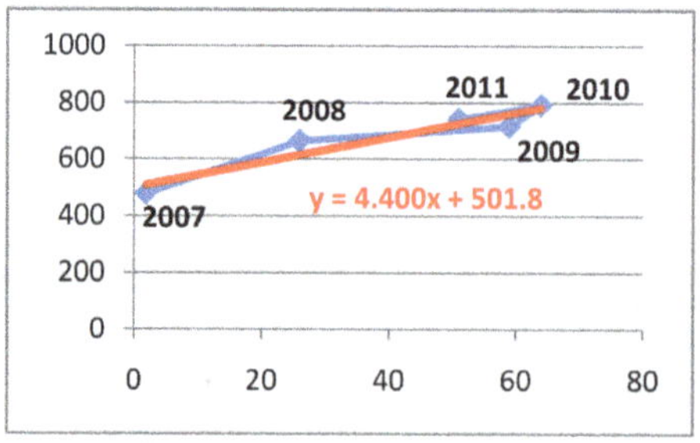

Figure 19. The relationship between cases referred in the prosecutor's office and the foreign direct investments.

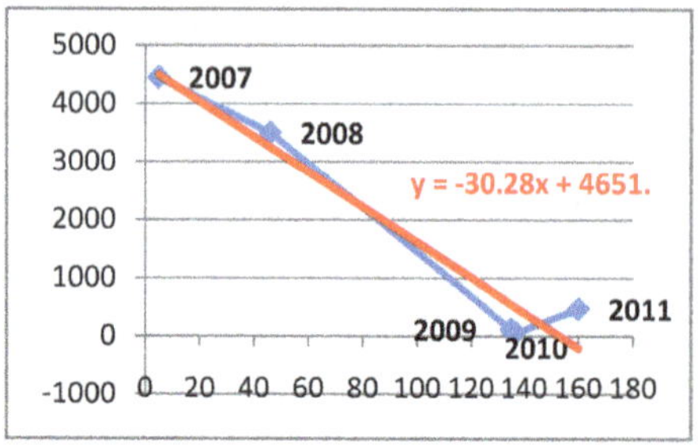

Figure 18. Relationship between the cases reported in the police and the investments in Albania.

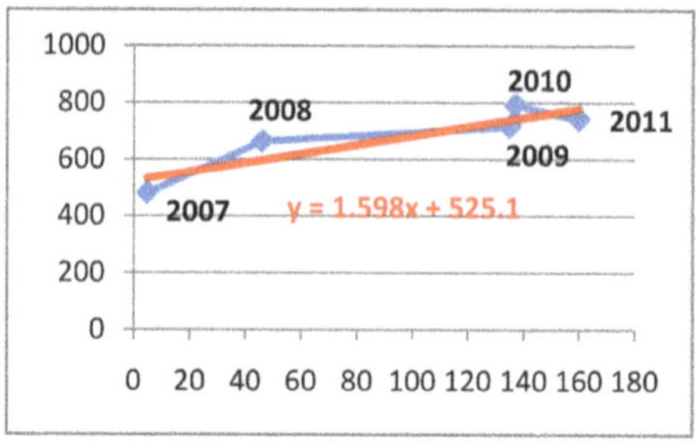

Figure 20. The relationship between cases reported in the police and the foreign direct investments in Albania.

The simple linear regression equation is

$$y = \beta_0 + \beta_1 x + \varepsilon ,$$

where ε denotes the random error term, see Bolton and David (2002).

Assume that the mean $E(\varepsilon) = 0$ and the variance $V(\varepsilon) = \sigma^2$. An estimation of the unknown variance σ^2 is S^2. If the random error ε is normally distributed, then both β_0 and β_1 are normally distributed. Furthermore, the random variable S^2 is independent of both β_0 and β_1.

Compute: $\sum_{k=1}^{5} x_k^2 = 10858$,

$$C_{00} = \frac{\sum_{k=1}^{5} x_k^2}{5S_{xx}} = 0.805,$$

$$C_{11} = \frac{1}{S_{xx}} = 0.00037,$$

and the critical value of t-distribution

$$t_c = t_{0.05}(3) = 3.182.$$

A 95% confidence interval for β_0 is

$$\beta_0 + t_{0.05}(3) S \sqrt{C_{00}}.$$

A 95% confidence interval for β_1 is

$$\beta_1 + t_{0.05}(3) S \sqrt{C_{11}}..$$

We obtain the following results:

$P(-14.49 < \beta_0 < 29.05) = 95\%$,

$P(-0.5251 < \beta_1 < 0.4081) = 95\%$.

Given the significance level $\alpha = 0.05$

Test the hypothesis

H_0: $\beta_1 = 0$,

H_a: $\beta_1 \neq 0$. (two - tailed test)

The appropriate test statistics is "Student's" t-distribution

$$t = \frac{\beta_1}{S\sqrt{C_{11}}}$$

The observed value of the test statistics is

$$t = \frac{-0.0585}{7.260.01923} = -0.402$$

and the critical value of the test statistics is $t_c = 3.182$.

Decision Rule:

$$|t| = 0.402 < t_c = 3.182$$

Accept the null hypothesis H_0: $\beta_1 = 0$ at the 95% confidence level.

A 95% confidence interval for the expected value

$$E(y) = \beta_0 + \beta_1 x^* ,$$

where x^* denotes a particular value of x, is given by the formula:

$$\beta_0 + \beta_1 x^* \pm t_c S \sqrt{\frac{1}{n} + \frac{(x^* - \overline{x})^2}{S_{xx}}}.$$

see [5].

We obtain:

$$7.2827 - 0.0585 x^* \pm 24.066 \sqrt{0.2 + \frac{(40.4 - x^*)^2}{2697.2}}$$

A 95% prediction interval for y when $x = x^*$ is given by the formula:

$$\beta_0 + \beta_1 x^* \pm t_c S \sqrt{1 + \frac{1}{n} + \frac{(x^* - \overline{x})^2}{S_{xx}}}.$$

see [5].

We obtain the following results:

$$7.2827 - 0.0585 x^* \pm 24.266 \sqrt{1.2 + \frac{(40.4 - x^*)^2}{2697.2}}.$$

Let consider the relationship between the annual number of cases reported in the police for money laundering (denoted by x) and the rate of annual real GDP growth (denoted by y) during the period 1 January 1997-31 December 2011 in Albania, see **Table 1**.

Given the data set containing $n = 5$ observations:

$x_1 = 5$, $x_2 = 46$, $x_3 = 135$, $x_4 = 137$, $x_5 = 160$,

$y_1 = 6$, $y_2 = 8$, $y_3 = 3.1$, $y_4 = 3.9$, $y_5 = 3.6$,

Compute:

$S_{xx} = 18077.2$,

$S_{xy} = -449.56$,

$S_{yy} = 16.748$,

$SSE = 466.308$,

$S = 12.467$, $\sum_{k=1}^{5} x_k^2 = 64735$,

$\beta_0 = 7.322335$, $\beta_1 = -0.02487$,

$C_{00} = 0.7162$, $\sqrt{C_{00}} = 0.8463$,

$C_{11} = 0.0000553$, $\sqrt{C_{11}} = 0.00744$,

$$t_c = 3.182.$$

A 95% confidence interval for β_0 is : (−26.248; 40.892).

A 95% confidence interval for β_1 is : (−0.320; 0.270).

Given the significance level $\alpha = 0.05$.

Test the hypothesis

H_0: $\beta_1 = 0$,

H_a: $\beta_1 \neq 0$. (two - tailed test)

The observed value of the test statistics is

$$t = \frac{\beta_1}{S\sqrt{C_{11}}} = -0.268$$

The critical value of the test statistics is

$$t_c = 3.182$$

Decision Rule

$$|t| = 0.268 < t_c = 3.182$$

Accept the null hypothesis H_0: $\beta_1 = 0$ at the 95% level of confidence.

A 95% confidence interval for the expected value

$$E(y) = \beta_0 + \beta_1 x^*,$$

where x^* denotes a particular value of x is

$$7.322 - 0.02487x^* \pm 39.671\sqrt{0.2 + \frac{(x^* - 6.6)^2}{18077.2}}.$$

A 95% prediction interval for y when $x = x^*$ is given by the formula

$$7.322 - 0.02487x^* \pm 39.671\sqrt{1.2 + \frac{(x^* - 6.6)^2}{18077.2}}.$$

Similarly, we can develop the confidence intervals and hypothesis testing for the parameters β_0 and β_1 of simple linear regression equation in all relationships considered in the paper.

Remark

The probability distribution for Pearson coefficient of correlation r is difficult to obtain in the small sample case. For large random samples this difficulty could be overcome by using the Fisher z - transformation, see [5].

4. Conclusions

By analyzing the official data available to us during the period of 2007-2011 for the Republic of Albania, it is possible to conclude that the economic growth is negatively correlated with the number of cases of money laundering, because as we have seen in the cases above mentioned the coefficient of correlation "r" is a negative number. However, further researches may be necessary to draw light on this topic, as we had no access to data regarding the exact amount of money laundering. The only information available to us was the number of cases of money laundering. With this information at our disposal, it is not possible to clearly state the percentage that illicit money occupies on the annual GDP growth rate over January 2007-December 2011. Actually, in other countries, these data are published and available for scientific studies, for instance, in America, Germany, and other 11 countries that have done similar researches. But, in Albania the amount of money laundering is not of public domain. Nevertheless, in our case, where the correlation between the two variables X and Y is negative, anti-money laundering policies discourage criminal activities.

Some surprising results of this study are:

1) The positive correlation between government expenditure and the number of cases referred in the prosecutor's office for money laundering ($r = 0.72$);

2) The positive correlation between government expenditure and the number of cases reported in police for money laundering ($r = 0.79$);

3) The positive correlation between export and the number of cases referred in the prosecutor's office for money laundering ($r = 0.17$);

4) The positive correlation between export and the number of cases reported in police for money laundering ($r = 0.50$);

5) The positive correlation between the number of cases referred in the prosecutor's office for money laundering and net export ($r = 0.89$);

6) The positive correlation between the number of cases reported in police for money laundering and net export ($r = 0.68$);

7) The positive correlation between the number of cases referred in the prosecutor's office for money laundering and annual unemployment rate ($r = 0.64$);

8) The positive correlation between the number of cases reported in the police and the annual unemployment rate ($r = 0.60$);

9) The positive correlation between the number of cases referred in the prosecutor's office for money laundering and the inflation rate ($r = 0.22$);

10) The positive correlation between the number of cases reported in the police and the inflation rate ($r = 0.136$);

11) The positive correlation between the number of cases referred in the prosecutor's office for money laundering and the foreign direct investments in Albania ($r = 0.95$);

12) The positive correlation between the number of cases reported in the police and the foreign direct investments in Albania ($r = 0.8941$).

Actually, it is not possible to give a scientific argument for these surprising results, because officially data for the exact amount of money laundering during the years 2007, 2008, 2009, 2010, and 2011 are not published in the Republic of Albania.

REFERENCES

[1] J. Ferweda, "The Economics of Crime and Money Laun-

dering: Does Anti-Money Laundering Policy Reduce Crime?" *Review of Law & Economics*, Vol. 5, No. 2, 2008. http://ideas.repec.org/p/use/tkiwps/0835.html

[2] F. G. Schneider, "Turnover of Organized Crime and Money Laundering: Some Preliminary Empirical Findings," *Public Choice*, Vol. 144, No. 3, 2010, pp. 473-486. http://EconPapers.repec.org/RePEc:kap:pubcho:v:144:y:2010:i:3:p:473-486

[3] R. J. Bolton and D. J. David, "Statistical Fraud Detection: A Review," *Institute of Mathematical Statistics*, Vol. 19, No. 3, 2002, pp. 235-249.

[4] I. Stancu and D. Rece, "The Relationship between Economic Growth and Money Laundering: A Linear Regression Model," *Theoretical and Applied Economics*, Vol. 9, No. 9, 2009, pp. 3-8. http://www.doaj.org/doaj?func=abstract&id=452320

[5] K. M. Ramachandran and C. P. Tsoksos, "Mathematical Statistics with Applications," Academic Press, London, 2009.

[6] Bank of Albania, "Eksport of Goods According to Partners," Bank of Albania, Republic of Albania, 2011.

[7] Bank of Albania, "Governments Expenditures," Bank of Albania, Republic of Albania, 2011.

[8] Bank of Albania, "Imports of Goods According to Partners," Bank of Albania, Republic of Albania, 2011.

[9] General Directorate for the Prevention of Money Laundering, "Albanian Financial Intelligence Unit (AFIU)," Shtepiabotuese "Botart", Republic of Albania, 2011.

Stochastic Oscillators with Quadratic Nonlinearity Using WHEP and HPM Methods

Amnah S. Al-Johani[1,2]
[1]Department of Applied Mathematics, College of Science, Northern Borders University, Arar, Saudi Arabia
[2]College of Home Economics, Northern Borders University, Arar, Saudi Arabia

ABSTRACT

In this paper, quadratic nonlinear oscillators under stochastic excitation are considered. The Wiener-Hermite expansion with perturbation (WHEP) method and the homotopy perturbation method (HPM) are used and compared. Different approximation orders are considered and statistical moments are computed in the two methods. The two methods show efficiency in estimating the stochastic response of the nonlinear differential equations.

Keywords: Nonlinear Stochastic Differential Equations; Wiener-Hermite Expansion; WHEP Technique; Homotopy Perturbation Method

1. Introduction

Quadrate oscillation arises through many applied models in applied sciences and engineering when studying oscillatory systems [1]. These systems can be exposed to a lot of uncertainties through the external forces, the damping coefficient, the frequency and/or the initial or boundary conditions. These input uncertainties cause the output solution process to be also uncertain. For most of the cases, getting the probability density function (p.d.f.) of the solution process may be impossible. So, developing approximate techniques through which approximate statistical moments can be obtained, is an important and necessary work.

Since Meecham and his co-workers [2] developed a theory of turbulence involving a truncated Wiener-Hermite expansion (WHE) of the velocity field, many authors studied problems concerning turbulence [3-8]. A lot of general applications in fluid mechanics were also studied in [9-11]. Scattering problems attracted the WHE applications through many authors [12-16]. The nonlinear oscillators were considered as an opened area for the applications of WHE as can be found in [17-23]. There are a lot of applications in boundary value problems [24, 25] and generally in different mathematical studies [26-29]. The WHE properties and description of its usage are given in [30].

In HPM technique [31-34], the response of nonlinear differential equations can be obtained analytically as a series solution. The basic idea of the homotopy method is to deform continuously a simple problem (and easy to solve) into the difficult problem under study [35]. The HPM method is a special case of homotopy analysis method (HAM) propounded by Liao in 1992 [36]. The HAM was systematically described in Liao's book in 2003 [37] and was applied by many authors in [38-41]. The HAM method possesses auxiliary parameters and functions which can control the convergence of the obtained series solution.

The stochastic oscillator with cubic nonlinearity (Duffing oscillator) was considered in [17,42]. The nonlinear term is due to the restoring nonlinear force. In some applications, the restoring force is quadratic and it is required to estimate the response in this case. The main goal of this paper is to consider the quadratic nonlinear oscillator under stochastic excitation. The WHEP and HPM methods are used and compared.

This paper is organized as follows. The problem formulation is outlined in Section 2. The WHEP technique is described and applied to the stochastic quadratic oscillator in Section 3. The HPM is outlined in Section 4 and applied also to the quadratic oscillator. A comparison between the two methods is shown in Section 5.

2. Problem Formulation

In this section, the following quadratic nonlinear oscillatory equation is considered:

$$\ddot{x}(t;\omega)+2w\zeta\dot{x}+w^2x+\varepsilon w^2x^2=F(t;\omega),\ t\in[0,T] \quad (1)$$

under stochastic excitation $F(t;\omega)$ with deterministic initial conditions

$$x(0)=x_0,\quad \dot{x}(0)=\dot{x}_0,$$

where

w: frequency of oscillation,

ζ : damping coefficient,

ε : deterministic nonlinearity scale,

$\omega\in(\Omega,\sigma,p)$: a triple probability space with Ω as the sample space, σ is a σ-algebra on events in Ω and P is a probability measure.

3. WHEP Technique

The application of the WHE aims at finding a truncated series solution to the solution process of differential equations. The truncated series composes of two major parts; the first is the Gaussian part which consists of the first two terms, while the rest of the series constitute the non-Gaussian part. In nonlinear cases, there exists always difficulties of solving the resultant set of deterministic integro-differential equations got from the applications of a set of comprehensive averages on the stochastic integro-differential equation obtained after the direct application of WHE. Many authors introduced different methods to face these obstacles. Among them, the WHEP technique was introduced in [22] using the perturbation technique to solve perturbed nonlinear problems.

The WHE method utilizes the Wiener-Hermite polynomials which are the elements of a complete set of statistically orthogonal random functions [30]. The Wiener-Hermite polynomial $H^{(i)}(t_1,t_2,\cdots,t_i)$ satisfies the following recurrence relation:

$$H^{(i)}(t_1,t_2,\cdots,t_i)=H^{(i-1)}(t_1,t_2,\cdots,t_{i-1})\cdot H^{(1)}(t_i)$$
$$-\sum_{m=1}^{i-1}H^{(i-2)}(t_{i_1},t_{i_2},\cdots,t_{i_{i-2}})\cdot\delta(t_{i-m}-t_i),\ i\geq 2 \quad (2)$$

where

$$H^{(0)}=1,\quad H^{(1)}(t)=n(t),\quad H^{(2)}(t_1,t_2)=H^{(1)}(t_1)\cdot H^{(1)}(t_2)-\delta(t_1-t_2),$$
$$H^{(3)}(t_1,t_2,t_3)=H^{(2)}(t_1,t_2)\cdot H^{(1)}(t_3)-H^{(1)}(t_1)\cdot\delta(t_2-t_3)-H^{(1)}(t_2)\cdot\delta(t_1-t_3), \quad (3)$$
$$H^{(4)}(t_1,t_2,t_3,t_4)=H^{(3)}(t_1,t_2,t_3)\cdot H^{(1)}(t_4)-H^{(2)}(t_1,t_2)\cdot\delta(t_3-t_4)-H^{(2)}(t_1,t_3)\cdot\delta(t_2-t_4)-H^{(2)}(t_2,t_3)\cdot\delta(t_1-t_4),$$

in which $n(t)$ is the white noise with the following statistical properties

$$En(t)=0,\quad En(t_1)\cdot n(t_2)=\delta(t_1-t_2), \quad (4)$$

where $\delta(.)$ is the Dirac delta function and E denotes the ensemble average operator.

The Wiener-Hermite set is a statistically orthogonal set, *i.e.*

$$EH^{(i)}\cdot H^{(j)}=0\quad \forall i\neq j. \quad (5)$$

The average of almost all H functions vanishes, particularly,

$$EH^{(i)}=0\ \text{ for } i\geq 1. \quad (6)$$

Due to the completeness of the Wiener-Hermite set, any random function $G(t;\omega)$ can be expanded as

$$G(t;\omega)=G^{(0)}(t)+\int_{-\infty}^{\infty}G^{(1)}(t;t_1)H^{(1)}(t_1)\mathrm{d}t_1+\int_{-\infty}^{\infty}\int_{-\infty}^{\infty}G^{(2)}(t;t_1,t_2)H^{(2)}(t_1,t_2)\mathrm{d}t_1\mathrm{d}t_2$$
$$+\int_{-\infty}^{\infty}\int_{-\infty}^{\infty}\int_{-\infty}^{\infty}G^{(3)}(t;t_1,t_2,t_3)H^{(3)}(t_1,t_2,t_3)\mathrm{d}t_1\mathrm{d}t_2\mathrm{d}t_3+\cdots \quad (7)$$

where the first two terms are the Gaussian part of $G(t;\omega)$. The rest of the terms in the expansion represent the non-Gaussian part of $G(t;\omega)$. The average of $G(t;\omega)$ is

$$\mu_G=EG(t;\omega)=G^{(0)}(t) \quad (8)$$

The covariance of $G(t;\omega)$ is

$$\mathrm{Cov}(G(t;\omega),G(\tau;\omega))=E(G(t;\omega)-\mu_G(t))(G(\tau;\omega)-\mu_G(\tau))$$
$$=\int_{-\infty}^{\infty}G^{(1)}(t;t_1)G^{(1)}(\tau,t_1)\mathrm{d}t_1+2\int_{-\infty}^{\infty}\int_{-\infty}^{\infty}G^{(2)}(t;t_1,t_2)G^{(2)}(\tau,t_1,t_2)\mathrm{d}t_1\mathrm{d}t_2 \quad (9)$$
$$+2\int_{-\infty}^{\infty}\int_{-\infty}^{\infty}\int_{-\infty}^{\infty}G^{(3)}(t;t_1,t_2,t_3)\left[G^{(3)}(\tau,t_1,t_2,t_3)+G^{(3)}(\tau,t_1,t_3,t_2)+G^{(3)}(\tau,t_2,t_3,t_1)\right]\mathrm{d}t_1\mathrm{d}t_2\mathrm{d}t_3+\cdots$$

The variance of $G(t;\omega)$ is

$$\operatorname{Var}G(t;\omega)=E\left(G(t;\omega)-\mu_G(t)\right)^2=\int_{-\infty}^{\infty}\left[G^{(1)}(t;t_1)\right]^2\mathrm{d}t_1+2\int_{-\infty}^{\infty}\int_{-\infty}^{\infty}\left[G^{(2)}(t;t_1,t_2)\right]^2\mathrm{d}t_1\mathrm{d}t_2$$
$$+2\int_{-\infty}^{\infty}\int_{-\infty}^{\infty}\int_{-\infty}^{\infty}\left[G^{(3)}(t;t_1,t_2,t_3)\right]^2\mathrm{d}t_1\mathrm{d}t_2\mathrm{d}t_3+2\int_{-\infty}^{\infty}\int_{-\infty}^{\infty}\int_{-\infty}^{\infty}\left[G^{(3)}(t,t_1,t_2,t_3)G^{(3)}(t,t_1,t_3,t_2)\right]\mathrm{d}t_1\mathrm{d}t_2\mathrm{d}t_3 \tag{10}$$
$$+2\int_{-\infty}^{\infty}\int_{-\infty}^{\infty}\int_{-\infty}^{\infty}\left[G^{(3)}(t,t_1,t_2,t_3)G^{(3)}(t,t_2,t_3,t_1)\right]\mathrm{d}t_1\mathrm{d}t_2\mathrm{d}t_3+\cdots.$$

The WHE method can be elementary used in solving stochastic differential equations by expanding the solution process as well as the stochastic input processes via the WHE. The resultant equation is more complex than the original one due to being a stochastic integro-differential equation. Taking a set of ensemble averages together with using the statistical properties of the WH polynomials, a set of deterministic integro-differential equations are obtained in the deterministic kernels $G^{(i)}(t;\omega), i=0,1,2,\cdots$. To obtain an approximate solutions for these deterministic kernels, one can use perturbation theory in the case of having a perturbed system depending on, say, ε. Expanding the kernels as a power series of ε, another set of simpler iterative equations in the kernel series components are obtained. This is the main algorithm of the WHEP technique. The technique was successfully applied to several nonlinear stochastic equations; see [20,22,23,25].

The WHEP technique can be applied on linear or nonlinear perturbed systems described by ordinary or partial differential equations. The solution can be modified in the sense that additional parts of the Wiener-Hermite expansion can always be taken into considerations and the required order of approximations can always be made. It can be even run through a package if it is coded in some sort of symbolic languages.

Case-Study

The quadratic nonlinear oscillatory problem, Equation (1) under stochastic excitation $F(t;\omega)$ with deterministic initial conditions is solved using WHEP technique. The solution process takes the following form:

$$x(t;\omega)=x^{(0)}(t)+\int_{-\infty}^{\infty}x^{(1)}(t;t_1)H^{(1)}(t_1)\mathrm{d}t_1$$
$$+\int_{-\infty}^{\infty}\int_{-\infty}^{\infty}x^{(2)}(t;t_1,t_2)H^{(2)}(t_1,t_2)\mathrm{d}t_1\mathrm{d}t_2 \tag{11}$$
$$+\int_{-\infty}^{\infty}\int_{-\infty}^{\infty}\int_{-\infty}^{\infty}x^{(3)}(t;t_1,t_2,t_3)H^{(3)}(t_1,t_2,t_3)\mathrm{d}t_1\mathrm{d}t_2\mathrm{d}t_3+\cdots$$

Applying the WHEP technique, the following equations in the deterministic kernels are obtained:

$$Lx^{(0)}(t)+\varepsilon w^2\left(x^{(0)}(t)\right)^2+\varepsilon w^2\int_{-\infty}^{\infty}\left(x^{(1)}(t;t_1)\right)^2\mathrm{d}t_1+2\varepsilon w^2\int_{-\infty}^{\infty}\int_{-\infty}^{\infty}\left(x^{(2)}(t;t_1,t_2)\right)^2\mathrm{d}t_1\mathrm{d}t_2$$
$$+2\varepsilon w^2\left[\int_{-\infty}^{\infty}\int_{-\infty}^{\infty}\int_{-\infty}^{\infty}\left(x^{(3)}(t;t_1,t_2,t_3)\right)^2\mathrm{d}t_1\mathrm{d}t_2\mathrm{d}t_3+\int_{-\infty}^{\infty}\int_{-\infty}^{\infty}\int_{-\infty}^{\infty}x^{(3)}(t;t_1,t_2,t_3)x^{(3)}(t,t_2,t_3,t_1)\mathrm{d}t_1\mathrm{d}t_2\mathrm{d}t_3\right. \tag{12}$$
$$\left.+\int_{-\infty}^{\infty}\int_{-\infty}^{\infty}\int_{-\infty}^{\infty}x^{(3)}(t;t_1,t_2,t_3)x^{(3)}(t,t_1,t_3,t_2)\mathrm{d}t_1\mathrm{d}t_2\mathrm{d}t_3\right]=F^{(0)}(t)$$

$$Lx^{(1)}(t,t_1)+2\varepsilon w^2x^{(0)}(t)x^{(1)}(t,t_1)+4\varepsilon w^2\int_{-\infty}^{\infty}x^{(1)}(t;t_2)x^{(2)}(t;t_1,t_2)\mathrm{d}t_2$$
$$+4\varepsilon w^2\int_{-\infty}^{\infty}\int_{-\infty}^{\infty}x^{(1)}(t,t_2)x^{(2)}(t;t_1,t_2)\mathrm{d}t_2+8\varepsilon w^2\left[\int_{-\infty}^{\infty}\int_{-\infty}^{\infty}x^{(2)}(t;t_2,t_3)x^{(3)}(t;t_1,t_2,t_3)\mathrm{d}t_2\mathrm{d}t_3\right. \tag{13}$$
$$\left.+4\varepsilon w^2\int_{-\infty}^{\infty}\int_{-\infty}^{\infty}x^{(2)}(t;t_2,t_3)x^{(3)}(t;t_2,t_3,t_1)\mathrm{d}t_2\mathrm{d}t_3\right]=F^{(1)}(t,t_1)$$

$$\begin{aligned}
&Lx^{(2)}(t,t_1,t_2)+\varepsilon w^2\Bigg[2x^{(0)}(t)x^{(2)}(t,t_1,t_2)+x^{(1)}(t,t_1)x^{(1)}(t,t_2)+4\int_{-\infty}^{\infty}x^{(2)}(t;t_1,t_3)x^{(2)}(t;t_2,t_3)\mathrm{d}t_3\\
&+2\int_{-\infty}^{\infty}x^{(1)}(t;t_3)x^{(3)}(t;t_1,t_3,t_2)\mathrm{d}t_3+2\int_{-\infty}^{\infty}x^{(1)}(t;t_3)x^{(3)}(t;t_2,t_3,t_1)\mathrm{d}t_3+2\int_{-\infty}^{\infty}x^{(1)}(t;t_3)x^{(3)}(t;t_1,t_2,t_3)\mathrm{d}t_3\\
&+2\int_{-\infty}^{\infty}\int_{-\infty}^{\infty}x^{(3)}(t;t_2,t_5,t_6)x^{(3)}(t;t_1,t_6,t_5)\mathrm{d}t_5\mathrm{d}t_6+4\int_{-\infty}^{\infty}\int_{-\infty}^{\infty}x^{(3)}(t;t_2,t_5,t_6)x^{(3)}(t;t_1,t_5,t_6)\mathrm{d}t_5\mathrm{d}t_6\\
&+3\int_{-\infty}^{\infty}\int_{-\infty}^{\infty}x^{(3)}(t;t_2,t_5,t_6)x^{(3)}(t;t_5,t_6,t_1)\mathrm{d}t_5\mathrm{d}t_6+2\int_{-\infty}^{\infty}\int_{-\infty}^{\infty}x^{(3)}(t;t_1,t_5,t_6)x^{(3)}(t;t_2,t_6,t_5)\mathrm{d}t_5\mathrm{d}t_6\\
&+3\int_{-\infty}^{\infty}\int_{-\infty}^{\infty}x^{(3)}(t;t_1,t_5,t_6)x^{(3)}(t;t_5,t_6,t_2)\mathrm{d}t_5\mathrm{d}t_6+\int_{-\infty}^{\infty}\int_{-\infty}^{\infty}x^{(3)}(t;t_1,t_6,t_5)x^{(3)}(t;t_5,t_6,t_2)\mathrm{d}t_5\mathrm{d}t_6\\
&+2\int_{-\infty}^{\infty}\int_{-\infty}^{\infty}x^{(3)}(t;t_5,t_6,t_2)x^{(3)}(t;t_5,t_6,t_1)\mathrm{d}t_5\mathrm{d}t_6+\int_{-\infty}^{\infty}\int_{-\infty}^{\infty}x^{(3)}(t;t_2,t_6,t_5)x^{(3)}(t;t_5,t_6,t_1)\mathrm{d}t_5\mathrm{d}t_6\Bigg]=F^{(2)}(t,t_1,t_2)
\end{aligned} \tag{14}$$

$$\begin{aligned}
&Lx^{(3)}(t,t_1,t_2,t_3)+Lx^{(3)}(t,t_1,t_3,t_2)+Lx^{(3)}(t,t_2,t_3,t_1)+\varepsilon w^2\Big[2x^{(0)}(t)\big[x^{(3)}(t,t_1,t_2,t_3)+x^{(3)}(t,t_1,t_3,t_2)+\big]x^{(3)}(t,t_2,t_3,t_1)\\
&+4\int_{-\infty}^{\infty}x^{(2)}(t;t_4,t_1)x^{(3)}(t;t_4,t_3,t_2)\mathrm{d}t_4+4\int_{-\infty}^{\infty}x^{(2)}(t;t_4,t_1)x^{(3)}(t;t_4,t_2,t_3)\mathrm{d}t_4+4\int_{-\infty}^{\infty}x^{(2)}(t;t_4,t_2)x^{(3)}(t;t_4,t_3,t_1)\mathrm{d}t_4\\
&+4\int_{-\infty}^{\infty}x^{(2)}(t;t_4,t_2)x^{(3)}(t;t_4,t_1,t_3)\mathrm{d}t_4+4\int_{-\infty}^{\infty}x^{(2)}(t;t_4,t_3)x^{(3)}(t;t_4,t_1,t_2)\mathrm{d}t_4+4\int_{-\infty}^{\infty}x^{(2)}(t;t_4,t_3)x^{(3)}(t;t_4,t_2,t_1)\mathrm{d}t_4\\
&+4\int_{-\infty}^{\infty}x^{(2)}(t;t_4,t_1)x^{(3)}(t;t_3,t_2,t_4)\mathrm{d}t_4+4\int_{-\infty}^{\infty}x^{(2)}(t;t_4,t_2)x^{(3)}(t;t_3,t_1,t_4)\mathrm{d}t_4+4\int_{-\infty}^{\infty}x^{(2)}(t;t_4,t_3)x^{(3)}(t;t_1,t_2,t_4)\mathrm{d}t_4\Big]\\
&=G^{(3)}(t,t_1,t_2,t_3)+G^{(3)}(t,t_1,t_3,t_2)+G^{(3)}(t,t_2,t_3,t_1)
\end{aligned} \tag{15}$$

Let us take the simple case of evaluating the only Gaussian part (first order approximation) of the solution process of the previous case study, mainly

$$x(t;\omega)=x^{(0)}(t)+\int_{-\infty}^{\infty}x^{(1)}(t;t_1)H^{(1)}(t_1)\mathrm{d}t_1\,. \tag{16}$$

In this case, the governing equations are

$$Lx^{(0)}(t)+\varepsilon w^2\left[\left[x^{(0)}(t)\right]^2+\int_{-\infty}^{\infty}\left[x^{(1)}(t;t_1)\right]^2\mathrm{d}t_1\right]=G^{(0)}(t) \tag{17}$$

$$Lx^{(1)}(t,t_1)+\varepsilon w^2 2x^{(0)}(t)x^{(1)}(t,t_1)=G^{(1)}(t,t_1) \tag{18}$$

The ensemble average is

$$\mu_x(t)=x^{(0)}(t) \tag{19}$$

and the variance is

$$\sigma_x^2(t)=\int_{-\infty}^{\infty}\left[x^{(1)}(t;t_1)\right]^2\mathrm{d}t_1 \tag{20}$$

It has to be noticed that all the previous equations are deterministic linear ones in the general form $\ddot{x}+2w\xi\dot{x}+w^2x=F(t)$ with deterministic initial conditions $x(0)=x_0,\dot{x}(0)=\dot{x}_0$. It has the general solution

$$x(t)=x_0\phi_1(t)+\dot{x}_0\phi_2+\int_0^t h(t-s)F(s)\mathrm{d}s \tag{21}$$

In which we have

$$h(t)=\frac{1}{w\sqrt{1-\xi^2}}\mathrm{e}^{-w\xi t}\sin w\sqrt{1-\xi^2}\,t,$$

$$\phi_1(t)=\frac{\xi+\sqrt{\xi^2-1}}{2\sqrt{\xi^2-1}}\mathrm{e}^{mt}+\frac{-\xi+\sqrt{\xi^2-1}}{2\sqrt{\xi^2-1}}\mathrm{e}^{qt},$$

$$\phi_2(t)=\frac{1}{2w\sqrt{\xi^2-1}}\left[\mathrm{e}^{mt}-\mathrm{e}^{qt}\right],$$

where

$$m=-w\xi+w\sqrt{\xi^2-1},\quad q=-w\xi-w\sqrt{\xi^2-1}.$$

When adding the first term in the non-Gaussian part (the second approximation) of the solution process of the previous case study, mainly

$$\begin{aligned}
x(t;\omega)&=x^{(0)}(t)+\int_{-\infty}^{\infty}x^{(1)}(t;t_1)H^{(1)}(t_1)\mathrm{d}t_1\\
&+\int_{-\infty}^{\infty}\int_{-\infty}^{\infty}x^{(2)}(t;t_1,t_2)H^{(2)}(t_1,t_2)\mathrm{d}t_1\mathrm{d}t_2
\end{aligned}, \tag{22}$$

the governing equations become

$$Lx^{(0)}(t)+\varepsilon w^2\left[\left[x^{(0)}(t)\right]^2+\int_{-\infty}^{\infty}\left[x^{(1)}(t;t_1)\right]^2 \mathrm{d}t_1\right. \quad (23)$$
$$\left.+2\varepsilon w^2\int_{-\infty}^{\infty}\int_{-\infty}^{\infty}\left[x^{(2)}(t;t_1,t_2)\right]^2 \mathrm{d}t_1\mathrm{d}t_2\right]=G^{(0)}(t)$$

$$Lx^{(1)}(t,t_1)+2\varepsilon w^2 x^{(0)}(t)x^{(1)}(t,t_1) \quad (24)$$
$$+4\varepsilon w^2\int_{-\infty}^{\infty}x^{(1)}(t;t_2)x^{(2)}(t;t_1,t_2)\mathrm{d}t_2=G^{(1)}(t,t_1)$$

$$Lx^{(2)}(t,t_1,t_2)$$
$$+\varepsilon w^2\left[2x^{(0)}(t)x^{(2)}(t,t_1,t_2)+x^{(1)}(t,t_1)x^{(1)}(t,t_2)\right. \quad (25)$$
$$\left.+4\int_{-\infty}^{\infty}x^{(2)}(t;t_1,t_3)x^{(2)}(t;t_2,t_3)\mathrm{d}t_3\right]=G^{(2)}(t,t_1,t_2)$$

The ensemble average is still got by Equation (19) while the variance is got as

$$\sigma_x^2(t)=\int_{-\infty}^{\infty}\left[x^{(1)}(t;t_1)\right]^2\mathrm{d}t_1+2\int_{-\infty}^{\infty}\int_{-\infty}^{\infty}\left[x^{(2)}(t;t_1,t_2)\right]^2\mathrm{d}t_1\mathrm{d}t_2 \quad (26)$$

The WHEP technique uses the following expansion for its deterministic kernels as corrections made under each approximation order.

$$x^{(i)}(t)=x_0^{(i)}+\varepsilon x_1^{(i)}+\varepsilon^2 x_2^{(i)}+\varepsilon^3 x_3^{(i)}+\cdots,\ i=0,1,2,3,.. \quad (27)$$

Example:

Let us take $F(t;q)=\mathrm{e}^{-t}+\varepsilon\delta(t;\omega),\quad \varepsilon=0.3$ (28)

in the previous case-study and then solving using the WHEP technique. The following results are obtained, see **Figures 1-3**.

4. The Homotopy Perturbation Method (HPM)

In this technique, a parameter $p\in[0,1]$ is embedded in a homotopy function $v(r,p):\phi\times[0,1]\rightarrow\Re$ which satisfies

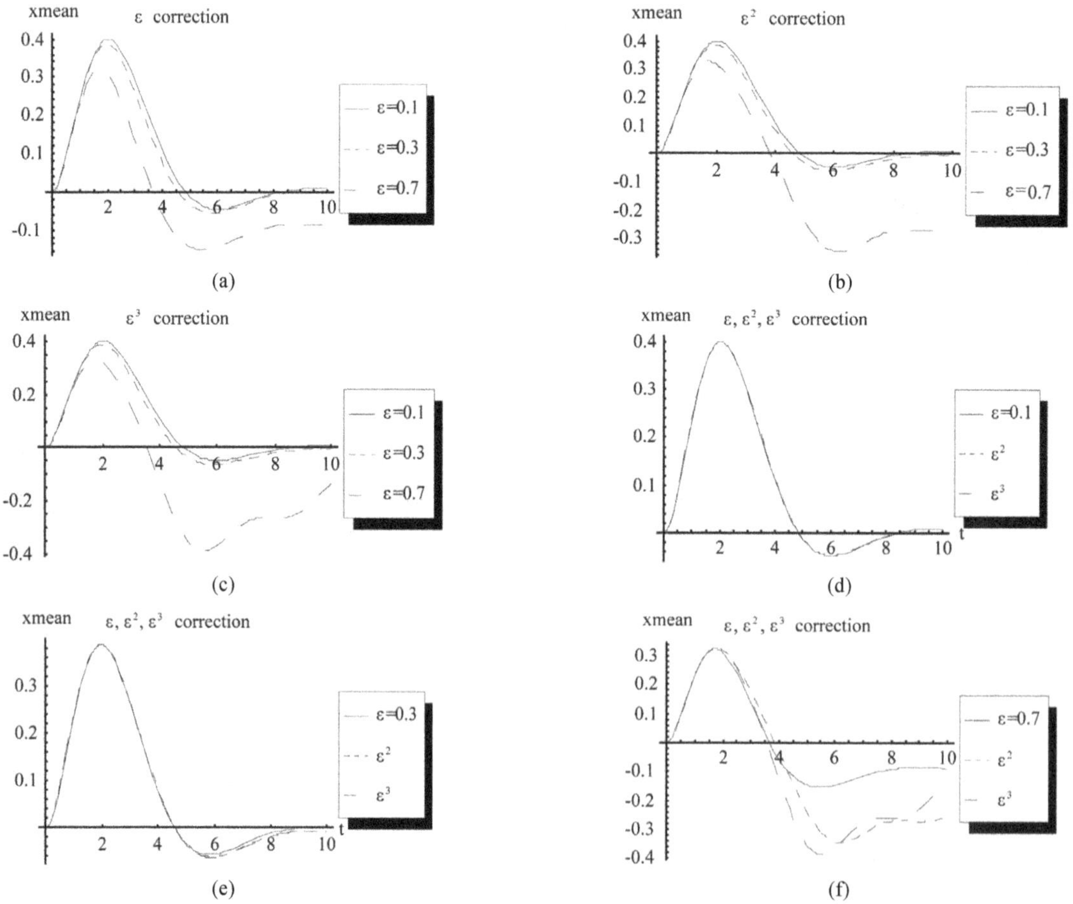

Figure 1. (a) The first order approximation of the mean at ε correction for different correction levels; (b) The first order approximation of the mean at ε^2 correction for different correction levels; (c) The first order approximation of the mean at ε^3 correction; (d) The first order approximation of the mean at ε, ε^2, ε^3 correction; (e) The first order approximation of the mean at ε, ε^2, ε^3 correction; (f) The first order approximation of the mean at ε, ε^2, ε^3 correction.

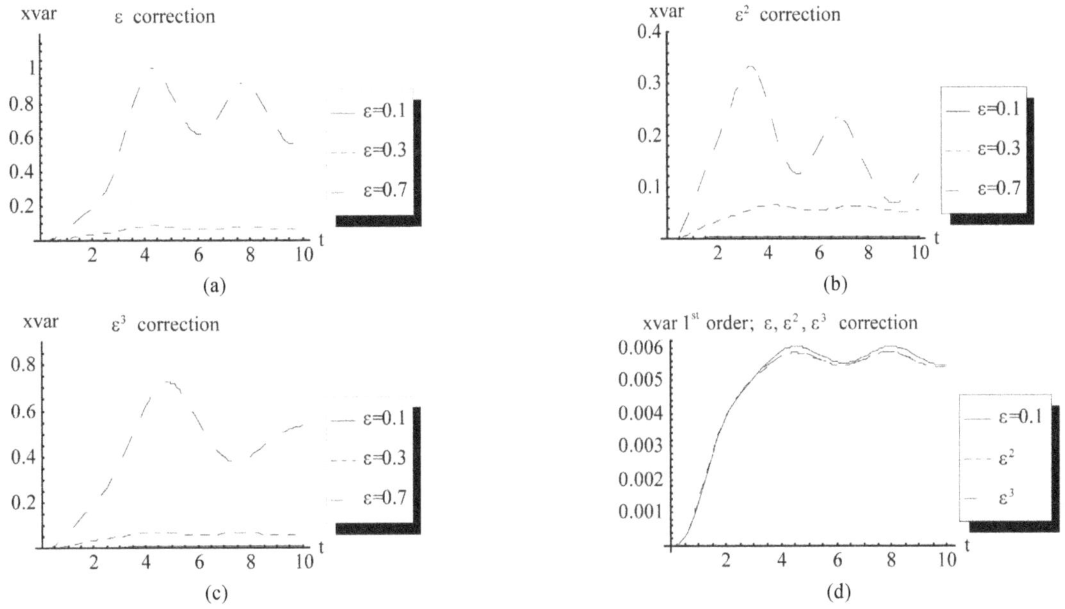

Figure 2. (a) The first order approximation of the variance at ε correction for different correction levels; (b) The first order approximation of the variance at ε^2 Correction for different correction levels; (c) The first order approximation of the variance at ε^3 correction for different correction levels; (d) The first order approximation of the variance at. ε, ε^2, ε^3 correction.

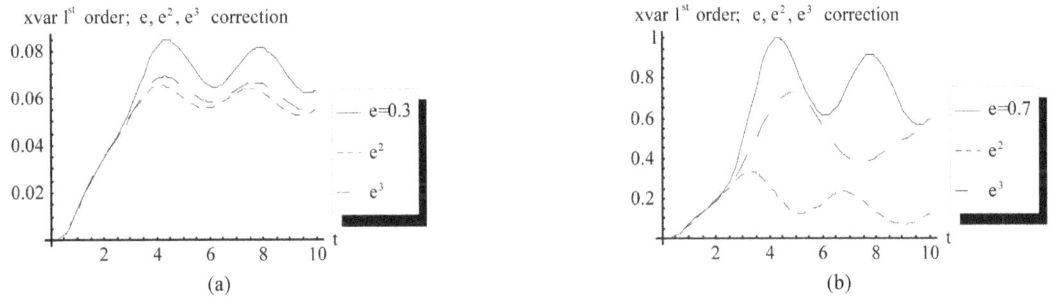

Figure 3. (a) The first order approximation of the variance at ε, ε^2, ε^3 correction; (b) The first order approximation of the variance at ε, ε^2, ε^3 correction.

$$H(v,p)=(1-p)\left[L(v)-L(u_0)\right]+p\left[A(v)-f(r)\right]=0 \tag{29}$$

where u_0 is an initial approximation to the solution of the equation

$$A(u)-f(r)=0,\ r\in\phi \tag{30}$$

with boundary conditions

$$B\left(u,\frac{\partial u}{\partial n}\right)=0,\ r\in\Gamma \tag{31}$$

in which A is a nonlinear differential operator which can be decompose into a linear operator L and a nonlinear operator N, B is a boundary operator, $f(r)$ is a known analytic function and Γ is the boundary of ϕ. The homotopy introduces a continuously deformed solution for the case of $p = 0$, $L(v)-L(u_0)=0$, to the case of $p = 1$, $A(v)-f(r)=0$, which is the original Equation (30). This is the basic idea of the homotopy method which is to deform continuously a simple problem (and easy to solve) into the difficult problem under study [35].

The basic assumption of the HPM method is that the solution of the original Equation (29) can be expanded as a power series in p as:

$$v=v_0+pv_1+p^2v_2+p^3v_3+\cdots \tag{32}$$

Now, setting $p = 1$, the approximate solution of Equation (23) is obtained as:

$$u=\lim_{p\to 1}v=v_0+v_1+v_2+v_3+\cdots \tag{33}$$

The rate of convergence of the method depends greatly on the initial approximation u_0.

The idea of the imbedded parameter can be utilized to solve nonlinear problems by imbedding this parameter to the problem and then forcing it to be unity in the obtained approximate solution if converge can be assured. A simple technique enables the extension of the applicability of the perturbation methods from small valued applications to general ones.

Example

Considering the same previous example of Sub-Section 3.1.1, one can get the following results w.r.t. homotopy perturbation:

$$A(x) = L(x) + \varepsilon w^2 x^2,$$

$$L(x) = \ddot{x} + 2w\varsigma\dot{x} + w^2 x,$$

$$N(x) = \varepsilon x^2,$$

$$f(r) = F(t;\omega).$$

The homotopy function takes the following form:

$$H(v,p) = (1-p)\left[L(v) - L(u_0)\right] + p\left[A(v) - f(r)\right] = 0$$

or equivalently,

$$L(v) - L(u_0) + p\left[L(u_0) + \varepsilon w^2 v^2 - F(t;\omega)\right] = 0. \quad (34)$$

Letting $v = v_0 + pv_1 + p^2 v_2 + p^3 v_3 + \cdots$, substituting in Equation (34) and equating the equal powers of p in both sides of the equation, one can get the following results:

1) $L(v_0) = L(y_0)$, in which one may consider the following simple solution:

$$v_0 = y_0, \quad y_0(0) = x_0, \dot{y}_0(0) = \dot{x}_0.$$

2) $L(v_1) = F(t;\omega) - L(v_0) - \varepsilon w^2 v_0^2, v_1(0) = 0, \dot{v}_1(0) = 0.$

3) $L(v_2) = -2\varepsilon w^2 v_0 v_1, v_2(0) = 0, \dot{v}_2(0) = 0$.

4) $L(v_3) = -\varepsilon w^2\left(v_1^2 + 2v_0 v_2\right), v_3(0) = 0, \dot{v}_3(0) = 0$.

5) $L(v_4) = -2\varepsilon_1\left(v_0 v_3 + v_1 v_2\right), v_4(0) = 0, \dot{v}_4(0) = 0$.

The approximate solution is

$$x(t;\omega) = \lim_{p\to 1} v = v_0 + v_1 + v_2 + v_3 + \cdots$$

which can be considered to any approximation order. One can notice that the algorithm of the solution is straight forward and that a lot of flexibilities can be made. For example, we have many choices in guessing the initial approximation together with its initial conditions. For zero initial conditions, we can choose $v_0 = 0$ which leads to:

$$\begin{aligned} x(t;\omega) &\cong x_5 = v_0 + v_1 + v_2 + v_3 + v_4 + v_5 \\ &= \int_0^t h(t-s)F(s;\omega)\mathrm{d}s - \varepsilon w^2 \int_0^t h(t-s)v_1^2(s;\omega)\mathrm{d}s \\ &\quad - 2\varepsilon w^2 \int_0^t h(t-s)v_1(s;\omega)v_3(s;\omega)\mathrm{d}s \end{aligned} \quad (35)$$

Figures 4-7 are obtained for $\varsigma = 0.5$: [42].

5. Comparisons between WHEP and HPM Methods

Figure [8] shows comparisons between the WHEP and HPM methods for different values of the nonlinearity strength, ε. As the nonlinearity strength increases, the deviation between the two methods is also increasing.

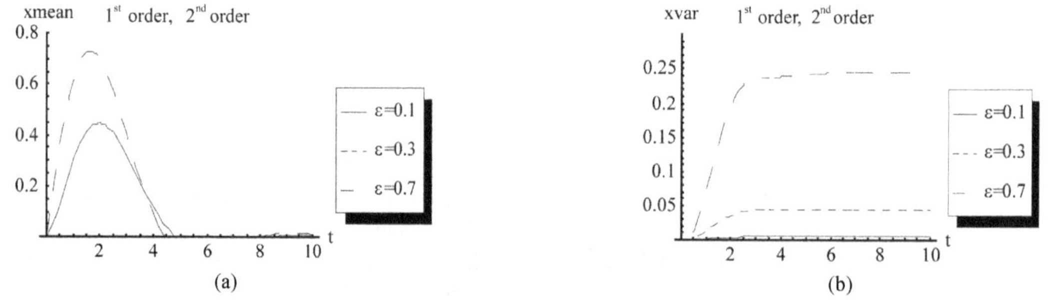

Figure 4. (a) The first and second order approximation of the mean for different correction levels; (b) The first and second order approximation of the variance at for different correction levels.

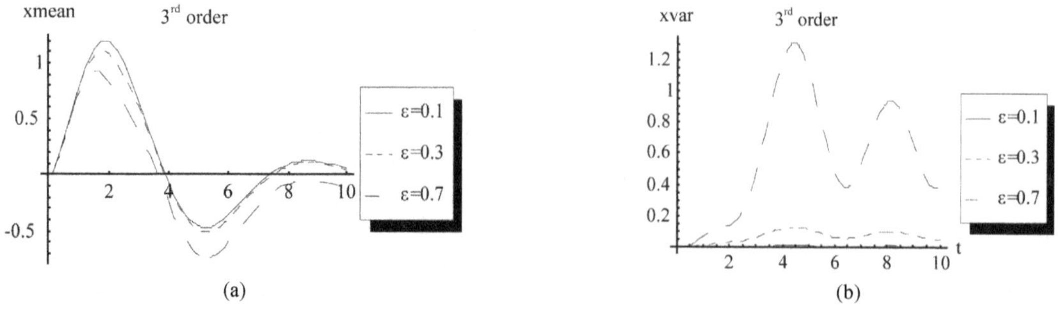

Figure 5. (a) The third order approximation of the mean for different correction levels; (b) The third order approximation of the variance for different correction levels.

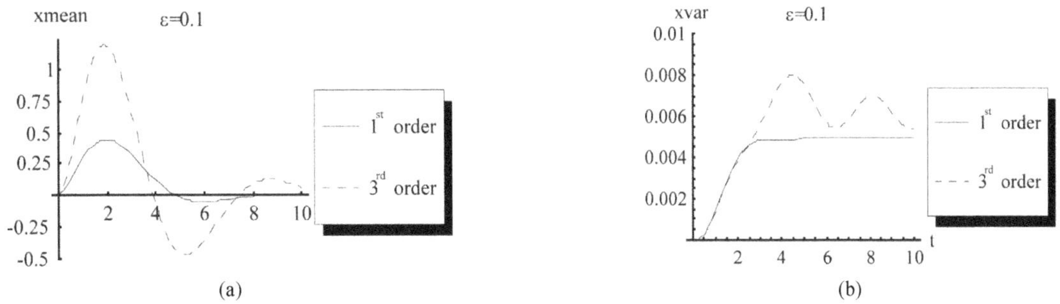

Figure 6. (a) A comparison between first, second order and the third order of the mean at $\varepsilon = 0.1$; (b) Comparison between first, second order and the, third order of the variance at $\varepsilon = 0.1$.

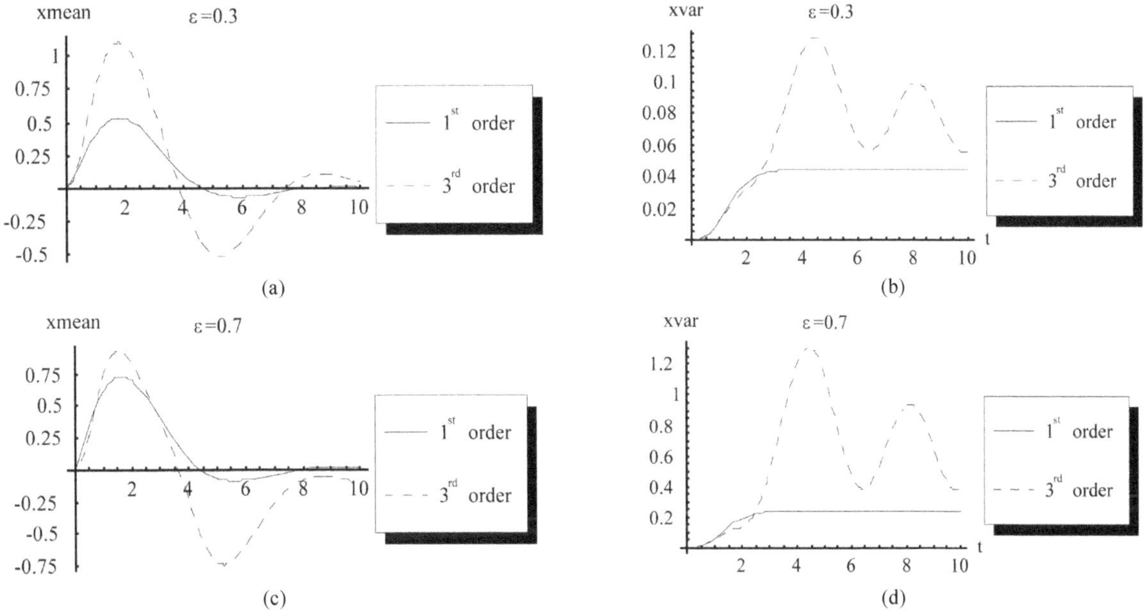

Figure 7. (a) A comparison between first, second order and the third order of the mean at $\varepsilon = 0.3$; (b) A comparison between first, second order and the, third order of the variance at $\varepsilon = 0.3$; (c) A comparison between first, second order and the o third order of the mean at $\varepsilon = 0.7$; (d) A comparison between first, second order and the third order of the variance at $\varepsilon = 0.7$.

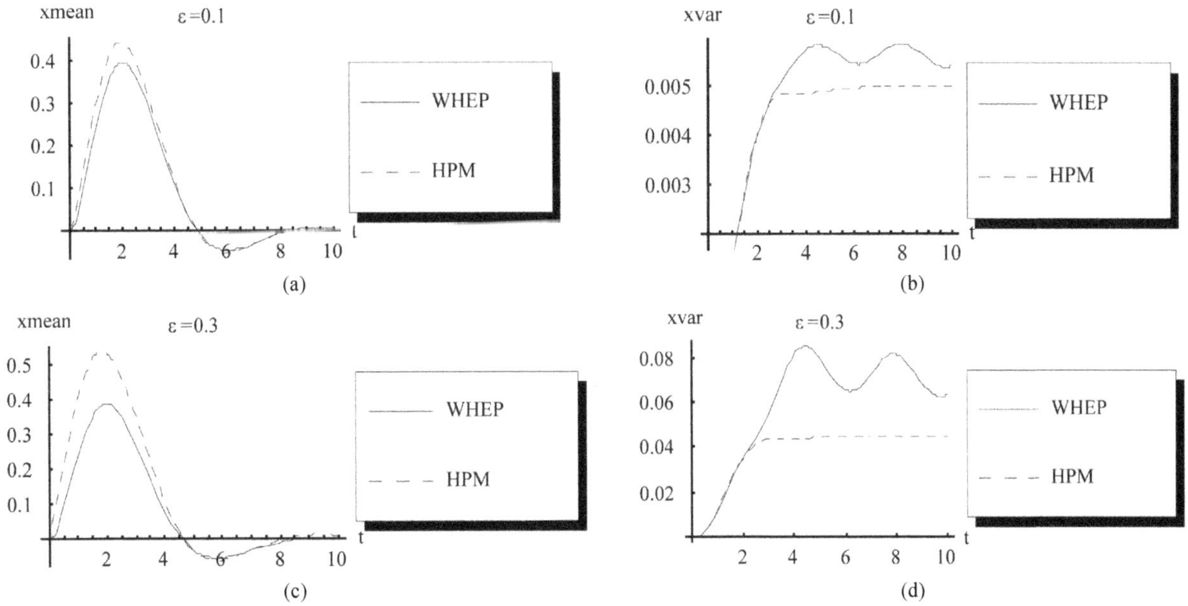

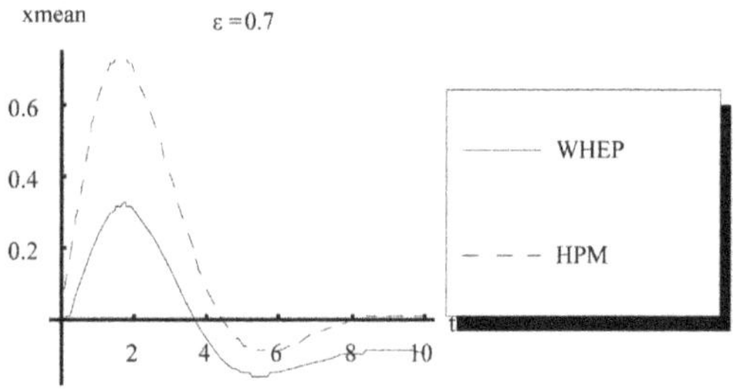

(e)

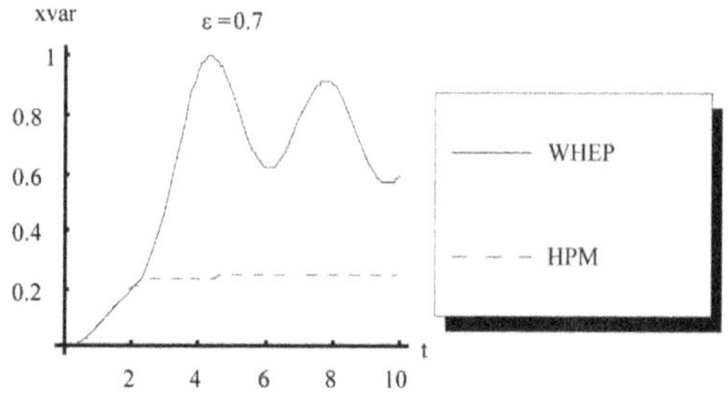

(f)

Figure 8. (a) A comparison between homotopy perturbation and Wiener-Hermite of the mean at $\varepsilon = 0.1$; (b) A comparison between homotopy perturbation and Wiener-Hermite of the variance at $\varepsilon = 0.1$; (c) A comparison between homotopy perturbation and Wiener-Hermite of the mean at $\varepsilon = 0.3$; (d) A comparison between homotopy perturbation and Wiener-Hermite of the variance at $\varepsilon = 0.3$; (e) A comparison between homotopy perturbation and Wiener-Hermite of the mean at $\varepsilon = 0.7$; (f) A comparison between homotopy perturbation and Wiener-Hermite of the variance at $\varepsilon = 0.7$.

This is due to the convergence condition of the WHEP technique which depends on ε. For small values of ε, the WHEP technique converges but after a certain value of ε it will diverge. The HPM is more accurate for higher values of ε. The HPM has advantages when used in solving differential equations with large nonlinearities.

6. Conclusion

The quadratic nonlinear oscillator with stochastic excitation is considered. The solution was obtained using the WHEP technique with different orders and different number of corrections. The HPM is used also with different approximations. The WHEP technique is more efficient but it converges only for certain limit of the nonlinearity strength. The HPM is more difficult in the stochastic differential equations but it is more preferable for higher values of the nonlinearity strength. The two methods are shown to be efficient in estimating the stochastic response of the quadratic nonlinear oscillators.

REFERENCES

[1] A. Nayfeh, "Problems in Perturbation," John Wiley, New York, 1993.

[2] S. Crow and G. Canavan, "Relationship between a Wiener-Hermite Expansion and an Energy Cascade," *Journal of Fluid Mechanics*, Vol. 41, No. 2, 1970, pp. 387-403.

[3] P. Saffman, "Application of Wiener-Hermite Expansion to the Diffusion of a Passive Scalar in a Homogeneous Turbulent Flow," *Physics of Fluids*, Vol. 12, No. 9, 1969, pp. 1786-1798.

[4] W. Kahan and A. Siegel, "Cameron-Martin-Wiener Method in Turbulence and in Burger's Model: General Formulae and Application to Late Decay," *Journal of Fluid Mechanics*, Vol. 41, No. 3, 1970, pp. 593-618.

[5] J. Wang and S. Shu, "Wiener-Hermite Expansion and the Inertial Subrange of a Homogeneous Isotropic Turbulence," *Physics of Fluids*, Vol. 17, No. 6, 1974, pp. 1130-1134.

[6] H. Hogge and W. Meecham, "Wiener-Hermite Expansion Applied to Decaying Isotropic Turbulence Using a Renormalized Time-Dependent Base," *Journal of Fluid of Mechanics*, Vol. 85, No. 2, 1978, pp. 325-347.

[7] M. Doi and T. Imamura, "An Exact Gaussian Solution for Two-Dimensional Incompressible Inviscid Turbulent Flow," *Journal of the Physical Society of Japan*, Vol. 46, No. 4, 1979, pp. 1358-1359.

[8] R. Kambe, M. Doi and T. Imamura, "Turbulent Flows Near Flat Plates," *Journal of the Physical Society of Japan*, Vol. 49, No. 2, 1980, pp. 763-778.

[9] A. J. Chorin, "Gaussian Fields and Random Flow," *Journal of Fluid of Mechanics*, Vol. 63, No. 1, 1974, pp. 21-32.

[10] Y. Kayanuma, "Stochastic Theory for Non-Adiabatic Crossing with Fluctuating Off-Diagonal Coupling," *Journal of the Physical Society of Japan*, Vol. 54, No. 5, 1985, pp. 2037-2046.

[11] M. Joelson and A. Ramamonjiarisoa, "Random Fields of Water Surface Waves Using Wiener-Hermite Functional Series Expansions," *Journal of Fluid of Mechanics*, Vol. 496, 2003, pp. 313-334.

[12] C. Eftimiu, "First-Order Wiener-Hermite Expansion in the Electromagnetic Scattering by Conducting Rough Surfaces," *Radio Science*, Vol. 23, No. 5, 1988, pp. 769-779.

[13] N. J. Gaol, "Scattering of a TM Plane Wave from Periodic Random Surfaces," *Waves Random Media*, Vol. 9, No. 11, 1999, pp. 53-67.

[14] Y. Tamura and J. Nakayama, "Enhanced Scattering from a Thin Film with One-Dimensional Disorder," *Waves in Random and Complex Media*, Vol. 15, No. 2, 2005, pp. 269-295.

[15] Y. Tamura and J. Nakayama, "TE Plane Wave Reflection and Transmission from One-Dimensional Random Slab," *IEICE Transactions on Electronics*, Vol. E88-C, No. 4, 2005, pp. 713-720.

[16] N. Skaropoulos and D. Chrissoulidis, "Rigorous Application of the Stochastic Functional Method to Plane Wave

Scattering from a Random Cylindrical Surface," *Journal of Mathematical Physics*, Vol. 40, No. 1, 1999, pp. 156-168.

[17] A. Jahedi and G. Ahmadi, "Application of Wiener-Hermite Expansion to Non-Stationary Random Vibration of a Duffing Oscillator," *Journal of Applied Mechanics, Transactions ASME*, Vol. 50, No. 2, 1983, pp. 436-442.

[18] I. I. Orabi and G. Ahmadi, "Functional Series Expansion Method for Response Analysis of Nonlinear Systems Subjected to Ransom Excitations," *International Journal of Nonlinear Mechanics*, Vol. 22, No. 6, 1987, pp. 451-465.

[19] I. I. Orabi, "Response of the Duffing Oscillator to a Non-Gaussian Random Excitation," *Journal of Applied Mechanics, Transaction of ASME*, Vol. 55, No. 3, 1988, pp. 740-743.

[20] E. Abdel Gawad, M. El-Tawil and M. A. Nassar, "Nonlinear Oscillatory Systems with Random Excitation," *Modeling, Simulation and Control B*, Vol. 23, No. 1, 1989, pp. 55-63.

[21] I. I. Orabi and G. Ahmadi, "New Approach for Response Analysis of Nonlinear Systems under Random Excitation," *American Society of Mechanical Engineers, Design Engineering Division (Publication) DE*, Vol. 37, 1991, pp. 147-151.

[22] E. Gawad and M. El-Tawil, "General Stochastic Oscillatory Systems," *Applied Mathematical Modelling*, Vol. 17, No. 6, 1993, pp. 329-335.

[23] M. El-Tawil and G. Mahmoud, "The Solvability of Parametrically Forced Oscillators Using WHEP Technique," *Mechanics and Mechanical Engineering*, Vol. 3, No. 2, 1999, pp. 181-188.

[24] Y. Tamura and J. Nakayama, "A Formula on the Hermite Expansion and Its Aoolication to a Random Boundary Value Problem," *IEICE Transactions on Electronics*, Vol. E86-C, No. 8, 2003, pp. 1743-1748.

[25] M. El-Tawil, "The Application of WHEP Technique on Stochastic Partial Differential Equations," *International Journal of Differential Equations and Applications*, Vol. 7, No. 3, 2003, pp. 325-337.

[26] Y. Kayanuma and K. Noba, "Wiener-Hermite Expansion Formalism for the Stochastic Model of a Driven Quantum System," *Chemical Physics*, Vol. 268, No. 1-3, 2001, pp. 177-188.

[27] O. Kenny and D. Nelson, "Time-Frequency Methods for Enhancing Speech," *Proceedings of SPIE—The International Society for Optical Engineering*, Vol. 3162, 1997, pp. 48-57.

[28] E. Isobe and S. Sato, "Wiener-Hermite Expansion of a Process Generated by an Ito Stochastic Differential Equations," *Journal of Applied Probability*, Vol. 20, No. 4, 1983, pp. 754-765.

[29] R. Rubinstein and M. Choudhari, "Uncertainty Quantification for Systems with Random Initial Conditions Using Wiener-Hermite Expansions," *Studies in Applied Mathematics*, Vol. 114, No. 2, 2005, pp. 167-188.

[30] T. Imamura, W. Meecham and A. Siegel, "Symbolic Calculus of the Wiener Process and Wiener-Hermite Functionals," *Journal of Mathematical Physics*, Vol. 6, No. 5, 1983, pp. 695-706.

[31] J. H. He, "Homotopy Perturbation Technique," *Computer Methods in Applied Mechanics and Engineering*, Vol. 178, 1999, pp. 257-292.

[32] J. H. He, "A Coupling Method of a Homotopy Technique and a Perturbation Technique for Nonlinear Problems," *International Journal of Nonlinear Mechanics*, Vol. 35, 2000, pp. 37-43.

[33] J. H. He, "Homotopy Perturbation Method: A New Nonlinear Analytical Technique," *Applied Mathematics and Computation*, Vol. 135, 2003, pp. 73-79.

[34] J. H. He, "The Homotopy Perturbation Method for Nonlinear Oscillators with Discontinuities," *Applied Mathematics and Computation*, Vol. 151, 2004, pp. 287-292.

[35] J. H. He, "Some Asymptotic Methods for Strongly Nonlinear Equations," *International Journal of Modern Physics B*, Vol. 20, No. 10, 2006, pp. 1141-1199.

[36] S. J. Liao, "On the Proposed Homotopy Analysis Techniques for Nonlinear Problems and Its Applications," Ph.D. Dissertation, Shanghai Jiao Tong University, 1992.

[37] S. J. Liao, "Beyond Perturbation: Introduction to the Homotopy Analysis Method," Chapman & Hall\CRC Press, Boca Raton, 2003.

[38] S. J. Liao, "An Approximate Solution Technique Which Does Not Depend upon Small Parameters: A Special Example," *International Journal of Nonlinear Mechanics*, Vol. 30, 1995, pp. 371-380.

[39] T. Hayat, M. Khan and S. Asghar, "Homotopy Analysis of MHD Flows of an Oldroyd 8-Constant Fluid," *Acta Mechanica*, Vol. 168, 2004, pp. 213-232.

[40] S. Asghar, M. Mudassar and M. Ayub, "Rotating Flow of a Third Fluid by Homotopy Analysis Method," *Applied Mathematics and Computation*, Vol. 165, 2005, pp. 213-221.

[41] S. P. Zhu, "A Closed Form Analytical Solution for the Valuation of Convertible Bonds with Constant Dividend Yield," *ANZIAM Journal*, Vol. 47, No. Part 4, 2006, pp. 477-494.

[42] M. A. El-Tawil and A. S. Al-Johani, "Approximate Solution of a Mixed Nonlinear Stochastic Oscillator," *Computers & Mathematics with Applications*, Vol. 58, No. 11-12, 2009, pp. 2236-2259.

12

Finite Element Analysis of the Ramberg-Osgood Bar

Dongming Wei[1], Mohamed B. M. Elgindi[2]
[1]Department of Mathematics, University of New Orleans, New Orleans, USA
[2]Texas A & M University-Qatar, Doha, Qatar

ABSTRACT

In this work, we present a priori error estimates of finite element approximations of the solution for the equilibrium equation of an axially loaded Ramberg-Osgood bar. The existence and uniqueness of the solution to the associated nonlinear two point boundary value problem is established and used as a foundation for the finite element analysis.

Keywords: Nonlinear Two Point Boundary Value Problem; Ramberg-Osgood Axial Bar; Existence and Uniqueness of Solutions; Finite Element Analysis; Convergence and a Priori Error Estimates

1. Introduction

The following Ramberg-Osgood stress strain equation

$$\varepsilon(x) = A\sigma(x) + B|\sigma(x)|^{q-2}\sigma(x), \qquad (1.1)$$

is accepted as the model for the material's constitutive equation in the stress analysis for a variety of industrial metals. Numerous data exist in literature that supports the use of (1.1) to represent the stress-strain relationship for aluminum and several other steel alloys exhibiting elastic-plastic behavior (see, for example, [1-4] and the references therein). In Equation (1.1), $\varepsilon(x)$ represents the axial strain, $\sigma(x)$ represents the axial stress, $0 < x < L$, $q \geq 2$ represents the material hardening index (where $q = 2$ describes the linear elastic material), the constants A, B and q are determined from the experimental values for the parameters E, σ_y, ε_y, ε_u, and ε_u by the formula

$$A = \frac{1}{E}, B = 0.002\left(\frac{1}{\sigma_y}\right)^{q-2}, q = 1 + \frac{\ln 20}{\ln(\sigma_u/\sigma_y)} \qquad (1.2)$$

where E is the Young's modulus, σ_y, ε_y are the material's yield stress and strain, σ_u, ε_u are the ultimate stress and the ultimate strain, and $L > 0$ stands for the length of the solid bar.

We observe that Equation (1.1) splits the strain into two parts: an elastic strain part with coefficient A and a nonlinear part with coefficient B. The linear part dominates for $\sigma < \sigma_y$, while the nonlinear part dominates for $\sigma > \sigma_y$. In many industrial applications, e.g., in light-weight ship deck titanium structures, welding-induced plastic zones play important roles in determining the structures' integrity (see [5,6]).

Figure 1 compares the stress-strain curves for Hooke's law, the double modulus, and Ramberg-Osgood law using material measured data. Among these models, the Ramberg-Osgood model appears to represent the material's behavior the best.

Table 1 gives experimental values of the material constants for some commonly used metals in industries.

Although (1.1) is widely used in industries for finite element analysis, no solvability and uniqueness or error analysis has been given in literature even for the following one-dimensional boundary value problem:

$$\begin{cases} \dfrac{d\sigma(x)}{dx} - c(x)u(x) + f(x) = 0, \ 0 < x < L, \\ \dfrac{du(x)}{dx} = A\sigma(x) + B|\sigma(x)|^{q-2}\sigma(x), \\ u(0) = 0, u'(L) = \beta \end{cases} \qquad (1.3)$$

where $c(x) \geq 0$ satisfies $c \in L^\infty(0,L)$ and $f(x) \in L^q(0,L)$. For simplicity, we consider only one boundary condition. Other Dirichlet type boundary conditions can be treated similarly.

We also consider the case when $c(x)u(x)$ is replaced by $\sum_{i=1}^{N} k_i u(x_i)\delta(x - x_i)$, where $\delta(x - x_i)$ is Dirac impulse functions, and k_i stands for concentrated elastic support constant at $0 \leq x_i \leq L$, for $i = 1, \cdots, N$.

In Section 2, we develop a week formulation of (1.3) subject to the given boundary condition and prove existence and uniqueness of the solution by using the theory

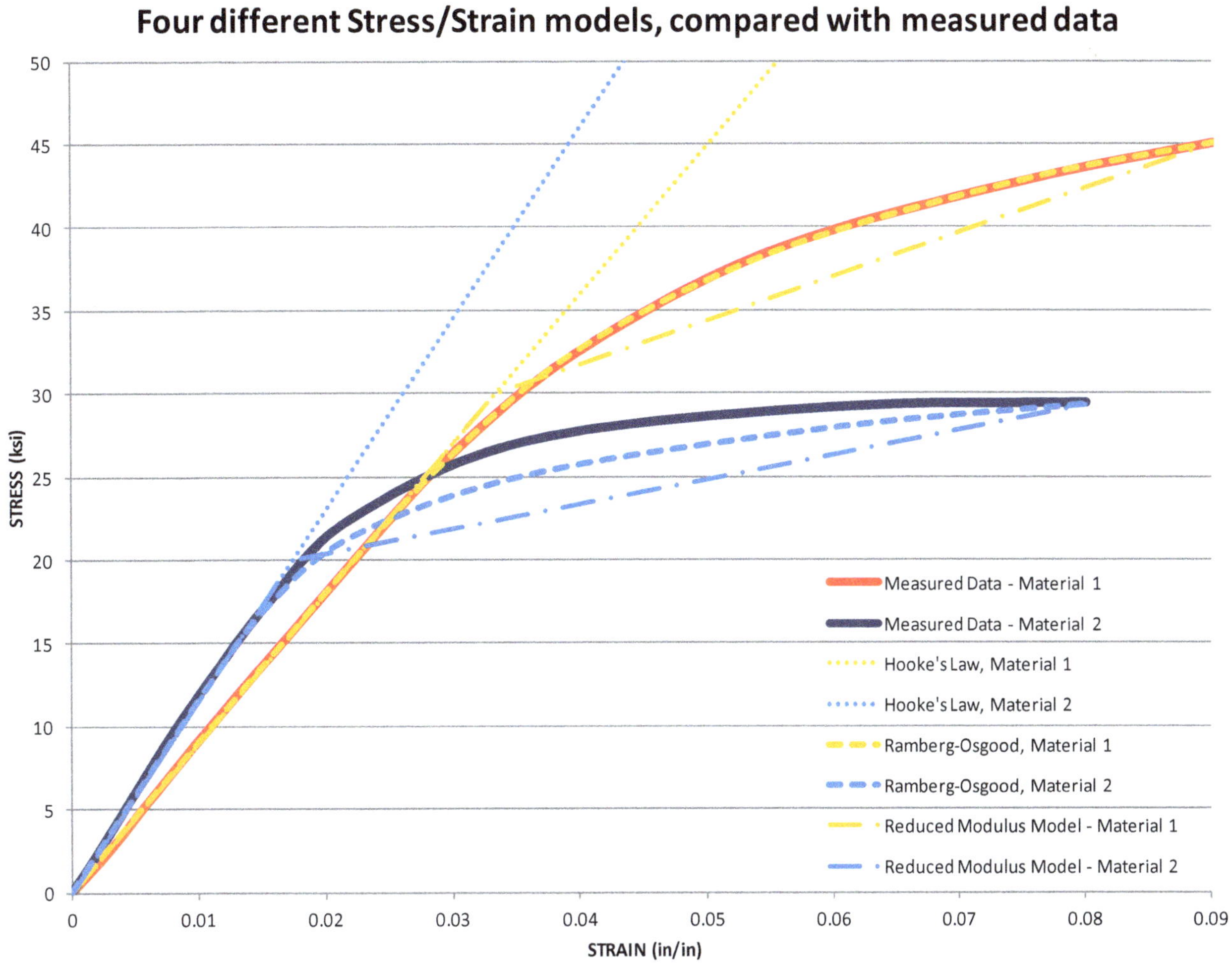

Figure 1. Ramburg-Osgood curves.

Table 1. Constants for Ramburg-Osgood materials.

Material	A	B	$q-1$
Inconel 718	3.33e−05	4.42e−71	32.00
5083 Aluminum	9.80e−05	2.50e−23	13.11
6061T6 Aluminum	1.00e−04	1.35e−58	34.44
304 Stainless Steel	3.57e−05	3.44e−13	6.32
304 L StainlessSteel	3.57e−05	2.24e−15	7.36

of perturbed convex variational problems in Sobolev spaces (see [7] for details.) We also prove that the solution is bounded in certain Sobolev norms. In Section 3, we derive an error estimates for the semi-discrete error between the week solution and the Galerkin's finite element solution of (1.3) for the standard conformal finite elements. The results of this section are based on the results in Section 2. We believe that the results established in these sections are novel and preliminary.

2. Existence and Uniqueness of Solutions

Let $W^{1,p}(0,L)$ and $W_0^{1,p}(0,L)$ be the standard Sobolev spaces, where $p=\dfrac{q}{q-1}$. Define $\phi(\sigma)=A\sigma+B|\sigma|^{q-2}\sigma$, where $A>0, B>0$, and $q>2$.

Observe that the mapping $\phi(\sigma)$ is one-to-one; however, its inverse cannot be written explicitly.

Since $\varepsilon=u'$, Equation (1.3) can be rewritten as:

$$\begin{cases} -\dfrac{\mathrm{d}\sigma(x)}{\mathrm{d}x}+c(x)u(x)=f(x), \ 0<x<L \\ \sigma=\phi^{-1}(u'), \qquad u(0)=0, u'(L)=\beta \end{cases} \tag{2.2}$$

Define the following space of admissible functions as

$$V\equiv\{u\in W^{1,p}(0,L)|u(0)=0,u'(L)=\beta\}. \tag{2.3}$$

The weak formulation of (2.2) can then be written as

Problem I: Find $u\in V$, such that

$$\int_0^L\phi^{-1}(u')v'\mathrm{d}x+\int_0^L cuv\mathrm{d}x=\int_0^l fv\mathrm{d}x \ \ \forall v\in W_0^{1,p}(0,L). \tag{2.4}$$

Let us define the operator:

$$a(u,v) \equiv \int_0^L \phi^{-1}(u')v'\mathrm{d}x + \int_0^L cuv\mathrm{d}x, \quad (2.5)$$

for $u,v \in V$. Then, $a(u,v)$ satisfies the following property:

$$a(u,u) = \int_0^L \phi^{-1}(u')u'\mathrm{d}x + \int_0^L cu^2\mathrm{d}x$$
$$= \int_0^L \phi^{-1}(u')\left[A\phi^{-1}(u') + B\left|\phi^{-1}(u')\right|^{q-2}\phi^{-1}(u')\right]\mathrm{d}x$$
$$+ \int_0^L cu^2\mathrm{d}x \ge \int_0^L \left[A\left|\phi^{-1}(u')\right|^2 + B\left|\phi^{-1}(u')\right|^q\right]\mathrm{d}x$$
$$= A\left\|\phi^{-1}(u')\right\|_{L^2}^2 + B\left\|\phi^{-1}(u')\right\|_{L^q}^q, \text{ for } u \in V. \quad (2.6)$$

Also, by the definition of ϕ, we have

$$\|u'\|_{L^q} = \left\|A\phi^{-1}(u') + B\left|\phi^{-1}(u')\right|^{q-2}\phi^{-1}(u')\right\|_{L^q}$$
$$\le A\left\|\phi^{-1}(u')\right\|_{L^q} + B\left\|\left|\phi^{-1}(u')\right|^{q-1}\right\|_{L^q} \quad (2.7)$$
$$\le A\left\|\phi^{-1}(u')\right\|_{L^q} + \bar{B}\left\|\phi^{-1}(u')\right\|_{L^q}^{q-1}$$

Lemma 2.1 For given positive constants A,B,q,L, there exists a constant C independent of the solutions $u(x) \in V$ of the BVP (1.3) such that $\left\|\phi^{-1}(u')\right\|_{L^2}^2 + \left\|\phi^{-1}(u')\right\|_{L^q}^q \le C$.

Proof: For a solution $u(x) \in V$, we can write: $v = u - u_b$, where $u_b \in V$ is a fixed function, so that $v \in W_0^{1,p}(0,L)$, and since: $\int_0^L \phi^{-1}(u')v'\mathrm{d}x + \int_0^L cuv\mathrm{d}x = \int_0^L fv\mathrm{d}x$, we get:

$$a(u,u) = \int_0^L \phi^{-1}(u')u'\mathrm{d}x + \int_0^L cu^2\mathrm{d}x$$
$$= \int_0^L fu\mathrm{d}x - \int_0^L fu_b\,\mathrm{d}x + \int_0^L cuu_b\mathrm{d}x \quad (2.8)$$
$$+ \int_0^L \phi^{-1}(u')u_b'\,\mathrm{d}x$$

Also, by (2.6) and (2.8), we have:

$$A\left\|\phi^{-1}(u')\right\|_{L^2}^2 + B\left\|\phi^{-1}(u')\right\|_{L^q}^q$$
$$\le \|f\|_{L^q}\|u_b\|_{L^p} + \|u\|_{L^p}\left(\|f\|_{L^q} + \|cu_b\|_{L^q}\right)$$
$$+ \left\|\phi^{-1}(u')\right\|_{L^q}\|u_b'\|_{L^p} \quad (2.9)$$
$$\le C_1 + C_2\|u\|_{L^p} + C_3\left\|\phi^{-1}(u')\right\|_{L^q},$$

where $C_1 = \|f\|_{L^q}\|u_b\|_{L^p}$, $C_2 = \left(\|f\|_{L^q} + \|cu_b\|_{L^q}\right)$, and $C_3 = \|u_b'\|_{L^p}$.

By the Sobolev inequality, we have (see e.g. [8,9]): $\|u\|_{L^p} \le C\|u'\|_{L^q}$, and therefore:

$$B\left\|\phi^{-1}(u')\right\|_{L^q}^q \le C_1 + C_2\|u'\|_{L^q} + C_3\left\|\phi^{-1}(u')\right\|_{L^q}.$$

Also, since by definition of ϕ,

$$\|u'\|_{L^q} = \left\|A\phi^{-1}(u') + B\left|\phi^{-1}(u')\right|^{q-2}\phi^{-1}(u')\right\|_{L^q}$$
$$\le A\left\|\phi^{-1}(u')\right\|_{L^q} + B\left\|\left|\phi^{-1}(u')\right|^{q-1}\right\|_{L^q}$$
$$\le A\left\|\phi^{-1}(u')\right\|_{L^q} + \bar{B}\left\|\phi^{-1}(u')\right\|_{L^q}^{q-1},$$

we have

$$\left\|\phi^{-1}(u')\right\|_{L^q}^q \le \bar{C}_1 + \bar{C}_2\left\|\phi^{-1}(u')\right\|_{L^q} + \bar{C}_3\left\|\phi^{-1}(u')\right\|_{L^q}^{q-1} \quad (2.10)$$

where $\bar{C}_i, i = 1,2,3$ are positive constants. From (2.10), we conclude that $\left\|\phi^{-1}(u')\right\|_{L^q}$ is bounded and that there exists a constant C such that $\left\|\phi^{-1}(u')\right\|_{L^q} \le C$, as $u(x)$ varies over the solution set of (1.3) in V. Therefore, the result of the lemma is follows.

Theorem 2.1 For a given $f \in L^q(0,L)$, $q \ge 2$, $A > 0$ & $B > 0$, problem (I) has a unique solution $u \in U$.

Proof:

The uniqueness follows from the following argument. Let u_1 and u_2 be two solutions of (2.4). Then (since $c(x) \ge 0$):

$$0 \ge \int_0^L \left(\phi^{-1}(u_1') - \phi^{-1}(u_2')\right)(u_1' - u_2')\mathrm{d}x,$$

which leads to:

$$0 \ge \int_0^L (\sigma_1 - \sigma_2)(\phi(\sigma_1) - \phi(\sigma_2))\mathrm{d}x$$
$$= A\int_0^L (\sigma_1 - \sigma_2)^2\,\mathrm{d}x$$
$$+ B\int_0^L \left(|\sigma_1|^{q-2}\sigma_1 - |\sigma_2|^{q-2}\sigma_2\right)(\sigma_1 - \sigma_2)\mathrm{d}x$$
$$\ge A\int_0^L (\sigma_1 - \sigma_2)^2\mathrm{d}x,$$

since $\sigma_1 = \phi^{-1}(u_1')$, $\sigma_2 = \phi^{-1}(u_2')$ and

$$\left(|\sigma_1|^{q-2}\sigma_1 - |\sigma_2|^{q-2}\sigma_2\right)(\sigma_1 - \sigma_2) \ge 0,$$

which is well-known [10,11].

Therefore, $\sigma_1 = \sigma_2$ and $u_1 = u_2$, and this establishes the uniqueness of the solution of (2.4).

For existence, we consider the variational formulation of (2.4) and define the total potential energy by:

$$J(u) = \frac{1}{2}\left[\int_0^L \sigma\varepsilon\mathrm{d}x + \int_0^L cu^2\mathrm{d}x\right] - \int_0^L fu\mathrm{d}x$$
$$= \frac{1}{2}\left[\int_0^L \phi^{-1}(u')u'\mathrm{d}x + \int_0^L cu^2\mathrm{d}x\right] - \int_0^L fu\mathrm{d}x$$

Let $\varphi(t)=\frac{1}{2}\phi^{-1}(t)t$, then $J(u)$ can be written as:

$$J(u)=\frac{1}{2}\left[\int_0^L \varphi(u')\mathrm{d}x+\int_0^L cu^2\mathrm{d}x\right]-\int_0^L fu\mathrm{d}x.$$

Also we have: $\varphi'(t)=\frac{1}{2}\phi^{-1}(t)+\frac{1}{2}\left[\phi^{-1}(t)\right]'t$.

Letting $t=\phi(y(t))=Ay(t)+B|y(t)|^{q-2}y(t)$.

Then $y=\phi^{-1}(t)$ and
$1=Ay'(t)+(q-1)B|y|^{q-2}y'(t)$.
Therefore, we get

$$\left[\phi^{-1}(t)\right]'=y'(t)=\frac{1}{A+(q-1)B|y|^{q-2}},\text{ and}$$

$$\varphi'(t)=\frac{1}{2}\left[y(t)+\frac{t}{A+(q-1)B|y|^{q-2}}\right]$$
$$=\frac{1}{2}\left[\phi^{-1}(t)+\phi^{-1}(t)\right]=\phi^{-1}(t).$$

Now the first variation of J can be expressed as:

$$\frac{\mathrm{d}}{\mathrm{d}\varepsilon}J(u+\varepsilon v)\bigg|_{\varepsilon=0}=\int_0^L\left[\varphi'(u')v'+cuv\right]\mathrm{d}x-\int_0^L fv\mathrm{d}x$$
$$=\int_0^L\left[\phi^{-1}(u')v'+cuv\right]\mathrm{d}x-\int_0^L fv\mathrm{d}x.$$

However:

$$\varphi''(t)$$
$$=\frac{1}{2}\left[\frac{2}{A+(q-1)B|y|^{q-2}}+t\frac{(q-1)(q-2)B|y|^{q-4}yy'}{\left[A+(q-1)B|y|^{q-2}\right]^2}\right]$$
$$=\frac{1}{2}\left\{\frac{2}{\left[A+(q-1)B|y|^{q-2}\right]}+\frac{(q-1)(q-2)B|y|^{q-2}}{\left[A+(q-1)B|y|^{q-2}\right]^2}\right\}$$
$$=\frac{1}{2}\frac{\left[A+q(q-1)B|y|^{q-2}\right]}{\left[A+(q-1)B|y|^{q-2}\right]^2}\geq 0.$$

We rewrite the total energy function as
$J(u)=F_1(u)+F_2(u)-F(u)$, where
$F_1(u)=\frac{1}{2}\int_0^L\varphi(u')\mathrm{d}x$, $F_2(u)=\frac{1}{2}\int_0^L cu^2\mathrm{d}x$, and

$F(u)=\int_0^L f(x)u(x)\mathrm{d}x$. Then weak formulation (2.4) is equivalent to $\underset{u\in V}{\text{Min}}\,J(u)$.

Since $\varphi''(t)\geq 0$, $F_1:V\to R$ is convex , and since $c\in L^\infty(0,L)$, $F_2:V\to R$ is weekly sequentially continuous (since $\{u_n\}$ converges weekly in V implies that $\{u_n\}$ converges strongly in $L^q(0,L)$.) Also (2.6) and (2.7) imply the coercivity of $J(u)$, see, e.g., [9-11]. Therefore, $J(u)$ satisfies the conditions of the theorem of 42.7, pp. 225-226, in [9], and the existence of a weak solution follows.

We now consider the second case when the term $c(x)u(x)$ is replaced by $\sum_{i=1}^N k_i u(x_i)\delta(x-x_i)$.

In this case, $F_2(u)=\frac{1}{2}\sum_{i=1}^N k_i\left[u(x_i)\right]^2$ and we only need show that it is weakly sequentially continuous. Suppose that $\{u_n\}$ converges weekly in V, then for a $v\in W^{-1,q'}(0,L)$,

$$\lim_{k\to\infty}\int_0^L(v'u_k'+vu_k)\mathrm{d}x=\int_0^L(v'u'+vu)\mathrm{d}x$$

and $\lim_{k\to\infty}\int_0^L vu_k\mathrm{d}x=\int_0^L vu\mathrm{d}x$. Therefore, since

$$u_k(x_i)=\int_0^{x_i}u_k'(x)\mathrm{d}x=\int_0^L v'u_k'(x)\mathrm{d}x,$$

where

$$v(x)=\begin{cases}x, & 0\leq x\leq x_i\\ x_i, & x_i<x\leq L\end{cases}$$

We have

$$\lim_{k\to\infty}u_k(x_i)=\lim_{k\to\infty}\int_0^L v'u_k'(x)\mathrm{d}x=\int_0^L v'u'(x)\mathrm{d}x$$
$$=\int_0^{x_i}u'(x)\mathrm{d}x=u(x_i)$$

$$\lim_{k\to\infty}F_2(u_k)=\lim_{k\to\infty}\sum_{i=1}^N k_i\left[u_k(x_i)\right]^2$$
$$=\sum_{i=1}^N k_i\left[u(x_i)\right]^2=F_2(u).$$

Therefore, Theorem 2.1 holds with the same conditions for the case when $c(x)u(x)$ is replaced by $\sum_{i=1}^N k_i u(x_i)\delta(x-x_i)$.

3. Finite Element Error Estimates

Let $V_h\equiv S_h^k(0,L)\subset W^{1,q}(0,L)$ be a standard conformal finite element space of order k (See [12-15]) satisfying the interpolation property:

$$\|v-\Pi_h v\|_{1,p}\leq C(v)h^k,\forall v\in W^{1,p}(0,L),\qquad(3.1)$$

where C is a positive constant depending only on v and L, $\Pi_h v$ is the finite element interpolation of v, k is the polynomial degree for the interpolation shape functions, and h the mesh size, $\|v-\Pi_h v\|_{1,p}$ the $W^{1,p}(0, L)$ norm.

The corresponding finite element Galerkin's finite element approximation problem for (2.1) is:

Problem II:

Find $u_h \in V_h \equiv \{_h \in S_h^k(0,L) | v_h(0) = A, v_h'(L) = B\}$, such that

$$\int_0^L \phi^{-1}(u_h')v_h'\mathrm{d}x + \int_0^L u_h v_h dx v_h' \mathrm{d}x = \int_0^L f v_h \mathrm{d}x, \tag{3.2}$$

$$\forall v_h \in V_0 \equiv \{v_h \in S_h^k(0,L) | v_h(0) = 0, v_h'(L) = 0\}.$$

Theorem 3.1 Problem II has a unique solution.

Proof: The proof is similar to the proof of Theorem 2.1.

Lemma 3.1

For given positive constants A, B, q, L, there exists a constant C independent of the solutions $u_h \in V_h$ of Problem II such that $\|\phi^{-1}(u_h')\|_{L^q} \le C$.

Proof:

The proof is similar to that of **Lemma 2.1**.

To derive finite element error estimates, let u denotes the exact solution of Problem I and u_h the finite element solution of Problem II.

Then

$$\begin{aligned} & a(u, u-u_h) - a(u_h, u-u_h) \\ & = a(u, u-\Pi_h u) - a(u_h, u-\Pi_h u) \\ & = \int_0^L (\phi^{-1}(u') - \phi^{-1}(u_h'))(u' - \Pi_h u')\mathrm{d}x \\ & \quad + \int_0^L c(u-u_h)(u-\Pi_h u)\mathrm{d}x \\ & \le \|\phi^{-1}(u') - \phi^{-1}(u_h')\|_{L^q} \|u' - \Pi_h u'\|_{L^p} \\ & \quad + \|c(u-u_h)\|_{L^q} \|u - \Pi_h u\|_{L^p} \end{aligned} \tag{3.3}$$

Let $\sigma \equiv \phi^{-1}(u')$, and $\sigma \equiv \phi^{-1}(u')$, . Also

$$\begin{aligned} & a(u, u-u_h) - a(u_h, u-u_h) \\ & = \int_0^L (\phi^{-1}(u') - \phi^{-1}(u_h'))(u' - u_h')\mathrm{d}x + \int_0^L c(u-u_h)^2 \mathrm{d}x \\ & = \int_0^L (\sigma - \sigma_h)(\phi(\sigma) - \phi(\sigma_h))\mathrm{d}x + \int_0^L c(u-u_h)^2 \mathrm{d}x \\ & = A\int_0^L (\sigma - \sigma_h)^2 \mathrm{d}x + B\int_0^L (|\sigma|^{q-2}\sigma - |\sigma_h|^{q-2}\sigma_h)(\sigma - \sigma_h)\mathrm{d}x \\ & \quad + \int_0^L c(u-u_h)^2 \mathrm{d}x \\ & \ge A\int_0^L |\sigma - \sigma_h|^2 \mathrm{d}x + \int_0^L c(u-u_h)^2 \mathrm{d}x. \end{aligned} \tag{3.4}$$

As a result of (3.3) and (3.4), we get:

$$\begin{aligned} & \|c(u-u_h)\|_{L^2}^2 + \|\sigma - \sigma_h\|_{L^2}^2 \\ & \le \frac{1}{A}\Big[\|\phi^{-1}(u') - \phi^{-1}(u_h')\|_{L^q} \|u' - (\Pi_h u)'\|_{L^p} \\ & \quad + \|c(u-u_h)\|_{L^q} \|u - \Pi_h u\|_{L^p} \Big] \end{aligned} \tag{3.5}$$

By **Lemma 2.1**, **Lemma 3.1**, and (3.4), we get the following error estimates:

$$\begin{aligned} & \|c(u-u_h)\|_{L^2}^2 + \|\sigma - \sigma_h\|_{L^2}^2 \\ & \le C(\|u' - \Pi_h u'\|_{L^p} + \|u - \Pi_h u\|_{L^p}) \\ & \le Ch^k \end{aligned} \tag{3.6}$$

Therefore, by (3.6), we have established the following convergence and error estimate result.

Theorem 2.3 For $c(x) = \sum_{i=1}^{N} k_i \delta(x - x_i)$, $k_i > 0$, or any $c(x) \ge 0$, let u and u_h be the unique solutions of Problems I and II, respectively, then

$$\|\sigma - \sigma_h\|_{L^2} \le Ch^{k/2}, \text{ and } \lim_{h\to 0} \|u' - u_h'\|_{L^p} = 0,$$

and if $c(x) \ge c_0$ for some $c_0 > 0$, or

$$c(x) = \sum_{i=1}^{N} k_i \delta(x - x_i), k_i > 0, \text{ then}$$

$$\|u - u_h\|_{L^2}^2 + \|\sigma - \sigma_h\|_{L^2} \le Ch^{k/2}, \text{ and } \lim_{h\to 0} \|u' - u_h'\|_{1,p} = 0,$$

in which $\sigma \equiv \phi^{-1}(u')$ and $\sigma_h \equiv \phi^{-1}(u_h')$ stand for the stresses.

Note that σ stands for the stress corresponding to the strain $\varepsilon = u'$.

4. Conclusion

In this work, we establish existence and uniqueness of the solution u of (2.4) in the Sobolev space U and its finite element solution u_h in a general finite element space $S_0^h(0,L)$ with elastic support for a class of load functions *f*. We derive convergence and error estimates for the semi-discrete error $e_h(x) \equiv u(x) - u_h(x)$.

5. Acknowledgements

The research in this paper is a part of a research project funded by the Research office, Texas A & M University at Qatar.

REFERENCES

[1] W. R. Osgood and W. Ramberg, "Description of Stress-Strain Curves by Three Parameters," NACA Technical Note 902, National Bureau of Standards, Washington DC, 1943.

[2] L. A. James, "Ramberg-Osgood Strain-Harding Charac-

terization of an ASTM A302-B Steel," *Journal of Pressure Vessel Technology*, Vol. 117, No. 4, 1995, pp. 341-345.

[3] K. J. R. Rasmussen, "Full-Range Stress-Strain Curves for Stainless Steel Alloys," Research Report R811, University of Sydney, Department of Civil Engineering, 2001.

[4] V. N. Shlyannikov, "Elastic-Plastic Mixed-Mode Fracture Criteria and Parameters, Lecture Notes Applied Mechanics, Vol. 7," Springer, Berlin, 2002.

[5] P. Dong and L. DeCan, "Computational Assessment of Build Strategies for a Titanium Mid-Ship Section," *11th International Conference on Fast Sea Transportation, FAST*, Honolulu, 26-29 September 2011, pp. 540-546.

[6] P. Dong, "Computational Weld Modeling: A Enabler for Solving Complex Problems with Simple Solutions, Keynote Lecture," *Proceedings of the 5th IIW International Congress*, Sydney, 7-9 March 2007, pp. 79-84.

[7] E. Zeidler, "Nonlinear Functional Analysis and Its Applications, Variational Methods and Optimization, Vol. III," Springer Verlag, New York, 1986.

[8] R. A. Adams, "Sobolev Spaces, Pure and Applied Mathematics, Vol. 65," Academic Press, Inc., New York, San Francisco, London, 1975.

[9] V. G. Maz'Ja, "Sobolev Spaces," Springer-Verlag, New York, 1985.

[10] F. E. Browder, "Variational Methods for Non-Linear Elliptic Eigenvalue Problems," *Bulletin of the American Mathematical Society*, Vol. 71, 1965, pp. 176-183.

[11] R. Temam, "Mathematical Problems in Plasticity," Gauthier-Villars, Paris, 1985.

[12] G. Strang and G. J. Fix, "An Analysis of the Finite Element Method," Prentice-Hall, Inc., Englewood Cliffs, 1973.

[13] P. G. Ciarlet, "The Finite Element Method for Elliptic Problems," North- Holland, Amsterdam, 1978.

[14] J. T. Oden And G. F. Carey, "Finite Elements," Prentice-Hall, Englewood Cliffs, 1984.

[15] S. C. Brenner and L. R. Scott, "The Mathematical Theory of Finite Element Methods, Texts in Applied Mathematics, v. 15," 3rd Edition, Springer Verlag, New York, 2008.

13

Conjugate Effect of Radiation and Thermal Conductivity Variation on MHD Free Convection Flow for a Vertical Plate

Rowsanara Akhter[1*], Mohammad Mokaddes Ali[2], Babul Hossain[2], M. Sharif Uddin[1]
[1]Department of Mathematics, Jahangirnagar University, Dhaka, Bangladesh
[2]Department of Mathematics, Mawlana Bhashani Science and Technology University, Tangail, Bangladesh

ABSTRACT

A numerical investigation is performed to study the effect of thermal radiation on magnetohydrodynamic (MHD) free convection flow along a vertical flat plate in presence of variable thermal conductivity in this paper. The governing equations of the flow and the boundary conditions are transformed into dimensionless form using appropriate similarity transformations and then solved employing the implicit finite difference method with Keller-box scheme. Results for the details of the velocity profiles, temperature distributions as well as the skin friction, the rate of heat transfer and surface temperature distributions are shown graphically. Results reveal that the thermal radiation is more significant in MHD natural convection flow during thermal conductivity effect is considered. To illustrate the accuracy of the present results, the results for the local skin fraction and surface temperature distribution excluding the extension effects are compared with results of Merkin and Pop designed for the fixed value of Prandtl number and a good agreement were found.

Keywords: Radiation; MHD; Thermal Conductivity; Finite Difference Method

1. Introduction

The physical phenomenon of free convection flow is driven by temperature difference. Using these considerations, the temperature variation generates a density gradient which responsible for buoyancy forces. The buoyancy effects are important in free convection flow of viscous incompressible electrically conducting fluid. Many practical applications of free convection flow exist, for example in the heater and coolers of mechanical devices, in chemical industries, in nuclear power plants, in the formation of microstructures during the cooling of molten metal's, in fluid flows around heat-dissipation fins, and solar ponds etc. Moreover, MHD free convection flow is used frequently in the field of stellar and planetary magnetospheres, aeronautics, chemical engineering and electronics. Furthermore, most of the engineering processes are related with a high temperature, accordingly, radiation heat transfer is significant to design the relevant equipment of heat transfer process. In addition, radiation effects on MHD free convection flow and heat transfer are important in the context of space technology. Considering it's important applications in engineering and industrial fields, a number of theoretical and experimental work have been conducted extensively by many researchers. Among them, Soundalgekar and Takhar [1] studied the effect of radiation on MHD free convection flow of a gas past a sami-infinite vertical plate using the Cogley-vincenti-Giles equilibrium model (Cogley *et al.* [2]). Hossain and Takhar [3] employed implicit finite difference methods to analyze the effect radiation on mixed convection flow along a heated vertical flat plate with a uniform free stream and a uniform surface temperature. The effects of radiation and transverse magnetic field near stretching sheet were investigated by Ghaly [4] in the presence of a uniform free stream of constant velocity, temperature and concentration to show that radiation have significant influences on the velocity and temperature profiles. Abd El-Naby *et al.* [5] studied the radiation effects on MHD unsteady free convection flow over a vertical plate with variable surface temperature. Badruddin *et al.* [6] explored the effect of radiation and viscous dissipation on natural convection flow in a porous medium

*Corresponding author.

by imposing finite element method (FEM). Furthermore, it is also known that the physical property may change significantly with temperature. To obtain better prediction of the flow behavior, it is necessary to take into account this variation of thermal conductivity of fluid. Mishra *et al.* [7] employed alternating direction implicit scheme and collapsed dimension method to investigate the effect of temperature dependent thermal conductivity and radiation heat transfer on transient conduction for a 2-D rectangular enclosure containing an absorbing, emitting and scattering medium. Seddeek and Salama [8] applied perturbation technique and shooting method to analyze the effects of variable viscosity and thermal conductivity on MHD unsteady two-dimensional laminar flow of a viscous incompressible conducting fluid past a semi-infinite vertical porous moving plate considering variable suction. Sharma and Singh [9] obtained the analytical and numerical solutions of the effects of thermal conductivity on MHD steady free convection flow of a viscous incompressible electrically conducting liquid along an inclined isothermal non-conducting porous plate in presence of viscous dissipation and Ohmic heating. The effects of thermal conductivity on unsteady MHD free convective flow over an isothermal semi infinite vertical plate were studied by Loganathan *et al.* [10] using implicit finite-difference method of Crank-Nicholson type.

In this paper, the effects of radiation and variable thermal conductivity on free convection flow for a vertical flat plate in presence of transverse magnetic field are studied. The detail derivation of the governing equations for the flow and the parametric discussion depending on the numerical results of the present simulations are presented in the following section.

2. Governing Equations of the Flow

We consider the conduction inside a vertical heated flat plate and free convection flow of an incompressible, viscous and electrically conducting fluid along that vertical flat plate of length l and width b. The fluid properties are assumed to be constant and the temperature T_b of the outer surface of the plate is greater than ambient temperature T_∞ and a uniform magnetic field of strength H_0 is imposed along the $\bar{y}$-axis. Here the $\bar{x}$-axis is taken along the vertical flat plate in upward direction and also the $\bar{y}$-axis is normal to that plate. The effects of radiation from the heated plate and thermal conductivity variation within the two dimensional flow region are considered in this analysis. Moreover, thermal conductivity of the fluid is assumed as $k_f = k_\infty\{1+\gamma^*(T_f - T_\infty)\}$. The flow configuration and the coordinates system are shown in following **Figure 1**.

The governing equations of the flow under the Boussinesq approximations can be expressed within the usual boundary layer as follows:

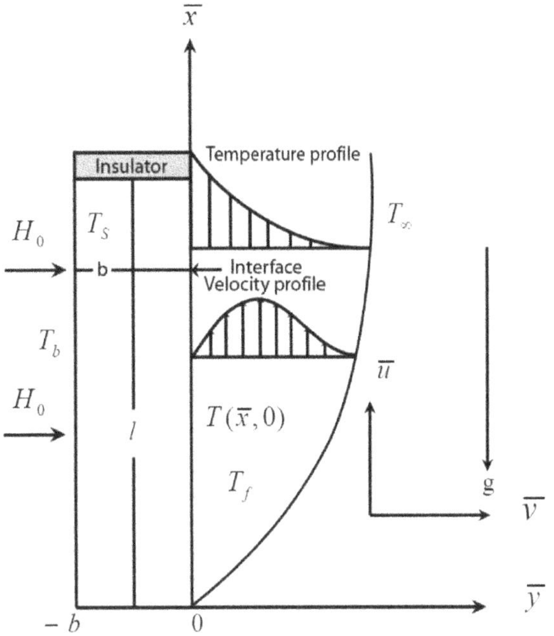

Figure 1. Physical model and coordinate system.

$$\frac{\partial \bar{u}}{\partial \bar{x}} + \frac{\partial \bar{v}}{\partial \bar{y}} = 0 \quad (1)$$

$$\bar{u}\frac{\partial \bar{u}}{\partial \bar{x}} + \bar{v}\frac{\partial \bar{u}}{\partial \bar{y}} = \nu\frac{\partial^2 \bar{u}}{\partial \bar{y}^2} + g\beta(T_f - T_\infty) - \frac{\sigma H_0^2 \bar{u}}{\rho} \quad (2)$$

$$\bar{u}\frac{\partial T_f}{\partial \bar{x}} + \bar{v}\frac{\partial T_f}{\partial \bar{y}} = \frac{1}{\rho C_p}\frac{\partial}{\partial \bar{y}}\left(k_f\frac{\partial T_f}{\partial \bar{y}}\right) - 4\Gamma(T_f - T_b) \quad (3)$$

where $\Gamma = \int_0^\infty K_{\lambda b}\left(\frac{\partial e_{b\lambda}}{\partial T_f}\right)_b \mathrm{d}\lambda$, $K_{\lambda b} = K_\lambda(T_b)$ is the mean absorption coefficient [5], $e_{b\lambda}$ is the Plank's function. The boundary conditions based on conduction are:

$$\left.\begin{aligned} &\bar{u} = \bar{v} = 0,\ T_f = T(\bar{x},0), \\ &\frac{\partial T_f}{\partial \bar{y}} = \frac{k_s}{bk_f}(T_f - T_b) \text{ at } \bar{y} = 0,\ \bar{x} > 0 \\ &\bar{u} \to 0, T_f \to T_\infty \text{ at } \bar{y} \to \infty,\ \bar{x} > 0 \end{aligned}\right\} \quad (4)$$

Equations (1) to (3) are nonlinear dimensional partial differential equations and these equations can be made non-dimensional by using the following dimensionless variables:

$$\left.\begin{aligned} &x = \frac{\bar{x}}{l}, y = \frac{\bar{y}}{l}Gr^{1/4},\ u = \frac{\bar{u}l}{\nu}Gr^{-1/2}, \\ &v = \frac{\bar{v}l}{\nu}Gr^{-1/4},\ \theta = \frac{T_f - T_\infty}{T_b - T_\infty}, \\ &Gr = \frac{g\beta l^3(T_b - T_\infty)}{\nu^2} \end{aligned}\right\} \quad (5)$$

Therefore, the dimensionless governing equations are:

$$\frac{\partial u}{\partial x}+\frac{\partial v}{\partial y}=0, \tag{6}$$

$$u\frac{\partial u}{\partial x}+v\frac{\partial u}{\partial y}+Mu=\frac{\partial^2 u}{\partial y^2}+\theta, \tag{7}$$

$$u\frac{\partial \theta}{\partial x}+v\frac{\partial \theta}{\partial y}=\frac{1}{\Pr}\left\{(1+\gamma\theta)\frac{\partial^2 \theta}{\partial y^2}+\gamma\left(\frac{\partial \theta}{\partial y}\right)\right\} -Ra(\theta-1) \tag{8}$$

The corresponding boundary conditions are:

$$\left.\begin{array}{l} u=v=0, \theta-1=p\dfrac{\partial \theta}{\partial y} \quad \text{at } y=0, x>0 \\ u\to 0, \theta\to 0 \quad \text{at } y\to\infty, x>0 \end{array}\right\} \tag{9}$$

Here $M=\frac{\sigma H_0{}^2 l^2}{\mu}Gr^{-1/2}$ is the magnetic parameter, $Ra=\frac{4\Gamma l^2}{\nu}Gr^{-1/2}$ is the radiation parameter, $\gamma=\gamma^*(T_b-T_\infty)$ is the thermal conductivity parameter, $\Pr=\frac{\mu C_p}{K_f}$ is the Prandtl number and $p=(k_f/k_s)(b/l)Gr^{1/4}$ is a conjugate conduction parameter. The value of the conjugate conduction parameter p depends on (b/l), (k_f/k_s) and Gr but each of which depends on the types of considered fluid and the solid. The steam function and similarity variable and the dimensionless temperature are considered in the following form to solve the governing equations:

$$\psi=x^{4/5}(1+x)^{-1/20}f(x,\eta),$$
$$\eta=yx^{-1/5}(1+x)^{-1/20}, \tag{10}$$
$$\theta=x^{1/5}(1+x)^{-1/5}h(x,\eta)$$

Using the above transformations, we obtain the following dimensionless governing equations:

$$f'''+\frac{16+15x}{20(1+x)}ff''-\frac{6+5x}{10(1+x)}f'^2 -Mx^{2/5}(1+x)^{1/10}f'+h=x\left(f'\frac{\partial f'}{\partial x}-f''\frac{\partial f}{\partial x}\right) \tag{11}$$

$$\frac{1}{\Pr}h''+\frac{\gamma}{\Pr}\left(\frac{x}{1+x}\right)hh''+\frac{\gamma}{\Pr}\left(\frac{x}{1+x}\right)(h')^2 +\frac{16+15x}{20(1+x)}fh'-\frac{1}{5(1+x)}f'h-Rax^{2/5}(1+x)^{1/10}h +Rax^{1/5}(1+x)^{3/10}=x\left(f'\frac{\partial h}{\partial x}-h'\frac{\partial f}{\partial x}\right) \tag{12}$$

The boundary condition (9) become

$$\left.\begin{array}{l} f(x,0)=f'(x,0)=0, \\ h'(x,0)=-(1+x)^{1/4}+x^{1/5}(1+x)^{1/20}h(x,0) \quad \text{at } y=0 \\ f'(x,\infty)\to 0, h(x,\infty)\to 0 \quad \text{at } y\to\infty. \end{array}\right\} \tag{13}$$

In practical point of view, it is important to calculate the values of the skin friction co-efficient in term of surface shear stress and the rate of heat transfer in term of the Nusselt number. These can be written in the dimensionless form as:

$$C_f=\frac{Gr^{-3/4}l^2}{\mu\nu}\tau_w \text{ and } Nu=\frac{Gr^{-1/4}l}{k_f(T_b-T_\infty)}q_w \tag{14}$$

where $\tau_w=\mu\left(\frac{\partial \bar{u}}{\partial \bar{y}}\right)_{\bar{y}=0}$ is the shearing stress and $q_w=-k_f\left(\frac{\partial T_f}{\partial \bar{y}}\right)_{\bar{y}=0}$ is the heat flux. Thus the local skin friction co-efficient and the local Nusselt number is obtained using the new variable systems that is describes in Equation (20) as follows:

$$C_{fx}=x^{2/5}(1+x)^{-3/20}f''(x,0) \text{ and}$$
$$Nu_x=-(1+x)^{-1/4}h(x,0) \tag{15}$$

The numerical value of the surface temperature distribution are obtained from the following relation

$$\theta(x,0)=x^{1/5}(1+x)^{-1/5}h(x,0) \tag{16}$$

We have discussed the velocity profiles and temperature distributions for various values of magnetic parameter, radiation parameter, thermal conductivity variation parameter and Prandtl number in the present investigation.

3. Method of Solution

The numerical solutions of this analysis are found by using implicit finite difference method with Keller-box [11] Scheme which is well documented by Cebeci and Bradshaw [12].

4. Comparison of the Results

The comparison of the skin friction coefficients (C_{fx}) and the surface temperature $(\theta(x,0))$ distribution between the present work and the work of Merin and Pop [13] is presented in following **Table 1**. We observed in this table, that the present analysis is an excellent agreement with the published work.

5. Results and Discussion

The main objective of the present work is to analyze

Table 1. Comparison of the present numerical results of the skin friction (C_{fx}) and surface temperature ($\theta(x,0)$) with Prandtl number Pr = 1.00, and p = 1.00.

$x^{\frac{1}{5}} = \xi$	Merin and Pop [13]		Present work (2013)	
	C_{fx}	$\theta(x,0)$	C_{fx}	$\theta(x,0)$
0.7	0.430	0.651	0.424	0.651
0.8	0.530	0.686	0.529	0.687
0.9	0.635	0.715	0.635	0.716
1.0	0.745	0.741	0.744	0.741
1.1	0.859	0.762	0.860	0.763
1.2	0.972	0.781	0.975	0.781

MHD free convection flow in presence of thermal conductivity variation with radiation effects. In this analysis, the numerical solutions are calculated from the transformed momentum and energy equations. The value of conjugate conduction parameter $p = 1.0$ is considered for the simulation of the present problem and the values of Prandtl number are considered 0.733, 0.930, 1.241 and 1.630 which corresponds to air, ammonia water and glycerin, respectively. The detailed numerical solutions have been obtained in terms of velocity, temperature, local skin friction, heat transfer rate and surface temperature for a wide range of values of the parameters as M = 0.10, 0.50, 0.70 and 1.00, Ra = 0.01, 0.03, 0.06 and 0.08, γ = 0.01, 0.10, 0.15 and 0.20 and then presented graphically in **Figures 2-11**, respectively.

The numerical values of velocity and temperature are obtained from the solution of the Equations (11) and (12) with the boundary condition (13) for different values of magnetic parameter M when Pr = 0.733, Ra = 0.01 and γ = 0.01 and are illustrated in **Figures 2(a)** and **(b)**, respectively. Here we observed that the velocity decreases for the increasing values of M. This is to be expected because, the magnetic field acting along the horizontal direction that introduces a retard force due to the interaction between applied magnetic field and fluid flow which acts against the fluid motion, as a result, the velocity of the fluid decreases. But near the surface of the plate the velocity increases and become maximum and then decrease and finally approaches to zero. Moreover, the velocity profiles meet together after certain position of η and cross the side. This is because, the gradient of decreasing of velocity decrease with the increasing value of magnetic parameter. In **Figure 2(b)**, the temperature within the boundary layer increases with the increasing values of magnetic parameter M due to the interaction of applied magnetic field and fluid motion that tends to heat the fluid. Furthermore, the temperature decreases monotonically with increasing of η for each value of M. Thus the magnetic field works to retard the fluid motion but increase the temperature within the thermal boundary layer region.

The variation of velocity and temperature for distinct values of the radiation parameter Ra together with an

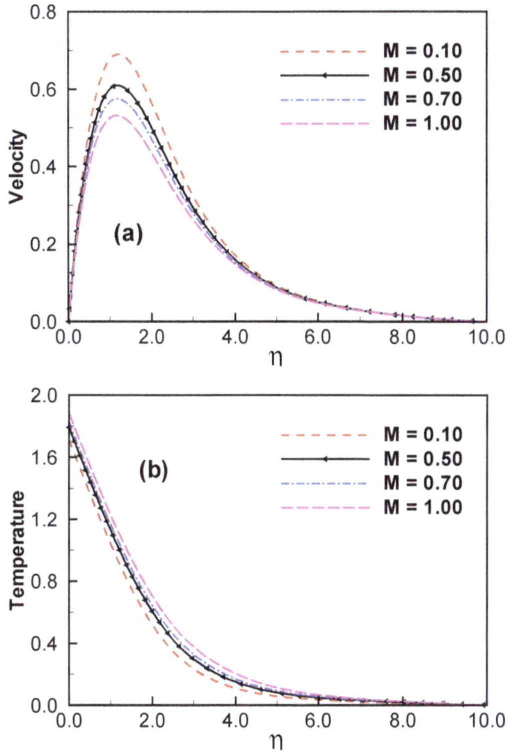

Figure 2. (a) Variation of velocity and (b) Variation of temperature against η for varying of M with Ra = 0.01 γ = 0.01, and Pr = 0.733.

individual certain value of Prandtl number, Pr, magnetic parameter, M and thermal conductivity variation parameter, γ are presented in **Figures 3(a)** and **(b)**, respectively. It can be seen that both the velocity and the temperature of the fluid increase within the velocity boundary layer and the temperature boundary layer, respectively for the increasing Ra due to the absorbsion of emitted heat from the heated plate that caused by the radiation effect. The trend is observed to shift upward and the peak velocity increases gradually with the increasing values of Ra. As the velocity and temperature of the fluid increases with the increasing value of radiation parameter as shown in **Figures 3(b)** and **(b)**, the thickness of the velocity and thermal boundary layer increase.

Figures 4(a) and **(b)** illustrate the effects of thermal conductivity variation parameter on velocity and temperature profiles associated with the certain value of M, Ra and Pr. Both figures reflect that the velocity and the temperature of the fluid increases with the increasing value of γ. The fact behind it's that the increasing value of thermal conductivity increases the energy transfer ability. It is also seen that near the surface of the plate the velocity becomes maximum with increasing of γ then after the peak position start to decrease and finally approaches to zero. On the other hand, the maximum values of temperature are occurred on the surface of the plate for each

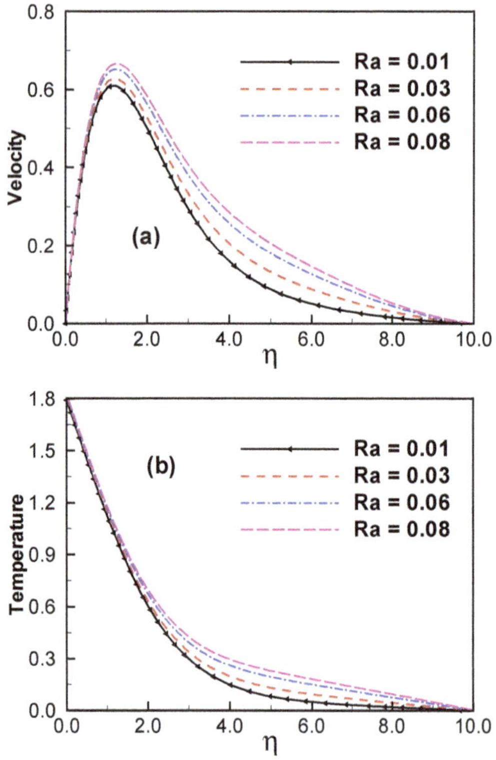

Figure 3. (a) Variation of velocity and (b) Variation of temperature against η for varying of Ra with M = 0.50 γ = 0.01, and Pr = 0.733.

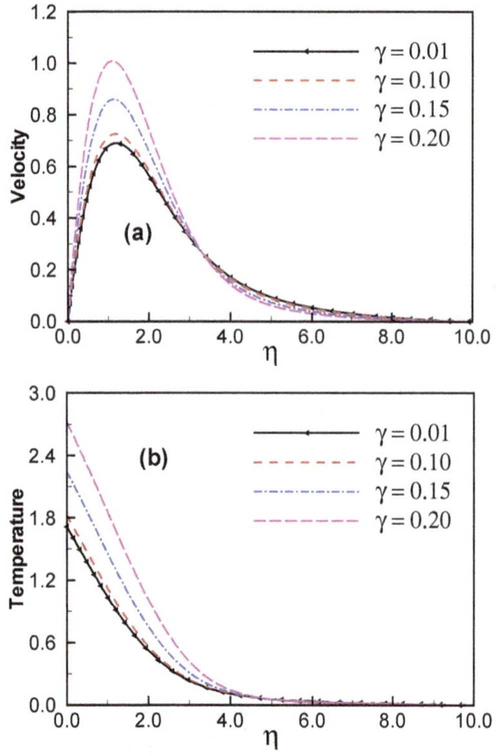

Figure 4. (a) Variation of velocity and (b) Variation of temperature against η for varying of γ with M = 0.50, Ra = 0.01 and Pr = 0.733.

value of γ and then turn to decrease asymptotically and finally approach to zero. These phenomenons are demonstrated in **Figures 4(a)** and **(b)**, respectively.

The effect of Pr on the velocity and temperature distributions is displayed in **Figures 5(a)** and **(b)**, respectively. From **Figure 5(a)** observed that the velocity of the fluid decrease as Pr increases. It is due to the fact that for increasing Pr, density of the fluid increases which creates a negative force to flow and then fluid does not move freely. Furthermore, **Figure 5(b)** shows that the temperature profiles for change in Pr from 0.733 to 1.630 and seen that the thermal boundary layer thickness decrease for increasing Pr, because of the increased Pr decrease the thermal diffusivity, which leads to the decrease of the energy transfer ability.

Figures 6(a) and **(b)**, respectively reveal that the skin friction coefficient and the heat transfer rate for some selected values of M with Pr = 0.733, Ra = 0.01, and $\gamma = 0.01$. The increased value of M leads to decrease the skin friction along the plate due to the fact the effect of magnetic field parameter opposes the fluid flow. Increasing fluid temperature for increasing M decrease the rate of heat transfer from the plate to fluid. This is because, the increased temperature reduces the temperature difference between the heated plate and fluid within the boundary layer.

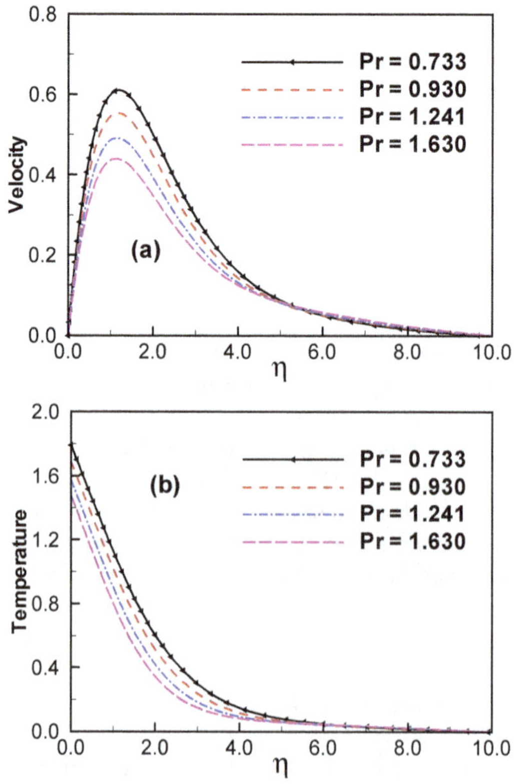

Figure 5. (a) Variation of velocity and (b) Variation of temperature against η for varying of Pr with, M = 0.50, Ra = 0.01 and γ = 0.01.

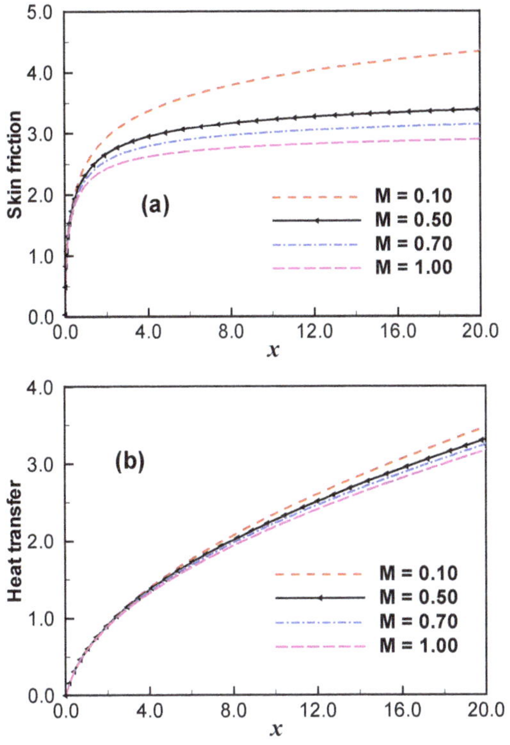

Figure 6. (a) Variation of skin friction and (b) Variation of heat transfer against x for varying of M with Ra = 0.01, γ = 0.01 and Pr = 0.733.

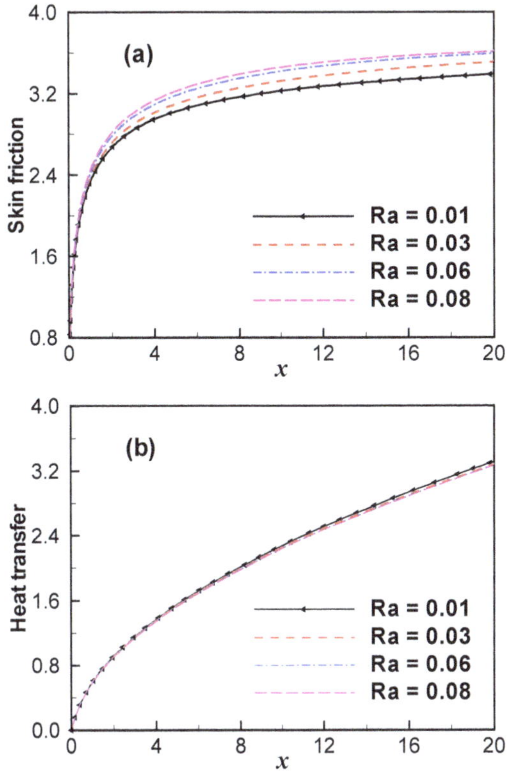

Figure 7. (a) Variation of skin friction and (b) Variation of heat transfer against x for varying of Ra with M = 0.50, γ = 0.01 and Pr = 0.733.

Figures 7(a) and **(b)** plotted the numerical values of the local skin friction coefficient (C_{fx}) and heat transfer rate (Nu_x) for different values of Ra associated with the distinct values of controlling parameter. As the radiation effect increases the fluid motion as well as temperature within the boundary layer which are mentioned earlier in **Figures 3(a)** and **(b)**, respectively. Accordingly, the corresponding skin friction increases and heat transfer rate decreases with the increasing value of Ra along the x direction.

The variation of skin friction and heat transfer rate for the effect of conductivity variation parameter are depicted in **Figures 8(a)** and **(b)**, respectively. The increasing value of γ generates greater buoyancy force which therefore increases the friction between the inner surface of the vertical plate and moving fluid particles. Thus the skin friction increases for the greater value of γ that is demonstrate in **Figure 8(a)**. Moreover, an increase in the value of γ leads to increase the energy transfer ability within the flow region, as a result, the heat transfers rate increases with the increasing of γ.

The effects of Prandtl number on the skin friction (C_{fx}) and heat transfer rate (Nu_x) against x for the fixed values of M, Ra and γ are shown respectively in **Figures 9(a)** and **(b)**. The increased values of Pr decrease both the velocity and temperature of the fluid within the boundary layer, consequently, the related skin friction on the plate decreases but the heat transfer rate from heated plate to fluid increases that has been exposed in **Figures 9(a)** and

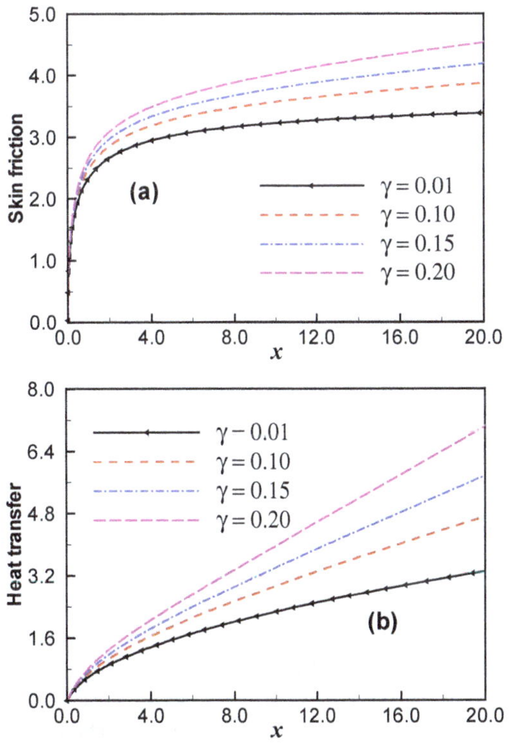

Figure 8. (a) Variation of skin friction and (b) Variation of heat transfer against x for varying of γ with M = 0.50, Ra = 0.01 and Pr = 0.733.

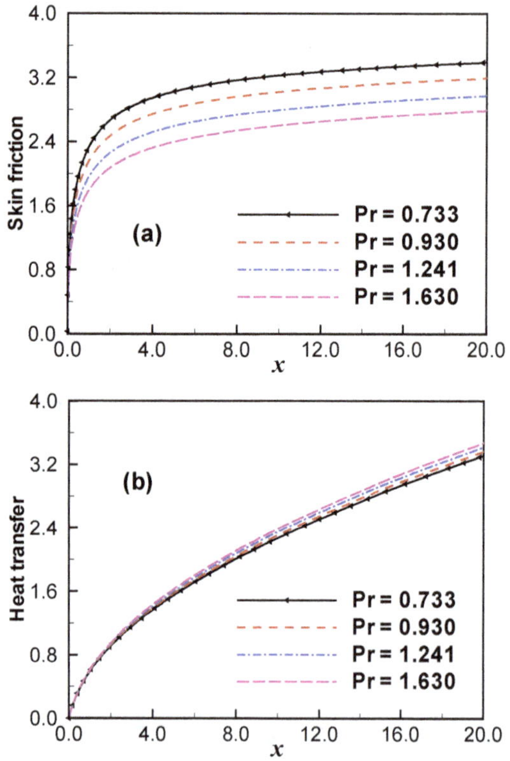

Figure 9. (a) Variation of skin friction and (b) Variation of heat transfer against x for varying of Pr with $M = 0.50$, $Ra = 0.01$ and $\gamma = 0.01$.

(b). Moreover, for a particular value of Pr, the local skin friction coefficient and heat transfer rate increase monotonically along the x direction.

The influence of magnetic field parameter, radiation parameter, conductivity variation parameter and Prandtl number on the interfacial temperature are depicted in **Figures 10** and **11**, respectively. The overall temperature profiles shift upward as well as the thermal boundary layer thickness increases with the increasing values of M, Ra, and γ observed in **Figures 2(b)**, **3(b)** and **4(b)**, respectively. Consequently, the surface temperature increase for enlarging values of M, Ra and γ, respectively. On the other hand, increasing Pr decreases the fluid temperature which results interfacial temperature decreases that is shown in **Figure 11(b)**.

6. Conclusions

In the present analysis, we have studied numerically the effects of radiation on MHD free convection flow under the influence of thermal conductivity variation for a vertical flat plate and the numerical solutions of the transformed governing equations associated with the specified boundary are obtained for different values of related physical parameters including magnetic parameter, radiation parameter, thermal conductivity variation parameter and Prandtl number. The particular conclusions in this

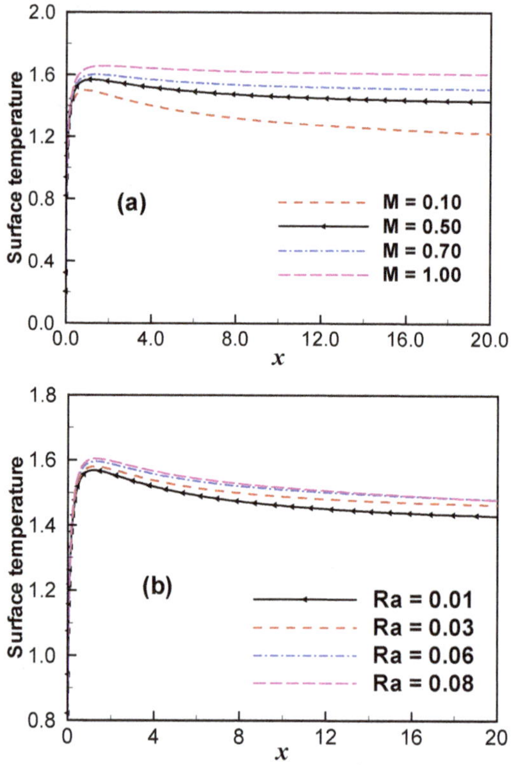

Figure 10. (a) Variation of surface temperature against x for varying M and (b) Variation of surface temperature against x for varying of Ra.

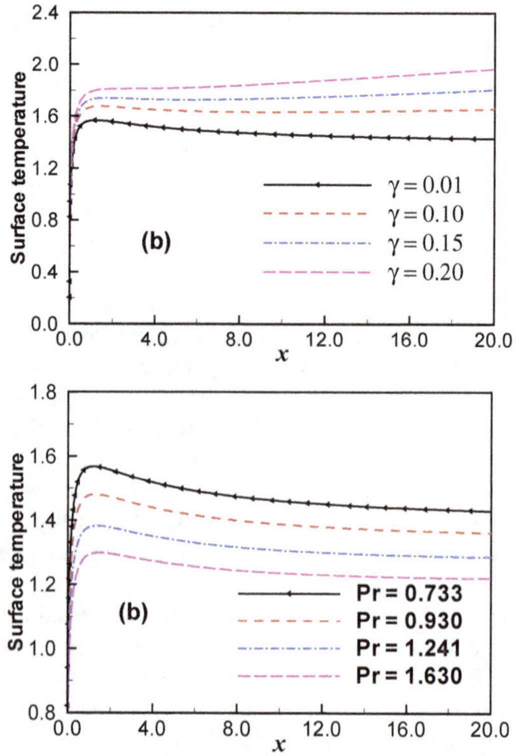

Figure 11. (a) Variation of surface temperature against x for varying γ and (b) Variation of surface temperature against x for varying of Pr.

study can be listed as follows:

The velocity of the fluid within the boundary layer and the skin friction at the interface increase for decreasing values of magnetic parameter, M, Prandtl number, Pr and increasing values of the radiation parameter, Ra and thermal conductivity variation parameter, γ.

The increasing value of M, Ra and γ leads to increase in the value of temperature within the thermal boundary layer as well as the surface temperature on the plate and the effect of M and Ra decrease heat transfer rate from plate to fluid within the boundary layer but opposite results hold for increasing of γ and Pr.

REFERENCES

[1] V. M. Soundalgekar and H. S. Takhar, "Radiative Convective Flow Past a Semi-Infinite Vertical Plate," *Modelling Measurement and Control*, Vol. 51, 1992, pp. 31-40.

[2] A. C. Cogley, W. G. Vincenti and S. E. Giles, "Differential Approximation for Radiative in a Non-Gray Gas Near Equilibrium," *American Institute of Aeronautics and Astronautics Journal*, Vol. 6, No. 3, 1968, pp. 551-553.

[3] M. A. Hossain and H. S. Takhar, "Radiation Effect on Mixed Convection along a Vertical Plate with Uniform Surface Temperature," *Heat and Mass Transfer*, Vol. 31, No. 4, 1996, pp. 243-248.

[4] A. Y. Ghaly, "Radiation Effects on a Certain MHD Free Convection Flow," *Chaos, Solitons & Fractals*, Vol. 13, No. 9, 2002, pp. 1843-1850.

[5] M. A. Abd El-Naby, E. M. E. Elsayed and N. Y. Abdelazem, "Finite Difference Solution of Radiation Effect on MHD Unsteady Free Convection Flow over a Vertical Plate Variable Surface Temperature," *Journal of Applied Mathematics*, Vol. 2003, No. 2, 2003, pp. 65-86.

[6] I. A. Badruddin, Z. A. Zainal, A. Zahid Khan and Z. Mallick, "Effect of Viscous Dissipation and Radiation on Natural Convection in a Porous Medium Embedded within Vertical Annulus," *International Journal of Thermal Science*, Vol. 46, No. 3, 2007, pp. 221-227.

[7] S. C. Mishra, P. Talukdar, D. Trimis and F. Durst, "Two-Dimensional Transient Conduction and Radiation Heat Transfer with Temperature Dependent Thermal Conductivity," *International Communications in Heat and Mass Transfer*, Vol. 32, No. 3-4, 2005, pp. 305-314.

[8] M. A. Seddeek and F. A. Salama, "The Effects of Temperatute Dependent Viscosity and Thermal Conductivity on Unsteady MHD Convective Heat Transfer Past a Semi-Infinite Vertical Porous Moving Plate with Variable Suction," *Computational Material Science*, Vol. 40, No. 2, 2007, pp. 186-192.

[9] P. R. Sharma and G. Singh, "Effects of Variable Thermal Conductivity, Viscous Dissipation on Steady MHD Natural Convection Flow of Low Prandtl Fluid on an Inclined Porous Plate with Ohmic Heating," *Meccanica*, Vol. 45, No. 2, 2010, pp. 237-247.

[10] P. Loganathan, P. Ganesan and D. Iranian, "Effects of Thermal Conductivity on Unsteady MHD Free Convective Flow over a Semi Infinite Vertical Plate," *International Journal of Engineering Science and Technology*, Vol. 2, 2010, pp. 6257-6268.

[11] H. B. Keller, "Numerical Methods in the Boundary Layer Theory," *Annual Review of Fluid Mechanics*, Vol. 10, 1978, pp. 417-433.

[12] T. Cebeci and P. Bradshow, "Physical and Computational Aspects of Convective Heat Transfer," Springer, New York, 1984.

[13] J. H. Merkin and I. Pop, "Conjugate Free Convection on a Vertical Surface," *International Journal of Heat and Mass Transfer*, Vol. 39, No. 7, 1996, pp. 1527-1534.

Nomenclature

C_{fx} : Local skin friction coefficient
C_p : Specific heat at constant pressure
f: Dimensionless stream function
g: Acceleration due to gravity
Gr : Grashof number
h: Dimensionless temperature
k_f , k_s : Fluid and solid thermal conductivities
M: Magnetic parameter
Nu_x: Local Nusselt number
p: Conjugate conduction parameter
Pr: Prandtl number
T_f : Temperature of the fluid
$\bar{u}$, $\bar{v}$: Velocity components
u, v: Dimensionless velocity components
$\bar{x}$, $\bar{y}$: Cartesian co-ordinates
x, y: Dimensionless Cartesian co-ordinate
β: Coefficient of thermal expansion
γ: Thermal conductivity variation parameter
η: Dimensionless similarity variable
θ: Dimensionless temperature
μ, ν: Dynamic and kinematic viscosities
ρ: Density of the fluid
σ: Electrical conductivity

14

Full-waveform Velocity Inversion Based on the Acoustic Wave Equation

Wensheng Zhang, Jia Luo
Institute of Computational Mathematics and Scientific/Engineering Computing, LSEC
Academy of Mathematics and Systems Science, Chinese Academy of Sciences, Beijing, P. R. China

ABSTRACT

Full-waveform velocity inversion based on the acoustic wave equation in the time domain is investigated in this paper. The inversion is the iterative minimization of the misfit between observed data and synthetic data obtained by a numerical solution of the wave equation. Two inversion algorithms in combination with the CG method and the BFGS method are described respectively. Numerical computations for two models including the benchmark Marmousi model with complex structure are implemented. The inversion results show that the BFGS-based algorithm behaves better in inversion than the CG-based algorithm does. Moreover, the good inversion result for Marmousi model with the BFGS-based algorithm suggests the quasi-Newton methods can provide an important tool for large-scale velocity inversion. More computations demonstrate the correctness and effectives of our inversion algorithms and code.

Keywords: Finite Difference; Acoustic Wave Equation; Full-waveform Inversion; CG Method; BFGS Method; Marmousi Model

1. Introduction

The velocity of underground media is an important parameter for resources exploration. It may indicate the structure and position of reservoir. Full-waveform inversion uses travel time, amplitude and phase information simultaneously to inverse media velocity. It is a powerful way in reconstructing complex velocity structures. The inversion can be performed in the time-space domain [1-4] or in the frequency-space domain [5-9]. The frequency-domain inversion approach is equivalent to the time-domain inversion approach if all of the frequency data components are used in the inversion process [9].

Full-waveform inversion can be formulated as a nonlinear optimization problem with an objective function consisting of the difference between observed and synthetic wave field in a suitable norm, usually L_2 norm. This choice results in least-squares formulation. However, solving full-waveform inversion problems using nonlinear optimization methods remains a challenge. This is mostly because of mathematical and numerical difficulties, which include multiple minima and ill-posedness. The objective function is relatively insensitive to wave number components of the model that are shorter than wavelength. The eigenvalues of the linearized forward operator corresponding to those components are nearly zero. Recently, the technique of multifrequency inversion of full-waveform has been investigated. For frequency domain inversion, the important issue is how to choose the various frequencies for inversion [10]. Moreover, the wave modeling in the frequency domain requires solving a large-scale system at each iteration which is time-consuming for large-scale problem. Comparing with the frequency-domain inversion methods, the time-domain methods have the advantage of high efficiency in forward modeling.

The algorithms used to solve the optimization problem include linearized approximation, the steepest descent algorithm, the conjugate gradient (CG) [11] and the quasi-Newton methods [12-15]. However, these methods behave differently as their own characters and the essential ill-posedness of the problem. In this paper, we implement full-waveform inversion with the CG method which is a typical gradient-based method, and with the BFGS method which is a representative of the quasi-Newton methods. As to the forward method, the finite-difference method is employed as its high efficiency. The discretization schemes including the absorbing boundary conditions and corner conditions are given.

2. Theoretical Methods

2.1. Finite Difference Simulation

The wave equations can be solved in either the time or the frequency domain. The usual numerical methods such as the finite difference method, the finite volume method and the finite element method can all be employed. Here we use the finite difference method for its high efficiency. The 2-D acoustic wave equation in the time domain can be written as

$$\frac{1}{v(x,z)^2}\frac{\partial^2 u}{\partial t^2}-(\frac{\partial^2 u}{\partial x^2}+\frac{\partial^2 u}{\partial z^2}) = f(t)\delta(x-x_s)\delta(z-z_s), \tag{1}$$

where $u(x,z,t)$ is the pressure, $v(x,z)$ is the velocity, $f(t)$ is the source time function, (x_s, z_s) is the source location. Usually the initial conditions are zero, i.e.

$$u(x,z,t=0)=0,\ u_t(x,z,t=0)=0. \tag{2}$$

Let $u^l_{n,m}$ be the pressure at the time $l\Delta t$ and at the spatial position $(n\Delta x, m\Delta z)$, and $v_{n,m}$ be the velocity at $(n\Delta x, m\Delta z)$, then the difference scheme of (1) is

$$\begin{aligned} u^{l+1}_{n,m} &= u^{l-1}_{n,m} - 2u^l_{n,m} \\ &+ v^2_{n,m}\Delta t^2 (\frac{u^l_{n+1,m}-2u^l_{n,m}+u^l_{n-1,m}}{\Delta x^2 r} \\ &+ \frac{u^l_{n+1,m}-2u^l_{n,m}+u^l_{n-1,m}}{\Delta x^2 r}) \\ &+ f(l\Delta t)\delta(n-n_s)\delta(m-m_s), \end{aligned} \tag{3}$$

with the initial conditions

$$u^{-1}_{n,m}=0,\ u^0_{n,m}=0 \tag{4}$$

The numerical stability condition for (3) and (4) is

$$\min(\Delta x,\Delta z) > \sqrt{2}\Delta t \max(v). \tag{5}$$

In the wave simulation, the unsuitable choice of time and spatial steps may cause serve dispersion and wave distortion. In order to suppress numerical dispersion, it is usually required that there are 10 sampling points in a wavelength for the second-order difference scheme, i.e.,

$$\min(\Delta x,\Delta z) < \frac{\min(v)}{10 f_{\max}}. \tag{6}$$

Obviously for the same frequency the numerical dispersion increases as the spatial steps rise. However, the dispersion can be depressed if the high-order accurate computational schemes are used.

In numerical computations, the computational domain is usually truncated to a finite region. So absorbing boundary conditions (ABCs) are required to absorb boundary reflections of outgoing waves. A number of ABCs have been introduced in wave simulation, for example, the paraxial approximation method [16] and the perfectly matched layer (PML) method [17]. The PML technique was proposed for electromagnetic fields equations, but can be applied to other wave equations and can fit irregular boundary.

Suppose the computational domain is a rectangular area

$$\Omega = \{(x,z), 0\le x\le X, 0\le z\le Z\}.$$

We set the boundary condition on surface ($z=0$) be the Dirichlet boundary condition as the data are gathered near the surface in our problem. For the boundary conditions of other three sides and four corners, we select the following Clayton and Engquist [16] conditions.

For the top boundary:

$$u(x,z=0,t)=0. \tag{7}$$

For the bottom boundary:

$$\left(\frac{\partial^2 u}{\partial z\partial t}+\frac{1}{v}\frac{\partial^2 u}{\partial t^2}-\frac{v}{2}\frac{\partial^2 u}{\partial x^2}\right)(x,z=Z,t)=0. \tag{8}$$

For the left boundary:

$$\left(\frac{\partial^2 u}{\partial x\partial t}-\frac{1}{v}\frac{\partial^2 u}{\partial t^2}+\frac{v}{2}\frac{\partial^2 u}{\partial z^2}\right)(x=0,z,t)=0. \tag{9}$$

For the right boundary:

$$\left(\frac{\partial^2 u}{\partial x\partial t}+\frac{1}{v}\frac{\partial^2 u}{\partial t^2}-\frac{v}{2}\frac{\partial^2 u}{\partial z^2}\right)(x=X,z,t)=0. \tag{10}$$

The absorbing conditions for four corners can be derived through the 45° rotation of the ABCs above. The results are the following:

For the left-upper corner:

$$\left(-\frac{\partial u}{\partial x}-\frac{\partial u}{\partial z}+\frac{\sqrt{2}}{v}\frac{\partial u}{\partial t}\right)=0. \tag{12}$$

For the right-upper corner:

$$\left(\frac{\partial u}{\partial x}-\frac{\partial u}{\partial z}+\frac{\sqrt{2}}{v}\frac{\partial u}{\partial t}\right)=0. \tag{13}$$

For the left-down corner:

$$\left(-\frac{\partial u}{\partial x}+\frac{\partial u}{\partial z}+\frac{\sqrt{2}}{v}\frac{\partial u}{\partial t}\right)=0. \tag{14}$$

For the right-down corner:

$$\left(\frac{\partial u}{\partial x}+\frac{\partial u}{\partial z}+\frac{\sqrt{2}}{v}\frac{\partial u}{\partial t}\right)=0. \tag{15}$$

The difference schemes for the absorbing boundary condition (7)-(10) can also be obtained by the second-order central finite difference method, i.e.,

$$u^{l+1}_{n,m}=0, \tag{16}$$

$$\begin{aligned} &\frac{\delta^z_+\delta^t_0 u^l_{n,m}}{2\Delta t\Delta z}+\frac{1}{2\Delta t^2 v_{i,j}}\delta^t_+\delta^t_-(u^l_{n,m}+u^l_{n,m-1}) \\ &-\frac{v_{i,j}}{4\Delta x^2}\delta^x_+\delta^x_-(u^{l+1}_{n,m-1}+u^{l-1}_{n,m})=0, \end{aligned} \tag{17}$$

$$\frac{\delta_+^x \delta_0^t u_{n,m}^l}{2\Delta t \Delta x} - \frac{1}{2\Delta t^2 v_{i,j}} \delta_+^t \delta_-^t (u_{n,m}^l + u_{n+1,m}^l) + \frac{v_{i,j}}{4\Delta z^2} \delta_+^z \delta_-^z (u_{n+1,m}^{l+1} + u_{n,m}^{l-1}) = 0, \quad (18)$$

$$\frac{\delta_+^x \delta_0^t u_{n,m}^l}{2\Delta t \Delta x} + \frac{1}{2\Delta t^2 v_{i,j}} \delta_+^t \delta_-^t (u_{n,m}^l + u_{n-1,m}^l) - \frac{v_{i,j}}{4\Delta z^2} \delta_+^z \delta_-^z (u_{n-1,m}^{l+1} + u_{n,m}^{l-1}) = 0, \quad (19)$$

where δ_+ and δ_- are the forward and backward difference operators with first-order accuracy respectively, δ_0 is the central difference operator with second-order accuracy, Δx and Δz are the spatial steps in x and z directions respectively. Similarly, the difference schemes for four corners conditions (12)-(15) can be obtained in the following:

For the left-upper corner: $(n,m) = (0,\ 1)$,

$$u_{n,m}^{l+1} = \frac{r_1 u_{n+1,m}^{l+1} + r_2 u_{n,m+1}^{l+1} + r_3 u_{n,m}^l}{r_1 + r_2 + r_3}. \quad (20)$$

For the right-upper corner: $(n,m) = (N_x - 1,\ 1)$,

$$u_{n,m}^{l+1} = \frac{r_1 u_{n-1,m}^{l+1} + r_2 u_{n,m+1}^{l+1} + r_3 u_{n,m}^l}{r_1 + r_2 + r_3}. \quad (21)$$

For the left-down corner:

$$(n,m) = (0, N_z - 1),\ (1, N_z - 1),\ (0, N_z - 2),$$
$$u_{n,m}^{l+1} = \frac{r_1 u_{n+1,m}^{l+1} + r_2 u_{n,m-1}^{l+1} + r_3 u_{n,m}^l}{r_1 + r_2 + r_3}. \quad (22)$$

For the right-down corner:

$$(n,m) = (N_x - 1, N_z - 1),\ (N_x - 2, N_z - 1),\ (N_x - 1, N_z - 2), \quad (23)$$
$$u_{n,m}^{l+1} = \frac{r_1 u_{n-1,m}^{l+1} + r_2 u_{n,m-1}^{l+1} + r_3 u_{n,m}^l}{r_1 + r_2 + r_3}.$$

where

$$r_1 = \frac{\Delta t}{\Delta x},\ r_2 = \frac{\Delta t}{\Delta z},\ r_3 = \frac{\sqrt{2}}{v_{n,m}}. \quad (24)$$

2.2. Two Full-waveform Inversion Algorithms

Full-waveform inversion is minimization of the following objective function

$$f(v) = \frac{1}{2} \| u_{obs} - u_{cal} \|^2 = \frac{1}{2} \| \Delta u \|^2, \quad (25)$$

where u_{obs} is the observed wave field and u_{cal} is the synthetic data. The functional (25) can be written the following discrimination form

$$f(v) = \sum_{n_s=0}^{N_s} \sum_{n_t=0}^{N_t} \sum_{nr=0}^{N_r} \{u_{obs}(r, n_t, n_s) - u_{cal}(v; n_r, m_r, n_t, n_s)\}^2, \quad (26)$$

where the number of spatial points is $M = N_x \times N_z$, which is also the dimension of unknown velocity v. Assume that the receiver number is N_r and the source number is N_s then the dimension of $u_{obs}(r, n_t, n_s)$ is $N_r \times N_s \times N_t$, here N_t is the time sampling points.

The optimization methods such as nonlinear conjugate gradient and BFGS can be implemented. The gradient-based methods are the commonly used methods in multidimensional unconstrained optimization. The CG method is one and it uses the conjugate directions instead of the negative gradient direction as search direction, which can be written as

$$p_k = \begin{cases} -g_k, & k = 0, \\ -g_k + \beta_k p_{k-1}, & k \ge 1. \end{cases} \quad (27)$$

Different choices of β_k lead to different conjugate gradient methods. For example, the FR method [11], the HS method [18] and the PRP method [19, 20]:

$$\beta_k^{FR} = \frac{\| g_{k+1} \|^2}{\| g_k \|^2},\ \beta_k^{HS} = \frac{g_{k+1}^T y_k}{p_k^T y_k}, \quad (28)$$

$$\beta_k^{PRP} = \frac{g_{k+1}^T y_k}{\| g_k \|^2}, \quad (29)$$

where $y_k = g_{k+1} - g_k$. Here we adopt the PRP method.

The partial derivative of $f(v)$ with respect to unknown velocity v is

$$\frac{\partial f(v)}{\partial v} = J^T \Delta u, \quad (30)$$

where J is the Fréchet derivative matrix with dimension $M \times (N_t \times N_r)$:

$$J = \left[\frac{\partial u_{cal}}{\partial v} \right] = \left[\frac{\partial u_{cal}}{\partial v_0}, \frac{\partial u_{cal}}{\partial v_1}, \ldots, \frac{\partial u_{cal}}{\partial v_{M-1}} \right]. \quad (31)$$

The gradient g is a vector with dimension M, which can be written as

$$g = -J^T \Delta u. \quad (32)$$

For the line search method we may use the Aimijo algorithm [21] or the strong Wolfe algorithm [22, 23]. The Aimijo algorithm requires

$$J(v + \alpha p) \le J(v) + b_1 \alpha p^T \nabla_J(v), \quad (33)$$

where $b_1 \in (0,1)$ and we select $b_1 = 0.001$ in computations. And the strong Wolfe algorithm requires the following condition besides condition (33) above

$$|p^T \nabla_J(v + \alpha p)| \le -b_2 \alpha p^T \nabla_J(v), \quad (34)$$

where $b_2 \in (b_1, 1)$ and we select $b_2 = 0.9$ in computations. Now we give the following full-waveform inversion algorithm combining with the CG method.

Algorithm 1:

1: Give the initial velocity model v_0 and the iterative

stopping condition of the gradient: $||g|| \le \varepsilon_{stop}$. Set iteration number $k = 0$;

2: Compute the objective function f_0 and the gradient g. If the gradient satisfies the stopping condition $||g|| \le \varepsilon_{stop}$ then stop the iteration. Otherwise set $p_0 = -g_0$;

3: Set $k = k+1$. If $k < k_{\max}$ then continue the iteration;

3.1: Use the strong Wolfe line search algorithm to get the step α_k;

3.2 : Update the velocity model: $v_{k+1} = v_k + \alpha_k p_k$;

3.3: Compute the objective function $f_{k+1}(v_{k+1})$ and the gradient g_{k+1};

3.4: If the gradient satisfies the stopping criterion then go out of the loop;

3.5: If the updating quantity of velocity satisfies the stopping condition then goes out of the loop;

3.6: Calculate, $s_k = v_{k+1} - v_k$, $y_k = g_{k+1} - g_k$;

3.7: Calculate β_k^{PRP} and update

$$p_{k+1}: \ p_{k+1} = -g_{k+1} + \beta_k^{PRP} p_k.$$

Set $k = k+1$, return step 3.

The algorithm above doesn't use the information of second-order derivate of objective function. The BFGS method is a quasi-Newton method which proposed dependently by Broyden, Fletcher and Goldfarb and Shanno [12-15]. The key of the quasi-Newton method is how to construct B_{k+1} satisfying the following so-called quasi-Newton condition:

$$B_{k+1}s_k = y_k, \tag{35}$$

where $s_k = v_{k+1} - v_k$ and $y_k = g_{k+1} - g_k$. The BFGS method modifies matrix B_{k+1} according to the following expression

$$B_{k+1} = B_k - \frac{B_k ss^T B_k}{s^T B_k s} + \frac{yy^T}{s^T y}, \tag{36}$$

and its inverse is given by

$$H_{k+1} = H_k + \frac{ss^T}{y^T s}\left[\frac{y^T H_k y}{y^T s} + 1\right] - \frac{1}{y^T s}\left[sy^T H_k + H_k y s^T\right], \tag{37}$$

where $H_k = B_k^{-1}$. If H_k is positive definite then H_{k+1} is positive definite as long as $y_k^T s > 0$. The following algorithm is the full-waveform inversion algorithm based on the BFGS method.

Algorithm 2:

1: Give the initial velocity model v_0 and the iterative terminal condition of the gradient: $||g|| \le \varepsilon_{stop}$. Set iteration number $k = 0$;

2: Compute the objective function f_0 and the gradient g. If the gradient satisfies the stopping rule $||g|| \le \varepsilon_{stop}$ then stop the iteration. Otherwise set $p_0 = -g_0$;

3: Set $k = k+1$. If $k < k_{\max}$ then continue the iteration;

3.1: Use the strong Wolfe line search algorithm to obtain the step α_k;

3.2 : Update the velocity model: $v_{k+1} = v_k + \alpha_k p_k$;

3.3: Compute the objective function $f_{k+1}(v_{k+1})$ and the gradient g_{k+1};

3.4: If the gradient satisfies the stopping condition, then go out of the loop;

3.5: If the updating quantity of velocity satisfies the stopping condition, then go out of the loop;

3.6: Calculate $s_k = v_{k+1} - v_k$, $y_k = g_{k+1} - g_k$;

3.7: Calculate variable scaling factor with

$$\tau_k = \frac{y_k^T s_k}{s_k^T B_k s_k} \text{ and let } B_k = \tau_k B_k;$$

3.8: Update B_{k+1} with (36).

3.9: Use the Cholesky method to solve the equation

$$B_{k+1} p_{k+1} = -g_{k+1} \text{ to get } p_{k+1}.$$

Set $k = k+1$, continue with step 3.

In step 3.7 above, the scaling factor τ_k is introduced which can improve convergence [24]. The CG method is only linearly convergent while the BFGS method is superlinear convergent but requires solving a system at each iteration.

3. Numerical Computations

First of all, we test the validity of the finite difference method for wave modeling. **Figure 1** is the wave filed received near the surface for a fixed source. The boundary reflections are very serious in **Figure 1** since only Dirichlet boundary conditions are used. **Figure 2** is the wave field corresponding to **Figure 1** but the ABCs are used. We can see the serious boundary reflections in **Figure 1** are eliminated obviously in **Figure 2**.

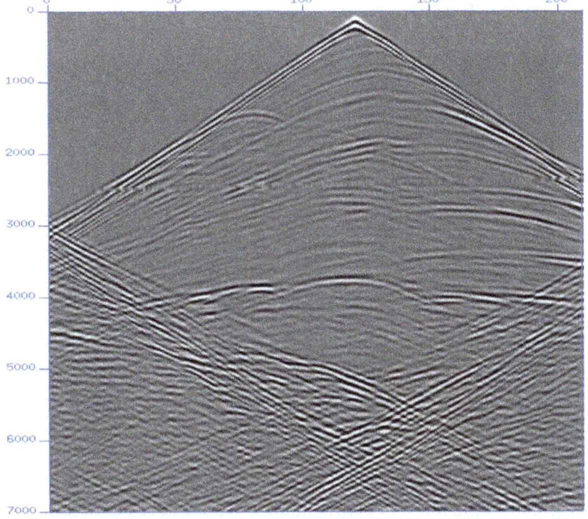

Figure 1. Wavefield received near the surface by the finite difference method with Dirichlet boundary conditions.

Next we test two inversion algorithms for a simple model shown in **Figure 3**. The model has a homogeneous background media with velocity 2000 *m/s* and two square areas with greater abnormal velocity. The left square media has velocity 2600 *m/s* and the right square media has the velocity 2100 *m/s*. The configuration of sources and receivers are shown in **Figure 4**. There are 16 sources totally and 31 receivers for each source. The source step is 10 *m* and the receiver step is 5 *m*. The source line and receiver line are 10 *m* and 5 *m* below the surface respectively. The spatial discretization points are $N_x = 17$ and $N_z = 33$. The spatial steps are $\Delta x = \Delta z = 5m$. The parameter ε_{stop} is 10^{-7}. The maximum number of iterations $k_{\max}$ is 200. The initial velocity v_0 is 2000*m/s*.

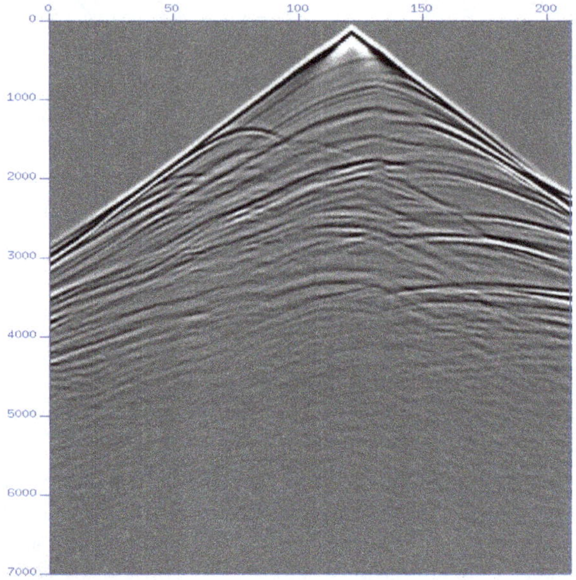

Figure 2. Wavefield received near the surface by the finite difference method with absorbing boundary conditions.

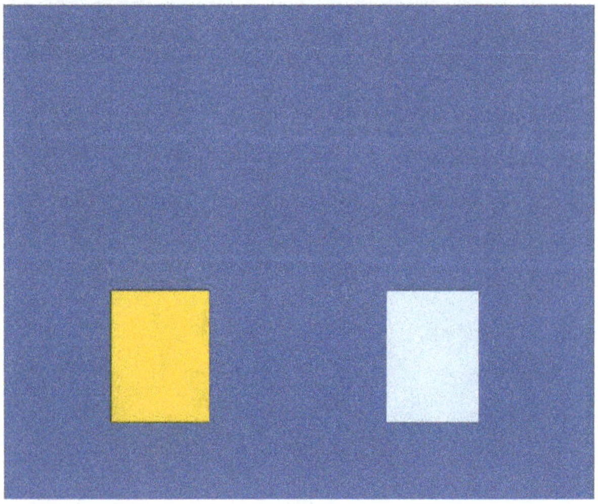

Figure 3. True model for testing two inversion algorithms.

The inversion results with the CG-based algorithm are shown in **Figures 5** and **6**. **Figure 5** is the inversion result after 20 iterations and **Figure 6** is that after 200 iterations. For the BFGS-based algorithm the inversion results are shown in **Figures 7-9**, which are the results after 20 iterations, 100 iterations and 200 iterations, respectively. Comparing **Figures 7-9** with **Figures 5** and **6**, we can see that the BFGS-based algorithm has higher inverse accuracy than the CG-based algorithm. In order to test the ability of the BFGS-based algorithm further, we inverse the Marmousi model next.

The Marmousi model is a benchmark model for testing the ability of migration or inversion methods [25]. The velocity model is shown in **Figure 10**. We can see it has many complex structures. The synthetic data is also generated by the finite difference code. The following computations parameters are adopted: $N_x = 493$ $N_z = 249$,

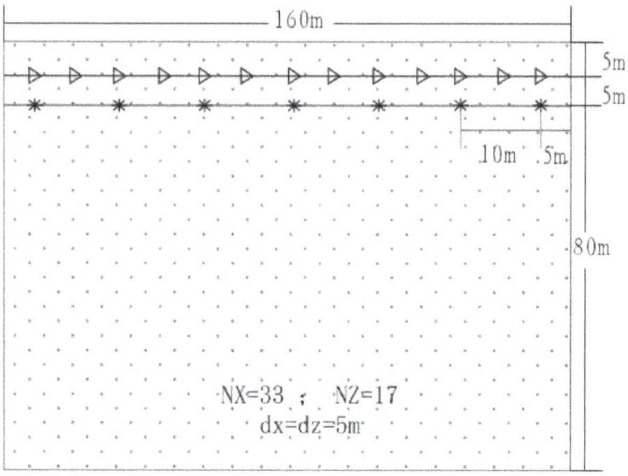

Figure 4. The configuration of sources and receivers for full-waveform inversion.

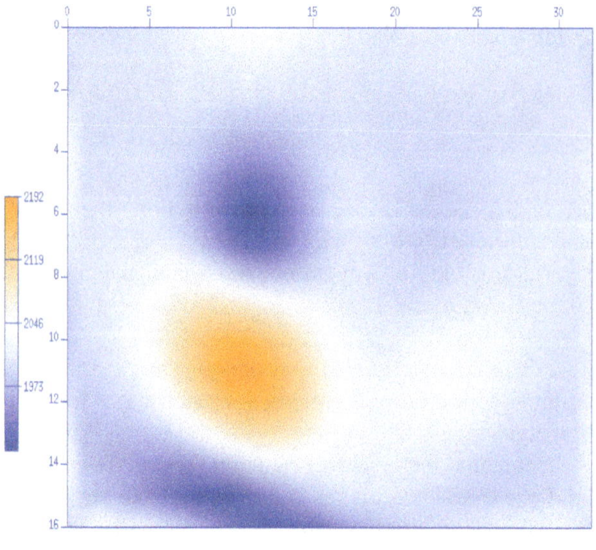

Figure 5. Inversion result after 20 iterations with the CG-based Algorithm 1.

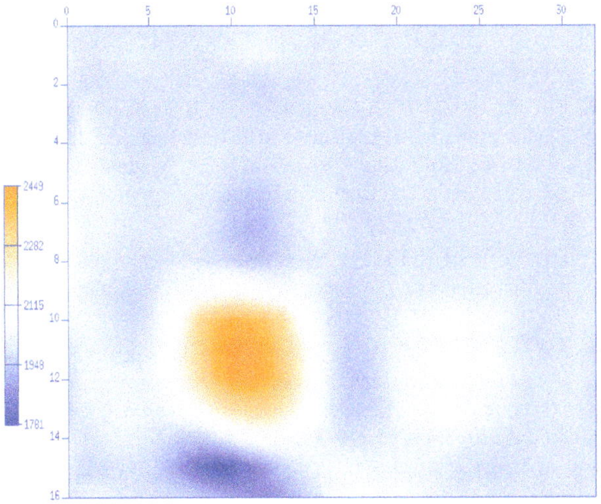

Figure 6. Inversion result after 200 iterations with the CG-based Algorithm 1.

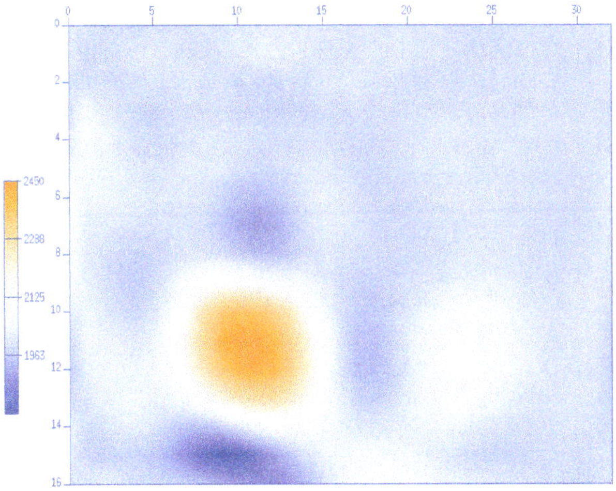

Figure 7. Inversion result after 20 iterations with the BFGS-based Algorithm 2.

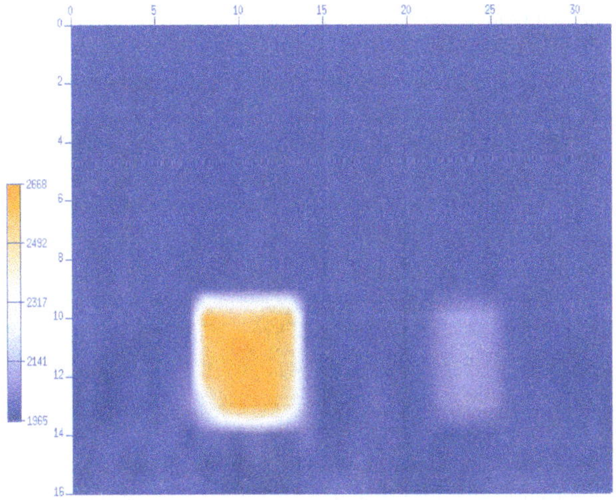

Figure 8. Inversion result after 100 iterations with the BFGS-based Algorithm 2.

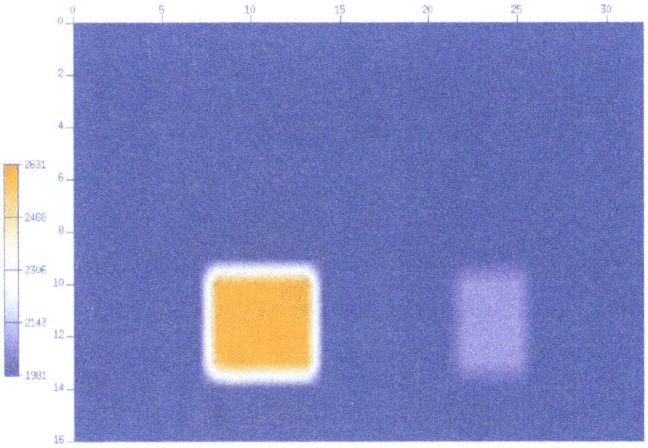

Figure 9. Inversion result after 200 iterations with the BFGS-based Algorithm 2.

Figure 10. Marmousi model with complex structures.

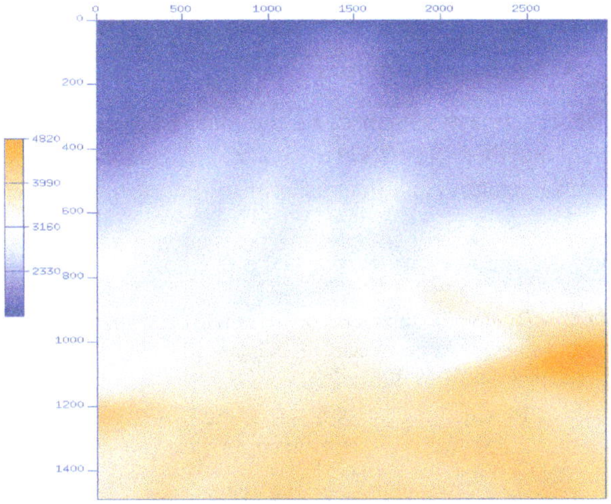

Figure 11. Initial velocity model for inversion.

$\Delta x = \Delta z = 6m$, $\Delta t = 0.6ms$, $N_t = 3501$. **Figure 11** is the initial model and **Figure 12** is the inversion result after 200 iterations. This result clearly shows that BFGS-based algorithm has obtained very good inversion accuracy.

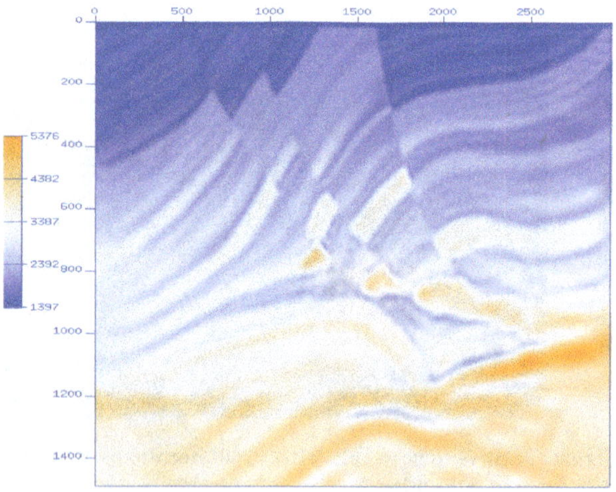

Figure 12. Inversion result after 200 iterations with the BFGS-based Algorithm 2.

4. Conclusions

We consider full-waveform inversion as an iterative minimization of the misfit between observed data and synthetic data. The algorithm is independent of the dimensionality of the problem. Two full-waveform inversion algorithms based on the CG method and BFGS method are proposed. Numerical computations for two models including complicated Marmousi model show that the BFGS-based method has higher inversion accuracy than the CG-based algorithm because the later uses the second-order derivative information in computations. The good inversion results for complicated Marmousi model show that the BFGS-based algorithm can be expected to inverse real data in the future.

5. Acknowledgements

This research is supported by the State Key project with grant number 2010CB731505 and the National Center for Mathematics and Interdisciplinary Sciences, CAS.

REFERENCES

[1] A. Tarantola, "Inversion of Seismic Reflection Data in the Acoustic Approximation", *Geophysics,* Vol. 49, No.8, 1984, pp. 1259-1266.

[2] C. Bunks, F. Saleck, S. Zaleski, and G. Chavent "Multiscale Seismic Waveform Inversion", *Geophysics*, Vol.60, No.5, 1995, pp.1457-1473.

[3] R. G. Pratt, C. Shin, and G. J. Hicks, "Gauss-Newton and Full Newton Methods in Frequency-space Seismic Waveform Inversion", *Ceophysical Journal International,* Vol.133, No.2, 1998, pp.341-362.

[4] C. Burs and O. Ghattas, "Algorithmic Strategies for Full Waveform Inversion: 1D experiments", *Geophysics*, Vol.74, No.6, 2009, pp.WC37-WC46

[5] R. G. Pratt, "Seismic Waveform Inversion in Frequency Domain Part I: Theory and Verification in a Physical Scale Model", *Geophysics*, Vol. 64, No.3, 1999, pp.888-901.

[6] Z. M. Song, P. R. Williamson, and R. G. Pratt, "Fre-Quency-domain Acoustic-wave Modeling and Inversion of Cross Hole data: Part 2—Inversion Method, Synthetic Experiments, and Real-data Results", *Geophysics*, Vol.60, No.3, 1995, pp.796-809.

[7] R. G. Pratt and M. H. Worthington, "The Application of Diffraction Tomography to Cross Hole Data", *Geophysics*, Vol.53, No.10, 1988, pp.1284-1294.

[8] C. Shin and W. Ha, "A Comparison between the Behavior of Objective Functions for Waveform Inversion in the Frequency and Laplace Domains", *Geophysics*, Vol.73, No.5, 2008, pp.VE119-VE133.

[9] W. Hu, A. Abubakar and T. M. Habashy, "Simultaneous Multifrequency Inversion of Full-waveform Seismic Data", *Geophysics*, Vol.74, No.2, 2009, pp.R1-R14.

[10] L. Sirgue and R. G. Pratt, "Efficient Waveform Inversion and Imaging: A Strategy for Selecting Temporal Frequency", *Geophysics*, Vol.69, No.1, 2004, pp.231-248.

[11] R. Fletcher and C. Reeves, "Function Minimization by Conjugate Gradients", *Computer Journal*, Vol.7, No.2, 1964, pp.149-154.

[12] C. G. Broyden, "The Convergence of a Class of Double-Rank Minimization Algorithms. 2. The New Algorithm", *J. of the Institute of Math. And its Appl.,* Vol. 6, 1970, pp.222-231.

[13] R. Fletcher, "A New Approach to Variable Metric Algorithms", *Computer Journal,* Vol.13, No.3, 1970, pp.317-322.

[14] D. Goldfarb, "A Family of Variable Metric Methods Derived by Variational Means", *Mathematics of Computation.*, Vol. 24, No. 109, 1970, pp. 23-26.

[15] D. F. Shanno, "Conditioning of Quasi-Newton Methods for Function Minimization", *Math. Comput.,* Vol.24, No.111, 1970, pp.647-656 .

[16] R. Clayton and B. Engquist, "Absorbing Boundary Conditions for Acoustic and Elastic Wave Equations", *Bulletin of the Seismological Society of America,* Vol.67, No.6, 1977, pp.1529-1540.

[17] J. P. Berenger, "A Perfectly Matched Layer for Absorbing of Electromagnetic Waves", *J. Comput. Phys.*, Vol. 114, No.2, 1994, pp.185-200.

[18] M. R. Hestenes and E. L. Stiefel, "Methods of Conjugate Gradients for Scaling Linear Systems", *J. Res. National Bureau Standards*, Vol. 49, No.6, 1952, pp.409-436.

[19] E. Polak and Ribiére, "Note Sur la Convergence de Directions Conjugate", *Rev. Francaise Informat Recherche Opertionelle*, 3e Année, 16, 1969, pp.35-43.

[20] B.T. Polyak, "The Conjugate Gradient Method in Extreme Problems", *USSR Comp. Math. and Math. Phys.*, Vol.9, No.4, 1969, pp. 94-112.

[21] L. Armijo, "Minimization of Functions Having Lipschitz Continuous First Partial Derivatives", *Pacific Journal of Mathematics,* Vol.16, No.1, 1966, pp.1-3.

[22] P. Wolfe, "Convergence Conditions for Ascent Methods",

SIAM Rev., Vol.11, No.2, 1969, pp.226-235.

[23] P. Wolfe, "Convergence Conditions for Ascent Methods II: Some Corrections", SIAM Rev., Vol.13, No.2, 1971, pp.185-188.

[24] J. Nocedal, Y. Yuan, "Analysis of a Self-scaling Quasi-Newton Method", *Math. Program,* Vol.61, No.1-3, 1993, pp.19-37.

[25] W. Zhang, "Imaging Methods and Computations Based on the Wave Equation", Beijing, Science Press, 2009.

Complete Solutions to Mixed Integer Programming

Ning Ruan
School of Science, Information Technology and Engineering, University of Ballarat, Ballarat, Australia

ABSTRACT

This paper considers a new canonical duality theory for solving mixed integer quadratic programming problem. It shows that this well-known NP-hard problem can be converted into concave maximization dual problems without duality gap. And the dual problems can be solved, under certain conditions, by polynomial algorithms.

Keywords: Duality Theory; Double Well; Global Optimization; Canonical Dual Transformation; Combinatorial Optimization; NP-hard Problems

1. Introduction

Mixed integer nonlinear programming refers to optimization problems which involve continuous and discrete variables [8]. In this paper, we consider the following constrained mixed integer quadratic programming:

$$(P_0)\ \ \min P(x,y) = f(x) + c^T y \qquad (1)$$

$$\text{s.t. } g(x) + w^T y \le 0,$$

$$-1 \le y \le 1,$$

$$x \in R^n, y \in R^n,$$

where, $f(x) = 1/2x^T Ax, g(x) = 1/2x^T Bx - bx - d$, c, w, b are given vectors, d is a given scalar, and $A, B \succ 0$, $c < 0$. X_a is a feasible space defined by

$$X_a = \{x \in R^n, y \in R^n | y_i \in \{-1,1\}, i = 1,\ldots,n\} \qquad (2)$$

Problem of the form (1) has a broad spectrum of applications, including process industry (process design [2, 13, 18], production planning [14], supply chain optimization [1,3], logistics and so on), management science (scheduling problem), financial (portfolio optimization problems [22]), engineering (network design [23]), machine learning (semi-supervised support vector machines), and computational chemistry /biology (solvent design problems).

Various methods have been proposed for solving mixed integer programming, such as branch and bound [4,5,19,21,24], cutting plane, branch and cut [16], branch and reduce, outer approximation [6,7,15], hybrid methods, and penalty method [17]. But the difficulty for developing an efficient method for such mixed integer programming lies not only on the nonlinearity of the functions involved, but also on existence of both discrete and continuous variables [20]. But if we introduce the canonical duality with some strategy, we can find global optima in polynomial time [10, 11, 12].

The rest of paper is arranged as follows. In section 2, we demonstrate how to rewrite the primal problem as a dual problem by using the canonical dual transformation. In section 3, optimality criterions for global solutions are discussed. Finally, in the last section, we present some conclusions.

2. Canonical Dual Transformation

Canonical duality theory [9] is a potentially powerful methodology which can be used to solve a large class of non-convex and discrete problems in nonlinear analysis, global optimization, and computational science.

Since $y \in \{-1,1\}^n$, one penalty term is added. Let a be a penalty factor, the original problem can be formulated

$$(P)\min P(x,y) = f(x) + c^T y + \frac{1}{2}a(y \circ y - e)^2$$

$$\text{s.t. } g(x) + w^T y \le 0$$

$$y \circ y - e = 0$$

$x \in R^n, y \in R^n$.

We choose the geometrically nonlinear operator

$$\xi = \Lambda(y) = y \circ y$$

then, the canonical function associated with this geometrical operator is

$$V(\xi) = \frac{1}{2}a(\xi - e)^2.$$

Let $\zeta \in R^n$ be the canonical dual variable corresponding to ξ, we have

$$\zeta = \nabla V(\xi) = a(\xi - e),$$

And the Legendre conjugates of the function $V(\xi)$ defined by

$$V^*(\zeta) = \{\xi^T\zeta - V(\xi) : \zeta = \nabla V(\xi)\} = \frac{1}{2}a^{-1}\zeta^T\zeta.$$

Thus the total complementarily function can be defined by

$$\begin{aligned}&\Xi(x,y,\varsigma,\sigma,\tau)\\&= f(x) + c^y + \xi^T\zeta - V(\xi) + \sigma(y \circ y - e)\\&\quad + \tau(g(x) + w^T y) = 0.\end{aligned}$$

By the criticality condition

$$\delta_y \Xi(x,y,\varsigma,\sigma,\tau) = 0,$$

We obtain

$$y = \frac{-(c+\tau w)}{2(\zeta+\sigma)}$$

Therefore, the canonical dual problem can be proposed as the following:

$$\begin{aligned}(P^d)\quad &\max P^d(\varsigma,\sigma,\tau)\\ &\text{s.t.}(\varsigma,\sigma,\tau) \in S_a\end{aligned} \tag{3}$$

and

$$\begin{aligned}&P^d(\varsigma,\sigma,\tau)\\&= -\frac{1}{2}\tau^2 b^T(A+\tau B)^{-1} b - \tau d\\&\quad -\frac{1}{4}\frac{(c+\tau w)^2}{(\varsigma+\sigma)} - \frac{1}{2a}\varsigma^2 - e(\zeta+\sigma),\end{aligned} \tag{4}$$

where $a = 10$, e is a vector with all its entry 1. Its dual feasible space S_a is defined as

$$S_a = \{\zeta \in R^n, \sigma \in R^n, \tau \in R | \tau \ge 0, \varsigma + \sigma \ne 0\}. \tag{5}$$

The notation $sta\{\cdot\}$ stands for finding all stationary points of $P^d(\varsigma,\sigma,\tau)$ over S_a. The following theorem shows that (P^d) is canonically (i.e., with zero duality gap) dual to the primal problem (P).

3. Global Optimality Condition

Theorem 1 The problem (P^d) is canonical dual to the primal problem (P) in the sense that if $(\bar\zeta,\bar\sigma,\bar\tau)$ is a KKT point of (P^d), then $(\bar x,\bar y)$ defined by

$$\bar x = \bar\tau(A+\bar\tau B)^{-1} b, \bar y = -\frac{c+\bar\tau w}{2(\bar\zeta+\bar\sigma)} \tag{6}$$

is a KKT point of (P), and

$$P(\bar x,\bar y) = P^d(\bar\zeta,\bar\sigma,\bar\tau) \tag{7}$$

Proof. By introducing a Lagrange multiplier

$$(\epsilon,\xi) \in R^n_- \times R^n_- \left(R^n_- := \{\epsilon \in R^n | \epsilon \le 0\}\right),$$

the Lagrangian $L: S_a \times R^n_- \times R^n_- \to R$ associated with the problem (P^d) is

$$L(\varsigma,\sigma,\tau,\epsilon,\xi) = P^d(\varsigma,\sigma,\tau) - \epsilon^T\sigma - \xi^T\tau.$$

The criticality conditions

$$\begin{aligned}&\nabla_\varsigma L(\bar\varsigma,\bar\sigma,\bar\tau,\epsilon,\xi) = 0,\\&\nabla_\sigma L(\bar\varsigma,\bar\sigma,\bar\tau,\epsilon,\xi) = 0,\\&\nabla_\tau L(\bar\varsigma,\bar\sigma,\bar\tau,\epsilon,\xi) = 0\end{aligned}$$

lead to

$$\bar\zeta = a(\bar y \circ \bar y - e), \tag{8}$$

$$\epsilon = \nabla_\sigma P^d(\bar\varsigma,\bar\sigma,\bar\tau) = \bar y \circ \bar y - \bar v, \tag{9}$$

$$\xi = \nabla_\tau P^d(\bar\varsigma,\bar\sigma,\bar\tau) = \bar v \circ \bar v - \bar v, \tag{10}$$

and the KKT conditions

$$0 < \bar\sigma \perp \epsilon = 0, \tag{11}$$

$$0 < \bar\tau \perp \xi = 0, \tag{12}$$

where $y = 1/2\mathrm{Diag}(\bar\varsigma+\bar\sigma)^{-1}(c/2 - \bar\tau w)$, the notation $s \circ t := (s_1t_1, s_2t_2, \ldots, s_nt_n)$ denotes the Hadamard product for any two vectors $s,t \in R^n$. This shows that if $(\bar\varsigma,\bar\sigma,\bar\tau)$ is a KKT point of the problem (P^d), and then $(\bar x,\bar y)$ is a KKT point of the primal problem (P).

By using the equations (6) we have

$$\partial_{\bar\varsigma} P^d = \frac{(c+\bar\tau w)^2}{4(\bar\varsigma+\bar\sigma)^2} - \frac{\bar\varsigma}{a} - e, \tag{13}$$

$$\partial_{\bar\sigma} P^d = \frac{(c+\bar\tau)^2}{4(\bar\varsigma+\bar\sigma)^2} = 0, \tag{14}$$

$$\partial_{\bar\tau} P^d = -\frac{(c+\bar\tau w)w}{2(\bar\varsigma+\bar\sigma)} \le 0, \tag{15}$$

and

$$\begin{aligned}&\bar\varsigma(\bar y \circ \bar y - e) = 0,\\&\bar\tau(g(\bar x) + w^T\bar y) - 0.\end{aligned} \tag{16}$$

So, in terms of

$$\bar x = \bar\tau(A+\bar\tau B)^{-1} b,\ \bar y = -\frac{c+\bar\tau w}{2(\bar\zeta+\bar\sigma)}, \tag{17}$$

we have

$$\begin{aligned}P^d(\bar\zeta,\bar\sigma,\bar\tau) &= \frac{1}{2}\bar x^T A\bar x + \tau\left(\frac{1}{2}\bar x^T B\bar x - \bar x^T b - d\right)\\&\quad -\frac{(c+\bar\tau w)^2}{4(\bar\varsigma+\bar\sigma)^2} - \frac{1}{2a}\bar\varsigma^2 - e(\bar\varsigma+\bar\sigma)\end{aligned}$$

$$= f(\bar{x}) + \tau g(\bar{x}) - \frac{(c+\bar{\tau}w)^2}{4(\bar{\varsigma}+\bar{\sigma})^2} - \frac{1}{2a}\bar{\varsigma}^2$$
$$-e(\bar{\varsigma}+\bar{\sigma})$$
$$= f(\bar{x}) + c^T\bar{y} + (\bar{y}\circ\bar{y})\bar{\varsigma} - \left(\frac{1}{2a}\bar{\varsigma}^2 + e\bar{\varsigma}\right)$$
$$+\sigma(\bar{y}\circ\bar{y} - e) + \bar{\tau}(g(\bar{x}) + w^T\bar{y}).$$

From (13), we have

$$\bar{y}\circ\bar{y} = \frac{\bar{\varsigma}}{a} + e. \tag{18}$$

Therefore,

$$(\bar{y}\circ\bar{y})\bar{\varsigma} - \left(\frac{1}{2a}\bar{\varsigma}^2 + e\bar{\varsigma}\right) = \frac{1}{2}a(\bar{y}\circ\bar{y} - e)^2. \tag{19}$$

Due to the fact that $y_i \in \{-1,1\}, i = 1,\ldots,n,$ we have

$$P^d(\bar{\varsigma},\bar{\sigma},\bar{\tau}) = P(\bar{x},\bar{y}). \tag{20}$$

This proves the theorem.

This theorem shows that there is no duality gap between the primal problem and its canonical dual. In order to identify the global minimize, we need to introduce a useful feasible space

$$S_a^+ = \{(\varsigma,\sigma,\tau) \in S_k | \varsigma + \sigma > 0\} \tag{21}$$

be a subset of S_a, and we have the following theorem.

Theorem 2 Suppose that the vector $(\bar{\varsigma},\bar{\sigma},\bar{\tau})$ is a critical point of the canonical dual function $P^d(\varsigma,\sigma,\tau)$. Let

$$\bar{x} = \bar{\tau}(A+\bar{\tau}B)^{-1}b,\ \bar{y} = -\frac{(c+\bar{\tau}w)}{2(\bar{\varsigma}+\bar{\sigma})} \tag{22}$$

If $(\bar{\varsigma},\bar{\sigma},\bar{\tau}) \in S_a^+$, then $(\bar{\varsigma},\bar{\sigma},\bar{\tau})$ is a global maximize of $P^d(\varsigma,\sigma,\tau)$ on S_a^+, the vector $(\bar{x},\bar{y})$ is a global minimize of $P(x,y)$ on R^n, and

$$P(\bar{x},\bar{y}) = \min_{(x,y)\in R^n\times R^n} \min_{(x,y)\in R^n\times R^n} P(x,y)$$
$$= \max_{(\varsigma,\sigma,\tau)\in S_a^+} \max_{(\varsigma,\sigma,\tau)\in S_a^+} P^d(\varsigma,\sigma,\tau) \tag{23}$$
$$= P^d(\bar{\varsigma},\bar{\sigma},\bar{\tau})$$

Proof. By Theorem 1 and the canonical duality theory, we know that vector $(\bar{\varsigma},\bar{\sigma},\bar{\tau}) \in S_a$ is a KKT point of the problem (P^d) if and only if $(\bar{x},\bar{y})$ defined by

$$\bar{x} = \bar{\tau}(A+\bar{\tau}B)^{-1}b,\ \bar{y} = -\frac{c+\bar{\tau}w}{2(\bar{\varsigma}+\bar{\sigma})}$$

is a critical point of the problem (P), and

$$P(\bar{x},\bar{y}) = P^d(\bar{\varsigma},\bar{\sigma},\bar{\tau})$$

By the fact that the canonical dual function $P^d(\varsigma,\sigma,\tau)$ is concave on S_a^+, the critical point $(\bar{\varsigma},\bar{\sigma},\bar{\tau})$ is a global maximize of $P^d(\varsigma,\sigma,\tau)$ over S_a^+, this proves the statement (23).

This theorem provides a sufficient condition for a global minimizer of the primal problem.

4. Conclusions

In this paper, the canonical duality theory has been applied to solve mixed integer programming problem. Theorems show that by the canonical dual transformation, primal problems can be converted into canonical dual problem. By the fact that the canonical dual function is concave on the dual feasible space, so the dual problem can be solved by well-developed deterministic optimization methods.

5. Acknowledgements

Dr. Ning Ruan was supported by a funding from the Australian Government under the Collaborative Research Networks (CRN) program.

REFERENCES

[1] K. Aardal," Capacited Facility Location: Separation Algorithms and Computational Experience," *Mathematical Programming,* Vol. 81, No. 2, 1998, pp. 149-175.

[2] A. Atamtürk, "Flow Pack Facets of the Single Node Fixed-charge Flow Polytope," *Operations Research Letters*, Vol. 29, No. 3, 2001, pp. 107-114.

[3] I. Barany, T. J. Van Roy and L. A. Wolsey, "Strong Formulations for Multi-item Capacitated Lot Sizing," *Management Science*, Vol. 30, 1984, pp. 1255-1261.

[4] P. Belotti, J. Lee, L. Liberti, F. Margot and A. Waechter, "Branching and Bounds Tightening Techniques for Non-convex MINLP," *Optimization Methods & Software*, Vol. 24, 2009, pp. 597-634.

[5] B. Borchers and J. E. Mitchell, "An Improved Branch and Bound Algorithm for Mixed Integer Nonlinear Programs," *Computer & Operations Research*, Vol. 21, No. 4, 1994, pp. 359-367.

[6] M. A. Duran and I. E. Grossmann, "An Outer-approximation Algorithm for a Class of Mixed-integer Nonlinear Programs," *Mathematical Programming*, Vol. 36, No. 3, 1986, pp. 307-339.

[7] R. Fletcher and S. Leyffer, "Solving Mixed Integer Nonlinear Programs by Outer Approximation," *Mathematical Programming*, Vol. 66, No. 1, 1994, pp. 327-349.

[8] C. A. Floudas, I. G. Akrotirianakis, S. Caratzoulas, C. A. Meyer and J. Kallrath, "Global Optimization in the 21st Century: Advances and Challenges," *Computers &*

Chemical Engineering, Vol. 29, 2005, pp. 1185-1202.

[9] D. Y. Gao, "Duality Principles in Nonconvex Systems: Theory, Methods and Applications," Kluwer Academic Publishers, Dordrecht/ Boston/ London, 2000.

[10] D. Y. Gao and N. Ruan, "Solutions to Quadratic Minimization Problems with Box and Integer Constraints," *Journal of Global Optimization*, Vol. 47, No. 3, 2010, pp. 463-484.

[11] D. Y. Gao, N. Ruan and H. D. Sherali, "Solutions and Optimality Criteria for Nonconvex Constrained Global Optimization Problems with Connections between Canonical and Lagrangian Duality," *Journal of Global Optimization*, Vol. 45, No. 3, 2009, pp. 473-497.

[12] D. Y. Gao, N. Ruan and H. D. Sherali, "Canonical Dual Solutions to Fixed Cost Quadratic Programs", In: A. Chinchuluun, P.M. Pardalos, R. Enkhbat and L. Tseveendorj, Eds., *Optimization and Optimal Control: Theory and Applications*, Springer, Vol. 39, 2010, pp. 139-156.

[13] F. Glover and H. D. Sherali, "Some Class of Valid Inequalities and Convex Hull Characterizations for Dynamic Fixed-charge Problems under Nested Condtraints," Vol. 40, No. 1, 2005, pp. 215-234.

[14] J. Kallrath, "Solving Planning and Design Problem in the Process Industry using Mixed-integer and Global Optimization," *Annals of Operations Research*, Vol. 140, No. 1, 2005, pp. 335-373.

[15] P. Kesavan, R. J. Allgor, E. P. Gatzke and P. I. Barton, "Outer Approximation Algorithms for Separable Nonconvex Mixed-integer Nonlinear Programs," *Mathematical Programming*, Vol. 1000, No. 3, 2004, pp. 517-535.

[16] P. Kesavan and P. I. Barton," Generalized Branch-and-cut Framework for Mixed-integer Nonlinear Optimization Problems," *Computers & Chemical Engineering*, Vol. 24, 2000, pp. 1361-1366.

[17] L. Grippo and S. Lucidi, "A Differentiable Exact Penalty Function for Bound Constrained Quadratic Programming Problems," *Optimization*, Vol. 22, No. 4, 1991, pp. 557-578.

[18] Z. Gu, G. L. Nemhauser and M. W. P. Savelsbergh, "Lifted Flow Cover Inequalities for Mixed 0-1 Integer Programs," *Math Program*, Vol. 85, No. 3, 1999, pp. 439-467.

[19] O. K. Gupta and A. Ravindran, "Branch and Bound Experiments in Convex Nonlinear Integer Programming," *Management Science*, 1985, pp. 1533-1546.

[20] P. Hansen, B. Jaumard, M. Ruiz and J. Xiong, "Global Minimization of Indefinite Quadratic Functions Subject to Box Constraints," *Naval Research Logistics*, Vol. 40, 1993, pp. 373-392.

[21] S. Leyffer, "Integrating SQP and Branch-and-bound for Mixed Integer Nonlinear Programming," *Computational Optimization and Applications,* Vol. 18, No. 3, 2001, pp. 295-309.

[22] X. Lin, C. A. Floudas and J. Kallarth, "Global Solution Approach for a Non-convex MINLP Problem in Product Portfolio Optimization.," *Journal of Global Optimization*, Vol. 32, No. 3, 2005, pp. 417-431.

[23] M. W. Padberg, T. J. Van Roy and L. A. Wolsey, "Valid Linear Inequalities for Fixed Charge Problems," *Operations Research*, Vol. 33, 1985, pp. 842-861.

[24] I. Quesada and I. E. Grossmann, "An IP/NLP based Branch and Bound Algorithm for Convex NINLP Optimization Problems," *Computers & Chemical Engineering*, Vol. 16, 1992, pp. 937-947.

Means of Choice for Interactive Management of Dynamic Geometry Problems Based on Instrumented Behaviour

Philippe R. Richard[1], Michel Gagnon[2], Josep Maria Fortuny[3], Nicolas Leduc[2], Michèle Tessier-Baillargeon[1]
[1]Département de didactique, Université de Montréal, Canada
[2]Département de génie informatique et génie logiciel, École Polytechnique de Montréal, Canada
[3]Departament de Didàctica de les Matemàtiques i de les Ciències Experimentals, Universitat Autònoma de Barcelona, Spain

ABSTRACT

Our paper presents a project that involves two research questions: does the choice of a related problem by the tutorial system allow the problem solving process which is blocked for the student to be restarted? What information about learning do related problems returned by the system provide us? We answer the first question according to the didactic engineering, whose mode of validation is internal and based on the confrontation between an *a priori* analysis and an *a posteriori* analysis that relies on data from experiments in schools. We consider the student as a subject whose adaptation processes are conditioned by the problem and the possible interactions with the computer environment, and also by his knowledge, usually implicit, of the institutional norms that condition his relationship with geometry. Choosing a set of good problems within the system is therefore an essential element of the learning model. Since the source of a problem depends on the student's actions with the computer tool, it is necessary to wait and see what are the related to problems that are returned to him before being able to identify patterns and assess the learning. With the simultaneity of collecting and analysing interactions in each class, we answer the second question according to a grounded theory analysis. By approaching the problems posed by the system and the designs in play at learning blockages, our analysis links the characteristics of problems to the design components in order to theorize on the decisional, epistemological, representational, didactic and instrumental aspects of the subject-milieu system in interaction.

Keywords: Didactics of Mathematics; Competencies; Geometric Thinking; Tutorial System; Related Problems; Dynamic Geometry; Instrumented Behavior; Cognitive Interactions; Conceptions; Mathematical Work Space; Means of Choice; Didactic Contract

1. Foreword

In the third year of secondary school, two students tried to solve a problem of proof at the interface of an interactive tutorial system. It was to compare the area of two triangles in a parallelogram and to proof the assumption made. After reading the statement and constructing or moving the elements of the figure in the dynamic geometry module (**Figure 1**), the students quickly agree on equal areas. They began to create a mathematical proof on the tutorial system interface and were therefore delighted to see that Prof. Turing, an artificial tutor agent, indicated with a smiley that their first intuition was well founded. Even though they were good students, they sometimes got stuck in their mathematical proof. Happily, with his messages, Prof. Turing was always successful in reviving the solution process. Without replacing the teacher, this tutor agent has 69,000 potential solutions "in mind" and was quickly able to target the solution envisaged by the students, thereby providing personalized support. Prof. Turing also knows how to recognize a student's persistent difficulty and can suggest that he get help from his teacher. Furthermore, once he arrives, the teacher sees what has happened from the messages received but instead of insisting on their meaning in the context of the problem he rather asks that a new problem be solved. The students launched on paper without too much difficulty then one said to his companion: "look, I've got it... look, this is why it works!" And the solution to the original problem is relaunched. The use of a related problem therefore is a means of choice for this didactic system. Can they be made available to Prof. Turing?

2. Introduction

According to the theory of didactic situations, we know that the only way to "do" mathematics is to try and solve some specific problems and in this regard asking new questions. The teacher must therefore not communicate knowledge but pass on the right problem. If this transfer happens, the student plays the game and ends up winning, while learning takes place. But what if the student rejects or avoids the problem, or doesn't solve it? The teacher then has the social obligation to help him [7]. Set in didactics of mathematics, our research project is based on three key concepts: on the necessity of seeking and resolving specific problems for learning geometry in high school, on the assistance that makes up a transfer of the "right problems" in a context of instrumented learning, and on the voluntary but surprising action of the teacher who chooses to set a problem as a message to help a student whose solving of an initial problem remains blocked.

Instrumented learning is based on the use by the student of a tutorial system created by our research team for learning geometry. This system supports the student in solving problems of proof, issuing messages as needed (verbal or iconic expressions) appropriate to the actions of the student in the internal logic of problems. During a validation phase of previous research (see next section), the introduction of a support structure that incorporates a set of related problems appeared necessary to acquire the means of choice that the teacher discusses with his students. Unlike existing approaches, these problems do not divide the original problem into sub-strategies. With a completely new approach, the new problems arise from the characteristics of relationships between problems and learning blockages, engendering new decision means for the tutorial system.

3. Research Program

In this section, key words are in bold.

Based on the **didactics of mathematics**, our project is a continuation of the project *a new approach to research on competential and instrumented learning of geometry in high school* (CRSH 410-2009-0179) and it renews the foundations laid down in the article *Didactic and theoretical-based perspectives in the experimental development of an intelligent tutorial system for the learning of geometry* [41]. These works were common to the design of geogebraTUTOR, a **tutorial system** which is intended to support the development of students' **mathematical competencies** [29, 46] and the **construction**

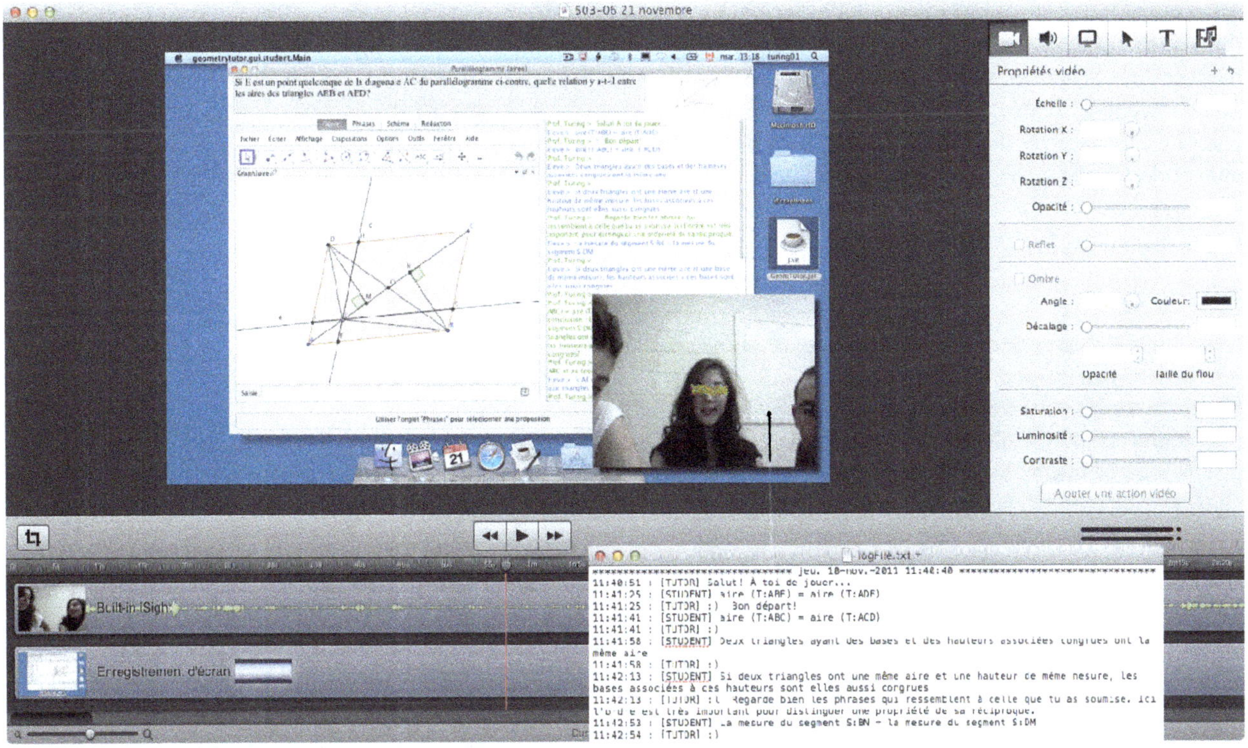

Figure 1. Analysis feature of interactions during solving of the parallelogram problem. In the background is the ScreenFlow software interface (recording sound, image and interaction on the screen) and in the foreground the log of the conversation between the student and the artificial tutor agent. On the student's screen, the "GeomTutor" Java applet launches the tutorial system and the "dictation" file saves the simultaneous recording of the teacher's intervention. The image used here shows the geometric (on the left) and discursive (on the right) modules but is hiding the modules for writing ("Statements" tab), structured arguments training ("Outline") and mathematical proof ("Writing"). For more information about the system's challenges, consult the video at http://www.matimtl.ca/evenements/evenement.jsp?id=106.

of geometric thinking [19, 22, 47]. It consists of two subsystems, Turing[1] (FQRSC 2005-AI-97435) and GeoGebra (http://www.geogebra.org/), a **dynamic geometry software** whose international influence is considerable in teaching mathematics and which includes a three dimensional geometry module in evolution [6]. By having to account for teacher intervention [49, 50], we have enriched our research program with assessment tools developed within Intergeo (http://i2geo.net), a consortium that manages a platform for sharing and assessment of the quality of resources to which our secondary mathematics teaching students have already contributed [51]. On the basis of these achievements, the current project still aims to improve learning but it now innovates by the original consideration of a structure of **related problems** which meets these student learning blockages, with a view to **instrumented behaviour** and whose reference geometry allows adaptation to actual class didactic contracts. Our reference to the decision-making theory of Schoenfeld [48] sheds light on both the resources, goals and orientations of the teacher intervention and the tutorial action, and the notions of **conceptions** [5] and **mathematical work space** [21] pose an epistemological, semiotic and instrumental view of cognitive interactions that emerge from the student **cognitive interaction** with the milieu. The notion of **means of choice** generalises the teacher's judgements and the decisions of the tutorial system when it returns a related problem following a learning blockage by the student, and that of a **didactic contract** designates the most frequently implicit expectations that there are for the student and the teacher respectively concerning mathematical knowledge.

4. Objectives

In seeking to better understand the means of choice for interactive management of dynamic geometry problems based on instrumented behaviour in high school, our research project has the objectives in the **Table 1**.

The idea of means of choice, as a voluntary action to consider one problem rather than another, transposes into decision making means within the tutorial system. We consider the instructional model (also known as instructional design) on two levels, that of didactics, the responsibility of active educational members and those in training, and that of the technology of computer programming and deployment, the responsibility of members with technological training (see the section *Research team, latest results and student training*). The achievement of our research objectives will have a direct affect on teacher training (**Table 2**).

[1] French acronym for TUtoRiel Intelligent en Géométrie and a nod to the engineer and mathematician Alan Mathison Turing.

5. Background

Well beyond the establishment of a simple online exerciser or a learning guide in a deterministic MOOC structure [16], our research project is based on a computerised environment for complex human learning which is in an international dynamic geometry movement where community development and sharing of reference activities are carried out among experts and teachers from different traditions. The idea of a tutorial system that supports students in problems of proofs is not new. Among first generation achievements, we can mention the Geometry Proofs Tutor [3], the Tigre-Mentoniezh project [32] and Géométrix (http://geometrix.free.fr/ by Jacques Gressier). All these systems are based on formal geometry models that, despite an evident advantage of computer programming, presuppose development in geometric thought

Table 1. The three general objectives.

Objective 1 Instructional model	• Design, index, implement and test a structure of related problems in a tutorial system (geogebraTUTOR) which is based on the means of usual choice for teacher intervention and the instrumented behaviour of the student during solving of fundamental problems involving proofs.
Objective 2 Interpretation and theorizing	• Interpret and theorize on the decisional, epistemological, representational, didactic and instrumental aspects of the subject-milieu system in interaction, with reference to the student's conceptions and the mathematical workspace.
Objective 3 Assessment and control	• Assess the consistency of the subject-milieu system in interaction, with reference to the development of mathematical competencies, construction of geometric thought and the student's learning in an instrumented perspective.

Table 2. The two training objectives.

Initial training	• Support, by trainees (university students), of a part of the cognitive, heuristic, semiotic and metamathematical means made available to students during simulated situations, to develop their ability to identify with what the student knows and, reciprocally, to test their teaching action.
Continuing education	Development of disciplinary competencies in geometry, as professional successor, critiques and interprets from its subjects or culture, in the exercise of his functions.

- Richard, Cobo, Fortuny and Hohenwarter [40]
- Trgalová, Richard and Soury-Lavergne [52]

by adherence to an axiomatic approach. Assisting student learning blockages is therefore formal. The same applies to second generation systems, although the interface, communication with the user and processing of significant actions are more developed. Among systems similar to geogebraTUTOR [42] should be mentioned the Advanced Geometry Tutor [28], the Baghera project [23], the Cabri-Euclide microworld [25] and the Geometry Explanation Tutor [1]. There is also the Andes Physics Tutor [54], for which some situations-problems are already premodeled in geometry.

Among the precursor achievements to our current system are AgentGeom [10] and Turing [39]. Unlike the systems which relate to formal axiomatic geometry, AgentGeom and Turing were developed on cognitive geometry models that lie between natural geometry and the axiomatic natural geometry of Kuzniak [20]. This allows full originality when considering situations-problems that bring together physical sciences to the process of discovery in mathematics [12], as Clairaut [9] appeared to desire in the Enlightenment by stating "this presumed induction carries its demonstration with it" (p. 64), following the representation on paper of a "dynamic" geometric figure [43]. In addition, although cognitive geometry is essential to considering reference geometry that is effectively practiced in the classroom, it allows for constitution of a structure of related problems that respond to informal learning blockages.

6. References Axes of the Conceptual Framework

6.1. Epistemological Axis

In the process of mathematical discovery, the epistemological dialectic of proofs and refutations of [24] considers the criticisms that arise with counter-examples in discussions between students and teacher. These criticisms are likely to require an adjustment of the conjecture, the proof or the counter-example itself, and also of knowledge and the problem. Although they can be seen as "breakpoints" [27] in solving a problem of proof, this is because like Lakatos we believe that the steps of proof are only summarised with formal or deductive approaches. In addition, we know that the supporting role of these criticisms can be incorporated into the steps of instrumented reasoning [18], while allowing the creation of a geometric workspace in continuity with the development of mathematical competencies inherited from elementary school [11]. Our desire to bring together the epistemology of mathematics into training programs, so the student can perform his work in geometry, is an attempt for subtle adaptation between an actual state of his mathematical competencies and the intrinsic requirement for performing geometry in class.

6.2. Semiotic Axis

When a study examines dynamic geometry and instrumented reasoning, questions on communications, processing of cognitive representations and objectivities of virtual representations are essential. The Duval's theory of language functions [13] sets out the conditions for learning based on the coordination of representation registers, of which the register of figures [14], and the functional-structural approach of Richard and Sierpinska [44] insists on the traditional semiotic means serving the quality of communications. When the student acts on a dynamic drawing, he is also acting on the system of representation, possibly for the communication of inductive reasoning [45]. This action can be the source of a learning blockage and, although the figural representations convey reasoning [36, 37] or simulate movement [2], they cannot generate it. In this context, a dynamic figure is also a kind of problem.

6.3. Situational Axis

In Brousseau's theory of didactical situations in mathematics [7], the main intervention of the teacher (arrow 1, **Figure 2**) occurs within a system which is itself in interaction, the student-milieu system (2), but with a didactic milieu which brings a tutor agent to him the role of arrows 1 and 2 is transposed with 6 and 7. According to [26], Brousseau will consider the subject-milieu interaction as the smallest unit of cognitive interactions. A state of equilibrium for this interaction defines a state of knowledge, the subject-milieu imbalance producing new knowledge (search for a new equilibrium)". However, the theory of didactic situations characterises each item of knowledge by situations that are specific to it, and the knowledge model of Balacheff and Margolinas [5] locates conceptions in the subject-milieu interaction, while first characterising a conception by the problems in which it is involved. If it results in a strong conceptual relationship between a moment of learning blockage and a related problem, it is also because a learning blockage is a breach of contract with what is expected in the context of the root problem and that the appearance of a new problem, in addition to relaunching the solving process, is not a sophisticated response suggestion which at its core considers the objective of the root problem.

6.4. Instrumental Axis

In Rabardel's theory of instrumentation [34], the instrument is the mode of action or thought constructed by the subject when he uses a tool; the manner in which the instrument is formed in the subject is the instrumental genesis. According to [21], instrumental genesis constitutes the geometric workspace and, when it is considered in the process of student-milieu interaction which creates its own space, instrumental genesis operates both during stages of discovery and validation [11]. Since the processes of instrumentation relate to the emergence and development of patterns of use and instrumented action, the progressive discovery of the tool's intrinsic properties by students, for whom the appearance of a related problem following a learning blockage is accompanied by the accommodation of their patterns and also changes in meaning of the instrument, results in association of the tool with new patterns [35]. The notions of conception and instrument occupy dual places in modelling of the subject-milieu [5].

6.5. Decisional Axis

The transposition of the teacher's intervention in the educational environment is necessarily accompanied by a transposition of means of the choice. Although these means can be interpreted in respect of a didactic contract, the implicitly shared significance that it supposes complicates understanding of the decisional process in all of the fundamental relationships. According to [7], the didactic contract is not really a true contract, since it is not explicit or voluntary and because neither the conditions for breach nor sanctions can be given in advance due to their didactic nature, which depends significantly on a knowledge of students that is as yet unknown. In Schoenfeld's decision making theory [48], if we know enough about what someone's resources, goals and orientations are, teacher or student, we can even come to understand, explain and model actions and decisions that seem unusual or abnormal. In the Introduction paragraph, the illustration of the sudden appearance of a new request for solving problem, while it was the first solving processed that was blocked, may surprise the observer but not the students, who are used to reacting to this type of requirement from their teacher.

7. Methodology

7.1. General Direction

As with our previous projects, we take advantage of a similar experimental effort to "verify a state or a change" and "develop accordingly". In fact, the illustration in the Foreword comes from an experiment (see the link below **Figure 1** and [50] which allowed, firstly, verification of the validity of a structure of messages from the tutor agent when two classes of high school students resolved problems of proof and, secondly, identification of the means of choice of their teachers with the use of related problems after indexation of conceptual, heuristic, semiotic and met mathematical criteria [42]. This requirement invites us to consider together two paradigms on the epistemological didactic level. On the conceptual side, the learning models that we claim identify with the use of our tutorial system based on the preceding axes. On the methodological side, the approach implemented combines the didactic engineering of Artigue [4] and the grounded theory analysis of Glaser and Strauss [17]. Since the functional integration of a structure of related problems in the instructional model is in line with our project (see *Acknowledgements*), we can achieve objectives 1 and 3 with a confirmatory model [53] of the hypothesis-deduction type [4]. But the intervention of these problems depends on the instrumented behaviour, and since they generate a non-deterministic learning itinerary (following related problems), the emergence of learning models adapted to the use of an advanced tutorial system commits us to achieving objective 2 using a comparative-inductive type model [17].

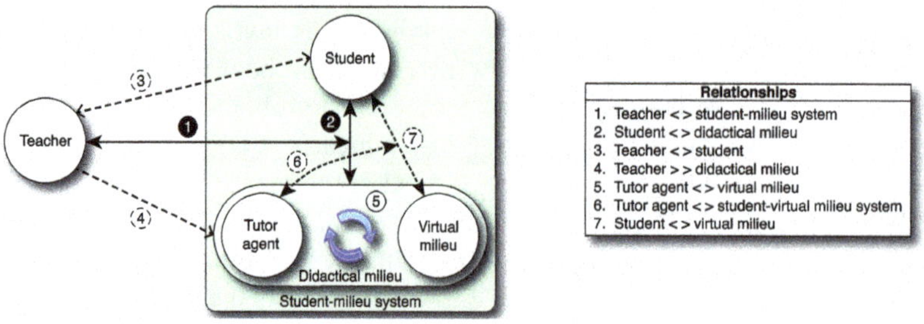

Figure 2. Fundamental relationships in the research system. It is an adaptation of [7] on the relationship of 1 to 4 between the teacher, the student and the milieu [41].

7.2. Description of Key Activities and Specific Procedures

Here is a brief description of phases of the research in relation to the objectives, for trimester Q.

Phase 1 ($Q_{1\text{-}2}$) ↓ Conception of the instructional model– preparation of objective 1 (1st part)

•From the 34 problems based on [33] and adaptation

of these to the curriculum in effect in the regions of the participating researchers, characterisation of problems according to the identity of problems [38] and indexation of them according to the above criteria.

•Implementation in geogebraTUTOR of a structure of related problems and testing experts on the functioning of the didactic milieu according to the learning model of [41].

Phase 2 (Q_{3-4}) ↓ 1st experiment in the schools – achievement of objective 1 (1st part)

To validate by cross-referencing primitive sources – or "triangulation" [15]:

•Collection of student-milieu interactions and teacher interventions in 3 classes of second cycle of high school in three different regions when, in their normal courses, students solve five problems of proof using the system interface (qualitative data, [30]).

•Request to teachers to reconstruct their interventions and compare the means of choice of the tutorial system with the didactic contract of the course [31, 8] during 50 minute explanatory discussions [55].

In addition to the log files, there will be "ScreenFlow" records (see **Figure 1**) for a sample of 24 volunteer students (4 teams of 2 per class), as well as audio recordings for the teachers of each class (intervention in class and explanatory discussions), in accordance with the ethical rules in force – ditto for entering and processing data in all phases.

Phase 3 (Q_{5-6}) ↓ Analysis and interpretation – preparation of objectives 2 and 3 (1st part)

•Analysis of times of learning blockage – or breakpoints – according to the knowledge model of Balacheff and Margolinas [5] by characterising conceptions **C** by a defining set of problems (P) for which solving tools (R) are provided based on representation systems (L) and a control structure (Σ) which permits judgements and decisions.

•Interpretation of times of learning blockage by joining the student's conceptions **C** = (P, R, L, Σ) with the characteristics of the related problems in play (of the system, phase 1; of teachers, phase 2).

Phase 4 (Q_{7-8}) ↓ Cross-referencing, validation and fine-tuning – preparation for objective 1 (2nd part) and achievement of objective 2 (1st part)

•Pooling of the results of phase 3 between researchers and identification of patterns.

•To improve the instructional model and better understand the common or invariant characteristics of the subject-milieu system in interaction, expert validation of the combination of phase 3.

•Optimisation of the articulation and continuity between the conception of the tutorial system and pursuit of the conception during solving by the students (or "conception in use", [34]).

Phase 5 (Q_{9-10}) ↓ 2nd experiment in the schools – preparation of objective 3 (2nd part) and achievement of objective 1 (2nd part)

•Resumption of the procedures for phase 2, by this time asking teachers to assess the solutions (without noting them) with a view to assessing competencies [38].

Phase 6 (Q_{11-12}) ↓ Modelling, synthesis, theorisation – achievement of objectives 2 (2nd part) and 3

•Modelling of an ontology on the basis of didactic contracts and instrumented behaviour.

•Summary of the non-deterministic learning itineraries and underlying means of choice.

•Theorisation on the fundamental relationships of the instrumented didactic situation (**Figure 2**).

8. Knowledge Mobilisation Plan

As a human sciences discipline, the teaching of mathematics has a scientific side and a professional side, including the initial and continuing training of teachers in the field. When a project involves not only teaching theories but also construction of computer environments for human learning, questions arise in a very practical way, which leads to seeking a functional modelling of knowledge by making distinctions that are useful feedback for the teaching of mathematics as a whole. Our plan for knowledge mobilisation, similar to that which is grounded in our research program, aims to continue the multidirectional exchange of knowledge between researchers, teachers and other persons involved in the world of teaching mathematics, in a collaborative spirit of sharing which includes quality, integration and popularisation. We can summarise the overall mobilisation plan in terms of *places*[2]:

•Publication in journals for the quality of the research (ESM, IJCML, ZDM, etc.), its integration (ADSC, REC, PME, etc.) and its popularisation (AMQ, PME, UNO, etc.);

•Organisation and participation in international symposia on teaching mathematics (EMF, ETM, CERME, etc.) as well as mathematics classroom technologies (INTERGEO, CADGME, E-LEARN, etc.);

•Initial and continuing training sessions for teachers (UDM, UAB, GRMS, etc.);

•Participation as appropriate in advisory bodies on training or technology programs (UDM, CCPÉ, MATI, etc.).

and of *means*, thanks to advanced technological skilled of team members:

•Community and sharing development platform for research quality (Turing, cK¢ wikibook, etc.), its integration (I2GÉO, NTLMP, etc.) and its popularisation

[2]We list acronyms and websites in the section References.

(GIC-IGC, GeoGebraTube, etc.).

We are also planning to organise a workshop during phase 3 in collaboration with other research groups, so that researchers and teachers associated with a project can share their experiences following the first experimental phase. The formula proposed is a symposium, like the one we organised most recently in Montreal http://turing.scedu.umontreal.ca/etm/, in the collaborative spirit typical of the *Congress of European Research in Mathematics Education* http://www.cerme8.metu.edu.tr, which balances quality (of seasoned researchers) and integration (of young researchers).

9. Results Expected

In the *Knowledge mobilisation plan* section, we insisted on the fact that the mutual contribution of didactics and computers generates research advantages and impacts within the university environment, with teachers and other stakeholders in the educational world, through interactions and increased access during the research itself.

By creating an "in use" tutorial system (within the meaning of [34]), our research approach is empirically based on the articulation and continuity between the institutional system design processes and the pursuit of the design in problem solving by the student. Since it was designed to produce a class of effects (support for learning through messages, problems and controls), implementation of the system, under the conditions provided for each phase of the project (see *Detailed description* section), allows updating of these effects following usage noted during experimental phases. In other words, if the cognitive outcome constitutes the design of the tutorial system, it is the source of its own existence by an expert anticipation of interactions of a of a changing student-milieu system. Unlike existing tutorial systems (see *Background* section), our choice of cognitive geometry is a significant mark of originality since it lets us both adapt the means of choice of the student's instrumented behaviour and integrate the specifics of authentic didactic contracts.

The idea of meeting a student's learning blockage by providing timely related problems to solve is an effective solution to one of the major difficulties in teaching: avoiding giving answers (discursive messages) at the same time as the questions (root problems). In this sense, our project theoretically answers the first didactic paradox of [7]: everything the teacher does to produce the expected behaviours by students tends to reduce the student's uncertainty and thereby deprives him of the necessary conditions for understanding and learning the intended concept; if the teacher tells or signifies what he wants from the student he can no longer get it other than as performance of an instruction and not by the exercise of his knowledge and judgement.

Apart from the institutional requirement for student training, the dissemination of knowledge and the influence on the community of researchers in the field, the integration of multidisciplinary doctoral research and the effective collaboration of the school institution remains a strategic advantage of the project, as is its influence on teachers practise and training. Whether first to improve students' geometrical skills, including deductive (reasoning, arguing), visual (observing, exploring), figural (modelling, conjecturing, defining) and operational (instrumentation, manipulation) skills, the potential for development of the tutorial system then allows the teacher to adapt a port of his pedagogical engineering according to the division of his responsibilities with those of the student. In initial training, these same arrangements sharpen students' abilities to simulate the effect of their teaching activities and to identify student's behaviour, since the anticipation of solutions up to planning (and not programming) learning itineraries adapted to the student.

The material benefits of our project are intended for public use in schools.

10. Conclusions: Four Centres of Originality

Although the current project is a continuation of the founding projects, it remains profoundly original in relation to it. We conclude by noting here four centres of originality.

A first centre relates to the organisation of a structure of related problems that responds to times of learning blockage by the student. Although desirable, this type of structure is unusual in math classes, since it is difficult to put in place in a paper-pencil environment and even if the use of related problems occasionally happens with some teachers, the choice of the problem remains limited by the environment. In addition, we know of no geometry tutorial system that integrates such organisation to restart a block problem solving process.

A second centre affects the joint consideration of the approaches to mathematical discovery and proof which, in the same context as instrumented learning, links the epistemology of mathematics with training programs. Attached to a reference geometry based first on the meaning of objects that it models and which makes as such an approximation possible, geometry becomes a means of learning and not longer its object, as is found with tutorial systems which develop geometric thinking by taking on an axiomatic approach.

A third centre relates to the functional modelling of knowledge in our tutorial system. Most of the time, a human learning computer environment is validated by comparing the results of a pre-test and a post-test, while requiring the user to comply with the system as it was designed. This attitude undoubtedly leads fairly quickly to concrete accomplishments, but it is necessary for these accomplishments to be effective learning aids. Although our tutorial system aims for effectiveness of the tutorial activity by first considering a modelling of human behaviour and designing a computer device which takes this model account, it is due to the structure of related problems is part of a learning model, distinct from the model of assessing mathematical competencies in an instrumented perspective.

A fourth centre looks at the non-deterministic character of the system and considers a large number of solutions. When a related problem is chosen, it is not so much because we know exactly why the student was blocked but because we suppose that the student knows what is expected of him and in return the teacher or the tutorial system knows the logic of the problem. The question of correlation between related a problem and a learning blockage is not deterministic since the system does not indicate how to proceed. It follows the student in his reasoning (by comparison with expert solutions generally in the range of 50-100,000), regardless if it belongs to a moment of discovery or proof and it invites him to remain in the logic of the situation, simultaneously offering personalised assistance based on the instrumented behaviour of each student in a single class.

11. Acknowledgements

The development of the research is made possible by a grant from the Conseil de recherches en sciences humaines (CRSH 410-2009-0179, Gouvernement du Canada) and a grant of the Secretaría de Estado de Investigación, Desarrollo e Innovación (EDU2011-23240, Gobierno de España).

REFERENCES

[1] V. Aleven, O. Popescu and K. R. Koedinger, "Towards Tutorial Dialog to Support Self-explanation : Adding natural Language Understanding to a Cognitive Tutor, dans Artificial Intelligence in Education : AI-ED in the Wired and Wireless Future (Éds. Moore, Redfield & Johnson)," IOS Press, Amsterdam, 2002, pp. 246-255.

[2] C. Alsina and R. Nelsen, "Math Made Visual, *Creating Images for Understanding Mathematics*", Washington : the Mathematical Association of America, 2006.

[3] J. R. Anderson, C. F. Boyle and Yost, "The Geometry Tutor", *The Journal of Mathematical Behavior*, 1986, pp. 5-20.

[4] M. Artigue, "Ingénierie didactique, *Recherches en didactique des mathématiques* **9.3**", 1990, pp. 281-308.

[5] N. Balacheff and C. Margolinas, "Modèle de connaissances pour le calcul de situations didactiques, dans *Balises pour la didactique des mathématique*", (Éds. Mercier & Margolinas), 2005, pp. 75-106. La pensée sauvage, Grenoble.
[Overview of the model in line http://ckc.imag.fr/].

[6] M. Blossier and P. R. Richard, "Modélisation instrumentée et conceptions a priori dans un espace de travail géométrique en évolution : un tour en géométrie dynamique tridimensionnelle", *Actes des journées mathématiques 2011 de l'École Normale Supérieure de Lyon (IFÉ 2011), Institut Français de l'Éducation* , 2011, pp. 93-101.

[7] G. Brousseau, "Théorie des situations didactiques", La Pensée Sauvage, Grenoble, 1998.

[8] F. Caron and S. R. de Cotret, "Un regard didactique sur l'évaluation en mathématiques : genèse d'une perspective" *Actes du Colloque 2007 du Groupe de didactique des mathématiques du Québec*, 2007, pp. 123-134.

[9] A. C. Clairaut (2006), Éléments de géométrie, reproduction en fac-similé de l'édition de Paris chez David fils, 1741: Élémens de géométrie, Gabay, Paris.

[10] P. Cobo, J. M. Fortuny, E. Puertas and P. R. Richard, Agentgeom : A multiagent system for pedagogical support in geometric proof problems, *International Journal of Computers for Mathematical Learning,* Vol. 12, No. 1, 2007, pp. 57-79.

[11] S. Coutat and P. R. Richard, "Les figures dynamiques dans un espace de travail mathématique pour l'apprentissage des propriétés géométriques", *Annales de didactique et de sciences cognitives,* Vol. 16, 2011, pp. 97-126.

[12] J. J. Dahan, "La démarche de découverte expérimentalement médiée par cabri-géomètre en mathématiques : un essai de formalisation à partir de l'analyse de démarches de résolutions de problèmes de boîtes noires", Thèse de Doctorat, Université Joseph Fourier Grenoble, 2005.

[13] R. Duval, "Sémiosis et pensée humaine : Registres sémiotiques et apprentissages intellectuels", Peter Lang, Berne, 1995.

[14] R. Duval, "Les conditions cognitives de l'apprentissage de la géométrie : développement de la visualisation, différenciation des raisonnements et coordination de leurs fonctionnements", *Annales de didactique et sciences cognitives*, Vol. 10, 2005, pp. 5-53.

[15] M. A. Eisenhart, "The Ethnographic Research Tradition and Mathematics Education Research", *Journal for Research in Mathematics Education,* Vol. 19, No. 2, 1988, pp. 99-114.

[16] G. Gadanidis and P. R. Richard, "Report of the Working Group in MOOCs and Online Mathematics Teaching and Learning", *Actes de la rencontre 2013 du Groupe canadien d'étude en didactique des mathématiques - Proceedings of the 2013 Canadian mathematics education study group conference*, Brock University, 2013.

[17] B. G. Glaser and A. L. Strauss, "The discovery of grounded theory : Strategies for qualitative research," Aldine de Gruyter, Hawthorne, 1967.

[18] K. F. Hollebrands, A. M. Conner and R. C. Smith, "The Nature of Arguments Provided by College Geometry Students With Access to Technology While Solving

Problems", *Journal for Research in Mathematics Education,* Vol. 41, No. 4, 2010, pp. 324-350.

[19] S. Johnston-Wilder and J. Mason, "Developing Thinking in Geometry", The Open University, 2005.

[20] A. Kuzniak, "Paradigmes et espaces de travail géométriques. Éléments d'un cadre théorique pour l'enseignement et la formation des enseignants en géométrie", *Revue canadienne de l'enseignement des sciences, des mathématiques et des technologies,* Vol. 6, No. 2, 2006, pp. 167-187.

[21] A. Kuzniak, "L'espace de travail mathématique et ses genèses", *Annales de didactique et de sciences cognitives* Vol. 16, 2011, pp. 9-24.

[22] A. Kuzniak, P. R. Richard and A. Gagatsis, "CERME7 Working Group 4 : Geometry teaching and learning", *Research in Mathematics Education,* Vol. 14, No. 2, 2012, pp. 191-192.

[23] Laboratoire Leibniz, "Baghera assessment project : Designing an hybrid and emergent educational society", dans *Rapport pour la commission européenne, Programme IST, Les Cahiers du Laboratoire Leibniz n° 81* (Éd. Soury-Lavergne). Grenoble, 2003.

[24] I. Lakatos, *Preuves et réfutations. Essai sur la logique de la découverte mathématique*. Hermann, Paris, 1984.

[25] V. Luengo, Some didactical and epistemological considerations in the design of educational software : The cabri-euclide example, *International Journal of Computers for Mathematical Learning,* Vol. 10, No. 1, 2005, pp. 1-29.

[26] C. Margolinas, Points de vue de l'élève et du professeur : essai de développement de la théorie des situations didactiques, Habilitation à diriger les recherches en sciences de l'éducation, Université de Provence, version électronique récupérée le 26 juillet 2010 à http://tel.archivesouvertes.fr/docs/00/42/96/95/PDF/HDR_Margolinas.pdf.

[27] J. Mason, *Researching Your Own Practice : The Discipline of Noticing*. Londres et New York : Routledge, 2005.

[28] N. Matsuda and K. VanLehn, "Advanced geometry tutor : An intelligent tutor that teaches proof-writing with construction", dans *The 12th International Conference on Artificial Intelligence in Education* (Éds. Looi, McCalla, Bredeweg & Breuker), 2005, pp. 443-450. IOS Press, Amsterdam.

[29] MÉLS (2001, 2006 et 2007), Programme de formation de l'école québécoise, éducation préscolaire, enseignement primaire (2001), enseignement secondaire 1er cycle (2006) et enseignement secondaire 2e cycle (2007). Publications du Gouvernement du Québec.

[30] P. Paillé and A. Mucchielli, "L'analyse qualitative en sciences humaines et sociales", Paris : Armand Colin, 2008.

[31] M. J. Perrin-Glorian and Y. Reuter, "Les méthodes de recherche en didactiques", Villeneuve d'Ascq, France : Presses Universitaires du Septentrion, 2006.

[32] D. Py, Aide à la démonstration en géométrie : le projet Mentoniezh, *Sciences et Techniques Educatives,* Vol. 3, No .2, 1996, pp. 227-256.

[33] D. Py, "Environnements interactifs d'apprentissage et démonstration en géométrie", Habilitation à diriger des recherches, Université de Rennes, 2001.

[34] P. Rabardel, "Les hommes et les technologies : Approche cognitive des instruments contemporains", Armand Colin, Paris, 1995.

[35] P. Rabardel and Pastré, "Modèles du sujet pour la conception : dialectiques activités développement", Toulouse : Octarès, 2005.

[36] P. R. Richard, "Raisonnement et stratégies de preuve dans l'enseignement des mathématiques", Peter Lang, Berne, 2004a.

[37] P. R. Richard, "L'inférence figurale : Un pas de raisonnement discursivo-graphique", *Educational Studies in Mathematics,* Vol. 57, No. 2, 2004b, pp. 229-263.

[38] P. R. Richard (2010a et b), "La geometría dinámica como herramienta para desarrollar competencias de modelización en el Bachillerato" (2010a), "La evaluación de competencias matemáticas : una apuesta de aprendizaje desde la elección de situaciones-problemas" (2010b), dans *Competencias matemáticas. Instrumentos para las ciencias sociales y naturales* (Éd. Chacón) 21-57 et 59-81. Publicaciones del Ministerio de Educación, Gobierno de España.

[39] P. R. Richard and J. M. Fortuny, "Amélioration des compétences argumentatives à l'aide d'un système tutoriel en classe de mathématique au secondaire", *Annales de didactique et de sciences cognitives,* Vol. 12, 2007, pp. 83-116.

[40] P. R. Richard, P. Cobo, J. M. Fortuny, M. Hohenwarter, "Training teachers to manage problem-solving classes with computer support", *Journal of Applied Computing,* Vol. 5, No. 1, 2009, pp. 38-50.

[41] P. R. Richard, J. M. Fortuny, M. Gagnon, N. Leduc, E. Puertas and M. Tessier- Baillargeon, "Didactic and theoretical-based perspectives in the experimental development of an intelligent tutorial system for the learning of geometry, dans Interoperable interactive geometry for Europe (Éds. Kortenkamp & Laborde)", *ZDM - The International Journal on Mathematics Education,* Vol. 43, 2011, pp. 425-439.

[42] P. R. Richard, J. M. Fortuny, M. Hohenwarter and M. Gagnon, "geogebraTUTOR : une nouvelle approche pour la recherche sur l'apprentissage compétentiel et instrumenté de la géométrie à l'école secondaire", *Actes de la World Conference on E-Learning in Corporate, Government, Healthcare, and Higher Education,* 2007.

[43] P. R. Richard, V. Meavilla and J. M. Fortuny, "Textos clásicos y geometría dinámica : estudio de un aporte mutuo para el aprendizaje de la geometría", *Revista Enseñanza de las Ciencias,* Vol. 28, No. 1, 2010, pp. 95-111.

[44] P. R. Richard and A. Sierpinska, "Étude fonctionnelle-structurelle de deux extraits de manuels anciens de

géométrie", In Lemoyne, G. et Sackur, C. (rédactrices invitées) *Le langage dans l'enseignement et l'apprentissage des mathématiques, Revue des sciences de l'éducation, Numéro thématique*, Vol. 30, No. 2, 2004, pp. 379-409.

[45] P. R. Richard, S. Coutat and C. Laborde, "L'apprentissage instrumenté de propriétés en géométrie : propédeutique à l'acquisition d'une compétence de démonstration", *Educational Studies in Mathematics* (sous presse), 2013.

[46] P. R. Richard, V. Freiman and D. H. Jarvis, "L'enseignement des mathématiques au Québec", *Actes de l'Espace mathématique francophone 2012, groupe spécial Comparaison de l'enseignement des mathématiques à travers les pays francophones : résultats, sens et usages*, Genève, 2013, pp. 7-30.

[47] P. R. Richard, E. Swoboda, M. Maschietto and J. Mithalal, "Introduction to the Geometrical Thinking Working Group", Actes du *Congress of European Research in Mathematics Education (CERME8)*, 2013, pp. 1-7.

[48] A. H. Schoenfeld, "How We Think - A Theory of Goal-Oriented Decision Making and its Educational Applications", New York : Routledge, 2011.

[49] M. Tessier-Baillargeon, V. Leduc and P. R. Richard, "Niveaux d'intervention enseignante pour le développement d'un système tutorial : une expérience didactique à l'école secondaire avec le système géogébraTUTOR", *Actes des journées mathématiques 2011 de l'École Normale Supérieure de Lyon (IFÉ 2011)*, Institut Français de l'Éducation, 2011, pp. 201-208.

[50] M. Tessier-Baillargeon, N. Leduc and P. R. Richard, "Développement expérimental d'un système tutoriel intelligent pour le développement de compétences en géométrie à l'école secondaire". Dans *Formation à la recherche en didactique des mathématiques* (Éds. Hitt & Cortés), pp. 101-113. Québec : Loze-Dion.

[51] J. Trgalová and P. R. Richard, "Analyse de ressources comme moyen de développement professionnel des enseignants", *Actes de l'Espace mathématique francophone 2012, groupe de travail Ressources et développement professionnel des enseignants,* 2012, pp. 12-23, Genève.

[52] J. Trgalová, P. R. Richard and S. Soury-Lavergne, "Évaluation de la qualité des ressources de géométrie dynamique : un outil pour le développement de compétences professionnelles des enseignants de mathématiques", *Actes des journées mathématiques 2011 de l'École Normale Supérieure de Lyon* (IFÉ 2011), Institut Français de l'Éducation, 2011, pp. 131-138, Lyon.

[53] J. M. Van der Maren, "Méthodes de recherche pour l'éducation", De Boeck Université, Bruxelles, 1996.

[54] K. Vanlehn, C. Lynch, K. Schulze, J. A. Shapiro, R. Shelby, L. Taylor, *et al.*, "The andes physics tutoring system : Lessons learned", *Int. J. Artif. Intell. Ed.*, Vol. 15, No. 3, 2005, pp. 147-204.

[55] P. Vermersch, *L'entretien d'explicitation en formation initiale et continue*, ESF, Paris,1994.

References to main websites

cK¢ wikibook	http://ckc.imag.fr/index.php/Main_Page
GeoGebraTube	http://www.geogebratube.org
GDM	http://turing.scedu.umontreal.ca/gdm/
GIC-IGC	http://www.geogebracanada.org/home
I2GÉO	http://i2geo.net (access to activities for registrants)
Turing	http://turing.scedu.umontreal.ca (access to Moodle for registrants)

Acronym reference

AC-GGB	Associació Catalana de geogebra
ADSC	Annales de didactique et de sciences cognitives
AMQ	Bulletin of the Association mathématique du Québec
CADGME	Computer Algebra and Dynamic Geometry Systems in Mathematics Education
CCPÉ	Advisory Committee on the curriculum of the Quebec Ministry of Education
CERME	Congress of European Research in Mathematics Education
CRSH	Conseil de recherche en sciences humaines du Canada
E-LEARN	World Conference on E-Learning in Corporate, Government, Healthcare & Higher Education
EMF	Espace mathématique francophone Symposium
ESM	Educational Studies in Mathematics – An International Journal
ETM	Symposium Espace de travail mathématique
GRMS	Group of leaders in high school mathematics
IJCML	International Journal of Computers for Mathematical Learning
INTERGEO	Interoperable Interactive Geometry Conference
MATI	Roland-Giguère training and learning technologies House
NTLMP	International Newsletter on the Teaching and Learning of Mathematical Proof
PME	Publicaciones del Ministerio de Educación de España
REC	Revista enseñanza de las ciencias – Investigación y experiencias didácticas
UAB	Universitat Autònoma de Barcelona
UDM	Université de Montréal
UNO	Revista Uno – Didáctica de las matemáticas
ZDM	Zentralblatt für Didaktik der Mathematik – The International Journal on Mathematics Education

17

Run-Up Flow of a Maxwell Fluid through a Parallel Plate Channel

Syed Yedulla Qadri[1], M. Veera Krishna[2]
[1]Department of Mathematics, St. Josephs PG College, Kurnool, India
[2]Department of Mathematics, Rayalaseema University, Kurnool, India

ABSTRACT

We consider the flow of an incompressible viscous Maxwell fluid between two parallel plates, initially induced by a constant pressure gradient. The pressure gradient is withdrawn and the upper plate moves with a uniform velocity while the lower plate continues to be at rest. The arising flow is referred to as run-up flow. The unsteady governing equations are solved as initial value problem using Laplace transform technique. The expression for velocity, shear stresses on both plates and discharge are obtained. The behavior of the velocity, shear stresses and mass flux has been discussed in detail with respect to variations in different governing flow parameters and is presented through graphs.

Keywords: Run-Up Flow; Maxwell's Fluid; Laplace Transforms; Reynolds Number and Parallel Plate Channels

1. Introduction

In some technological problems related to petroleum industry, Lubrication technology etc., the fluid flow experiences phenomenon viz. run-up which arises due to sudden withdrawal of the pressure gradient causing the flow while its boundaries instantaneously move from rest. Under this phenomenon, steady flow in the unperturbed state gains unsteadiness later. Many research workers have paid attention to the study of Maxwell fluids. In lubrication theory and in many physical situations where we come across slip flows, there arises a class of problems referred to as "run-up and spin-up flows". The growing importance of the use of non-Newtonian fluids in modern technology and industries has led various researchers to attempt diverse flow problems related to several non-Newtonian fluids. One such fluid that has attracted the attention of research workers in fluid mechanics during the last four decades is the Maxwell fluid. This theory has several industrial and scientific applications as well, which comprise pumping fluids such as synthetic fluids, polymer thickened oils, liquid crystal, animal blood, synovial fluid present in synovial joints and the theory of lubrication (Naduvinamani *et al.* [1-5], Lin and Hung [6]). Kazakia and Rivlin [7] initiated the study of these flows and later Rivlin [8-10] elaborately studied the run-up and spin-up flows of visco-elastic fluids between rigid parallel plates and in circular geometries. Ramacharyulu and Raju [11] investigated the run-up flow of a viscous incompressible fluid in a long circular cylinder of porous material. Ramakrishna [12] discussed the run-up and spin-up flows related to a dusty viscous fluid. Later, M. Devakar and T. K. V. Iyengar [14] examined the run-up flow of an incompressible couple stress fluid between two infinite rigid parallel plates. The flow was assumed to be initially induced by a constant pressure gradient between two infinite rigid parallel plates. After the steady state was attained, the pressure gradient was suddenly withdrawn and the parallel plates were set to move instantaneously with different velocities in the direction of the applied pressure gradient. The time dependence of the resultant flow was investigated. Sugunamma *et al.* [15] analyzed the start-up flow of an incompressible visco-elastic Rivlin-Ericksen fluid. The initial flow is assumed due to the movement of boundaries. At an instant of time t, the boundaries are suddenly brought to rest and the flow is maintained due to a prescribed pressure gradient. Veera Krishna *et al.* [16] discussed the hall current effects on unsteady MHD flow of rotating Maxwell fluid through a porous medium in a parallel plate channel. Raji Reddy and Sambasiva Rao [17] analyzed run-up flow of viscous incompressible fluid through a rectangular pipe, a pipe of equilateral triangular cross section, parallel plate channel and a cylinder. They solved them by using ADI numerical technique. Basha [18] extended the analysis of the same by considering visco-elastic Rivlin-Ericksen fluid between parallel plates subjected to a constant suction. Malleswari [19] discussed run-up flow of Rivlin-Ericksen fluid with porous lining. With the recent researches in non-Newtonian fluid flows cited earlier, we consider the flow of an incompressible Maxwell fluid between two parallel plates, initially induced by a constant pressure gradient. The pressure gradient is suddenly withdrawn while the plates are impulsively started simultaneously. The up-flow is referred to as run-up flow. In this present paper, we are studying this flow of a Maxwell fluid.

2. Formulation and Solution of the Problem

Consider the flow of an incompressible Maxwell fluid between two infinite rigid parallel plates $y = 0$ and $y = h$ along the direction of x-axis (**Figure 1**). Since the flow is along the x-direction, we take the velocity $q = (u(y,t),0,0)$, which satisfies the continuity equation.

We consider a Cartesian system $O(x,y)$ so that the fluid flow takes place within boundary plates $y = 0$ and $y = h$. The linear momentum equation governing the flow $u(y, t)$ is given by

$$\lambda\left(\frac{\partial^2 u}{\partial t^2}\right) + \frac{\partial u}{\partial t} = -\frac{1}{\rho}\left(\frac{\partial p}{\partial x} + \lambda\frac{\partial^2 p}{\partial t\partial x}\right) + \nu\left(\frac{\partial^2 u}{\partial y^2}\right) \quad (1)$$

where ρ the density, p is the pressure, μ is the coefficient of viscosity and λ is the relaxation time. The boundary conditions are

$$u = \begin{cases} U; y = h \\ 0; y = 0 \end{cases} (t \geq 0) \quad (2)$$

We consider the run-up flow of the Maxwell's fluid through the parallel plate channel. Initially the flow due to a prescribed pressure gradient with boundaries at rest and at the time $t > 0$, the pressure gradient is withdrawn and the upper plate moves with a uniform velocity while the lower plate continues to be at rest. The equation governing the initial flow is

$$\nu\left(\frac{\partial^2 u}{\partial y^2}\right) = \frac{1}{\rho}\left(\frac{\partial p}{\partial x}\right) \quad (3)$$

The corresponding boundary conditions are

$$u = \begin{cases} 0; y = h \\ 0; y = 0 \end{cases} (t \leq 0) \quad (4)$$

We introduce the non-dimensional variables

$$x^* = \frac{x}{h}, y^* = \frac{y}{h}, u^* = \frac{u}{U}, p^* = \frac{p}{\rho U^2}, t^* = \frac{tU}{h}$$

Using non-dimensional variables, the governing equation are (dropping the asterisk)

$$\alpha\left(\frac{\partial^2 u}{\partial t^2}\right) + \frac{\partial u}{\partial t} = \frac{1}{R}\left(\frac{\partial^2 u}{\partial y^2}\right) \quad (5)$$

$$\frac{\partial^2 u}{\partial y^2} = PR \quad (6)$$

where, $R = \frac{\rho U h}{\mu}$ is the Reynolds's number, $\alpha = \frac{\lambda U}{h}$ is the Maxwell's fluid parameter, $P = \frac{\partial p}{\partial x}$ is the constant of pressure gradient. Using the boundary conditions (4), the Equation (6) reduces to

$$u = \frac{PR}{2} y(y-1) \quad (7)$$

Applying the Laplace transform to the Equation (5) and using the Equation (7), we get

$$\bar{u}(y,s) = A\text{Cosh}(ay) + B\text{Sinh}(ay) - \frac{PR}{2}\frac{(y-y^2)}{s} + \frac{P}{s^2(s\alpha+1)} \quad (8)$$

Now the transformed boundary conditions are

$$\bar{u} = \begin{cases} \frac{1}{s}; y = 1 \\ 0; y = 0 \end{cases} \quad (9)$$

On solving Equations (8) using (9); we get

$$\bar{u}(y,s) = -\frac{P}{s^2(s\alpha+1)}\frac{\text{Sinh}[a(1-y)]}{\text{Sinh}(a)} + \frac{\text{Sinh}(ay)}{s\text{Sinh}(a)} - \frac{P}{s^2(s\alpha+1)}\frac{\text{Sinh}(ay)}{\text{Sinh}(a)} - \frac{PR}{2}\frac{(y-y^2)}{s} + \frac{P}{s^2(s\alpha+1)} \quad (10)$$

On taking the inverse Laplace transform [13] for the Equation (10); we get

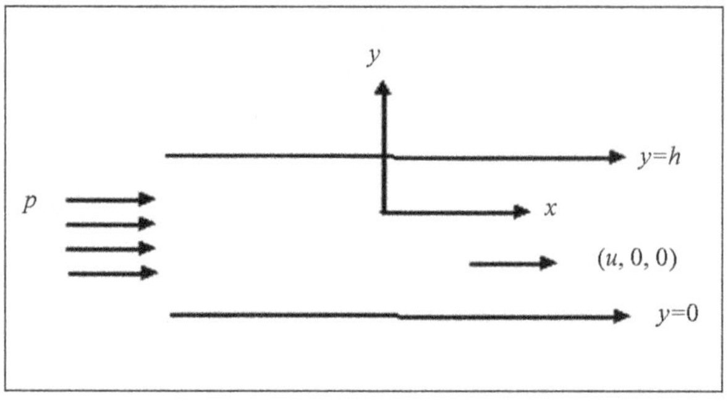

Figure 1. Configuration of the problem.

$$u(y,t) = y - \frac{PR}{2}(y - y^2)$$
$$+\sum_{n=1}^{\infty}(-1)^n e^{s_{n1}t}\left\{a_9\text{Sinh}(b_1 y) - a_3\text{Sinh}\left[b_1(1-y)\right]\right\}$$
$$+\sum_{n=1}^{\infty}(-1)^n e^{s_{n2}t}\left\{a_{10}\text{Sinh}(b_2 y) - a_6\text{Sinh}\left[b_2(1-y)\right]\right\} \quad (11)$$

The shear stresses on the upper plate and the lower plate are given by

$$\tau_L = \left(\frac{du}{dy}\right)_{y=0} = 1 - \frac{PR}{2}$$
$$+\sum_{n=1}^{\infty}(-1)^n e^{s_{n1}t}\left\{c_1 + c_2\text{Cosh}b_1\right\} \quad (12)$$
$$+\sum_{n=1}^{\infty}(-1)^n e^{s_{n2}t}\left\{c_3 + c_4\text{Cosh}b_2\right\}$$

and

$$\tau_U = \left(\frac{du}{dy}\right)_{y=1} = 1 + \frac{PR}{2}$$
$$+\sum_{n=1}^{\infty}(-1)^n e^{s_{n1}t}\left\{c_1\text{Cosh}b_1 + c_2\right\} \quad (13)$$
$$+\sum_{n=1}^{\infty}(-1)^n e^{s_{n2}t}\left\{c_3\text{Cosh}b_2 + c_4\right\}$$

The mass flux Q is given by

$$Q = \int_{y=0}^{1} u\mathrm{d}y = \frac{1}{2} - \frac{PR}{12}$$
$$+\sum_{n=1}^{\infty}(-1)^n e^{s_{n1}t}\left\{d_1\left[\text{Cosh}b_1 - 1\right] + d_2\left[1 - \text{Cosh}b_1\right]\right\}$$
$$+\sum_{n=1}^{\infty}(-1)^n e^{s_{n2}t}\left\{d_3\left[\text{Cosh}b_2 - 1\right] + d_4\left[1 - \text{Cosh}b_2\right]\right\} \quad (14)$$

3. Results and Discussion

The flow governed by the non-dimensional parameters R the Reynolds number, α the Maxwell fluid parameter. The velocity, the shear stresses on the plates and discharge between the plates are evaluated analytically and computationally discussed for different variations in the governing parameters R and α. It is interesting to note that the behavior of the flow very much depends on the pressure gradient, in accordance with the run-up flow, the initial steady flow is due to the prescribed pressure gradient, while the perturbed flow is due to the sudden movement of the boundary in absence of the pressure gradient. The flow in direction of the movement of the boundary may be considered as actual flow, where as the flow caused by the pressure gradient assumed to be against the direction of the boundary movement as the reversal flow. **Figures 2-9** represent the behaviour of the velocity component u for variations in low and high Reynolds number with Maxwell fluid parameter and for various values of time t. **Tables 1-3** show stresses on both boundaries and mass flux.

Figure 2 and **3** depict the variation of the velocity u slightly increases for increase in low Reynolds number R while it experience enhancement with increase in high

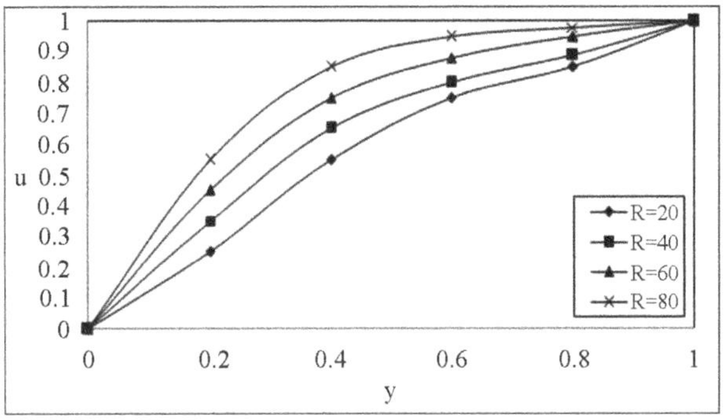

Figure 2. The velocity profile for u on low Reynolds number R with $P = 1, t = 0.1, \alpha = 0.25$.

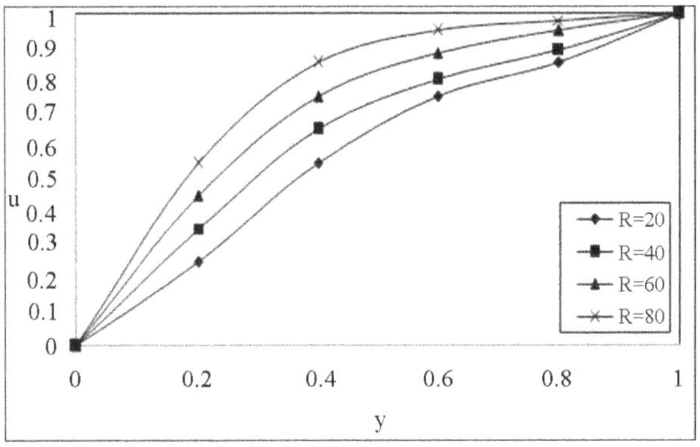

Figure 3. The velocity profile for u on high Reynolds number R with $P = 1, t = 0.1, \alpha = 0.25$.

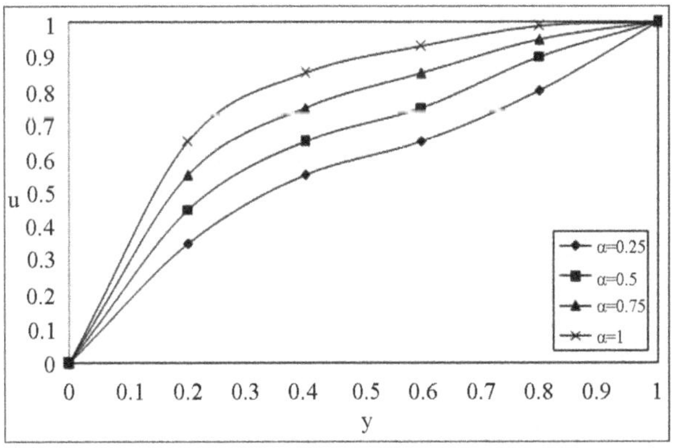

Figure 4. The velocity profile for u on the Maxwell fluid parameter α with low Reynolds number R = 20, P = 1, t = 0.1.

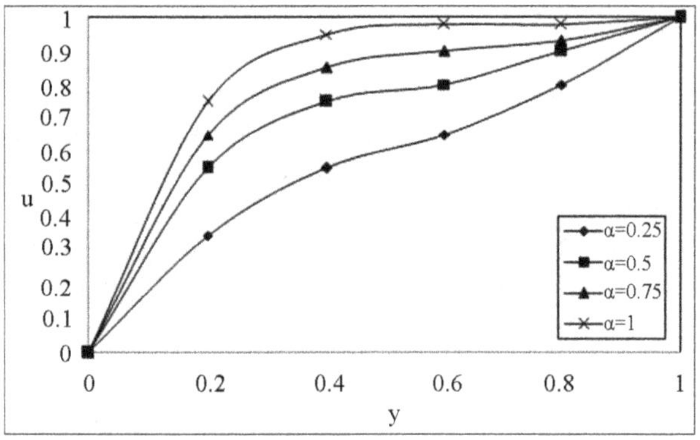

Figure 5. The velocity profile for *u* on the Maxwell fluid parameter α with high Reynolds number R = 20, P = 1, $t = 0.1$.

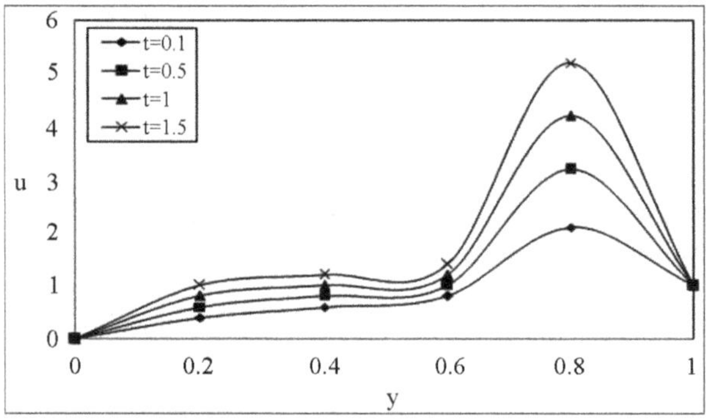

Figure 6. Time development of the velocity component *u* for $\alpha = 0.25$ with low Reynolds number $R = 20$, $P = 1$.

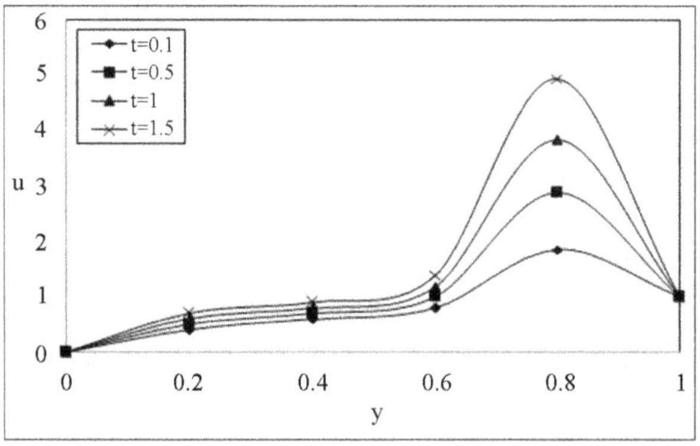

Figure 7. Time development of the velocity component *u* for $\alpha = 0.25$ with high Reynolds number $R = 20, P = 1$.

Reynolds number being the other parameter fixed. The velocity profiles (4 and 5) for u on the Maxwell fluid parameter α with low and high Reynolds number R = 20 and R = 1000 respectively. The similar behaviour is observed as earlier mentioned. We notice that the Maxwell fluid parameter with the Reynolds number affects the flow in the entire region. An interesting phenomenon is observed in **Figures 6** and 7, the magnitude of the velocity enhances with increase in time while fixing R and α. **Figures 8** and **9** indicates that for a fixed t, R and

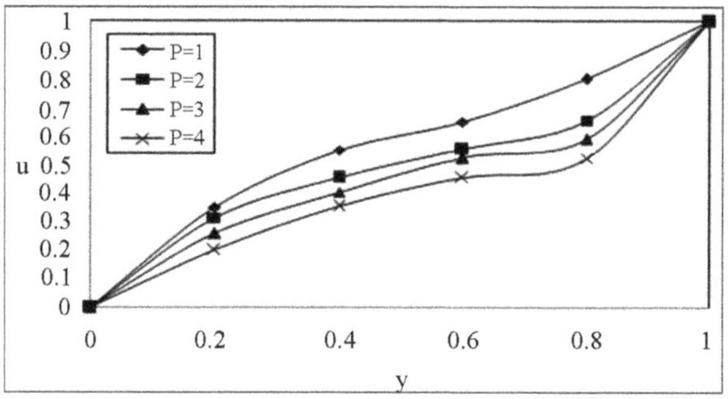

Figure 8. The velocity profile for *u* with pressure gradient *P* on low Reynolds number $R = 20$ with $t = 0.1, \alpha = 0.25$.

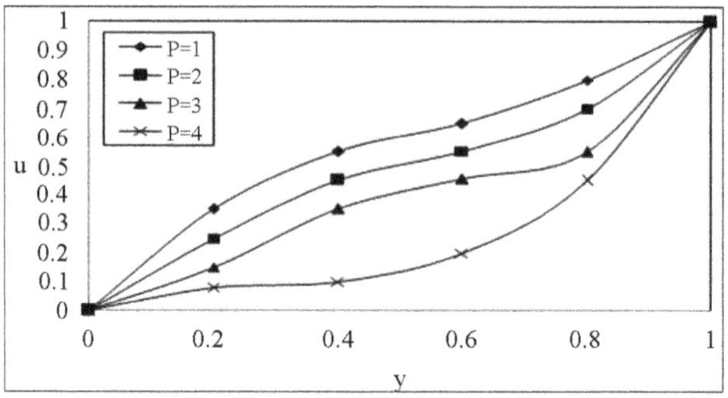

Figure 9. The velocity profile for *u* with pressure gradient *P* on low Reynolds number $R = 20$ with $t = 0.1, \alpha = 0.25$.

Table 1. The shear stresses on the upper plate.

R	I	II	III	IV
50	0.431432	0.511602	0.622056	0.736465
100	0.520844	0.623145	0.737066	0.833415
250	0.675562	0.746652	0.832455	0.911453
500	0.758568	0.833755	0.994585	1.833452
	I	II	III	IV
α	0.25	0.5	0.75	1.0

Table 2. The shear stresses on the lower plate.

R	I	II	III	IV
50	0.056752	0.074675	0.085326	0.094523
100	0.046675	0.052762	0.067745	0.073148
250	0.037562	0.047536	0.053742	0.052274
500	0.027458	0.037459	0.040742	0.052288
	I	II	III	IV
α	0.25	0.5	0.75	1.0

α, as P increases the velocity decreases for any y. This is in accordance with the fact that an increase in P implies a decrease in pressure which naturally results in a decrease of velocity.

The magnitude of the stresses on upper and lower plates enhance with increasing in α, and reduces in

Table 3. The Discharge between the plates.

R	I	II	III	IV
50	0.846752	0.931448	0.983246	1.132652
100	0.884675	0.966758	1.108321	1.234525
250	0.920554	1.103244	1.375265	1.578346
500	0.957562	1.346738	1.670812	1.846724
	I	II	III	IV
α	0.25	0.5	0.75	1.0

lower plate increases in upper plate with increasing *R* (**Tables 1** and **2**). The discharge between the plates enhances with increase in both *R* and α (**Table 3**).

4. Conclusion

The run-up flow of an incompressible Maxwell fluid between two infinite parallel plates is studied using Laplace transform technique. Analytical expressions for the fluid velocity field are obtained in Laplace transform domain. The magnitude of the velocity enhances with the increasing in both *R* and α and reduces with the increasing in *P*. The stresses on upper and lower plates enhance with the increasing in α, reduce in lower plate and increase in upper plate with increasing *R*. The discharge between the plates enhances with the increasing in both *R* and α.

5. Acknowledgements

The authors are thankful to Prof. R. Siva Prasad, Department of Mathematics, Sri Krishnadevaraya University, Anantapur, Andhra pradesh, India, and AJCM Journal for the support to develop this document.

REFERENCES

[1] N. B. Naduvinamani, P. S. Hiremath and G. Gurubasavaraj, "Squeeze Film Lubrication of a Short Porous Journal Bearing with Couple Stress Fluids," *Tribology International*, Vol. 34, No. 11, 2001, pp. 739-747.

[2] N. B. Naduvinamani, P. S. Hiremath and G. Gurubasavaraj, "Surface Roughness Effects in A Short Porous Journal Bearing with a Couple Stress Fluid," *Fluid Dynamics Residuals*, Vol. 31, No. 5-6, 2002, pp. 333-354.

[3] N. B. Naduvinamani, P. S. Hiremath and G. Gurubasavaraj, "Effects of Surface Roughness on the Couple Stress Squeeze Film between a Sphere and a Flat Plate," *Tribology International*, Vol. 38, No. 5, 2005, pp. 451-458.

[4] N. B. Naduvinamani, S. T. Fathima and P. S. Hiremath, "Hydro Dynamic Lubrication of Rough Slider Bearings with Couple Stress Fluids," *Tribology International*, Vol. 36, No. 12, 2003, pp. 949-959.

[5] N. B. Naduvinamani, S. T. Fathima and P. S. Hiremath, "Effect of Surface Roughness on Characteristics of Couple Stress Squeeze Film between Anisotropic Porous Rectangular Plates," *Fluid Dynamics Residuals*, Vol. 32, No. 5, 2003, pp. 217-231.

[6] J. R. Lin and C. R. Hung, "Combined Effects of Non-Newtonian Couple Stresses and Fluid Inertia on the Squeeze Film Characteristics between a Long Cylinder and an Infinite Plate," *Fluid Dynamics Residuals*, Vol. 39, No. 8, 2007, pp. 616-639.

[7] J. Y. Kazakia and R. S. Rivlin, "Run-Up and Spin-Up in a Visco-Elastic Fluid I," *Rheologica Acta*, Vol. 20, No. 2, 1981, pp. 111-127.

[8] R. S. Rivlin, "Run-Up and Spin-Up in a Visco-Elastic Fluid II," *Rheologica Acta*, Vol. 21, No. 2, 1982, pp. 107-111.

[9] R. S. Rivlin, "Run-Up and Spin-Up in a Visco-Elastic Fluid III," *Rheologica Acta*, Vol. 21, No. 3, 1982, pp. 213-222.

[10] R. S. Rivlin, "Run-Up and Spin-Up in a Visco-Elastic Fluid IV," *Rheologica Acta*, Vol. 22, No. 3, 1983, pp. 275-283.

[11] N. Ch. Pattabhi Ramacharyulu and K. A. Raju, "Run-Up in a Generalized Porous Medium," *Indian Journal of Pure Applied Mathematics*, Vol. 15, No. 6, 1984, pp. 665-670.

[12] D. Ramakrishna, "Some Problems in the Dynamics of Fluids with Particle Suspensions," Ph.D. Thesis, Kakatiya University, Warangal, 1986.

[13] G. Honig and U. Hirdes, "A Method for the Numerical In-version of Laplace Transforms," *Journal of Computational Applied Mathematics*, Vol. 10, No. 1, 1984, pp. 113-132.

[14] M. Diwakar and T. K. V. Iyengar, "Run-Up Flow of a Couple Stress Fluid between Parallel Plates," *Nonlinear Analysis: Modelling and Control*, Vol. 15, No. 1, 2010, pp. 29-37.

[15] V. Sugunamma, M. S. Latha and N. Sandeep, "Run-Up Flow of a Rivlin-Ericksen Fluid through a Porous Medium in a Channel," *International Journal of Mathematical Archive*, Vol. 2, No. 12, 2011, pp. 2625-2639.

[16] M. Veera Krishna, S. V. Suneetha and R. Siva Prasad, "Hall Current Effects on Unsteady MHD Flow of Rotating Maxwell Fluid through a Porous Medium," *Journal of Ultra Scientist of Physical Sciences*, Vol. 22, No. 1, 2010, pp. 133-144.

[17] R. Reddy Sheelam, "Computational Techniques in Transient Magneto Hydro Dynamics Dusty Viscous and Run-Up Flows," Ph.D. Thesis, Osmania University, Hyderabad, 1992, pp. 47-65.

[18] M. Basha, "Visco-Elastic Fluid Flow and the Heat Transfer through Porous Medium," Ph.D. Thesis, SriKrishnadevaraya University, Anantapur, 1994.

[19] D. Malleswari, "Unsteady Flow of a Rivlin-Ericksen Fluid through Plannar Channels with Porous Lining," Ph.D. Thesis, Sri Padmavathi Mahila Viswavidyalayam, Tirupati, 2010.

Appendix

$$a = \sqrt{R(s^2\alpha + s)}$$

$$s_{n_1} = -\frac{1}{2\alpha}\left[1 + \sqrt{1 - \frac{4\alpha n^2 \pi^2}{R}}\right]$$

$$s_{n_2} = -\frac{1}{2\alpha}\left[1 - \sqrt{1 - \frac{4\alpha n^2 \pi^2}{R}}\right]$$

$$s_{n_2} = -\frac{1}{2\alpha}\left[1 - \sqrt{1 - \frac{4\alpha n^2 \pi^2}{R}}\right]$$

$$a_1 = \frac{\sqrt{R}}{2}\frac{\left(1 + 2s_{n_1}\alpha\right)}{\sqrt{s_{n_1} + s_{n_1}^2\alpha}},$$

$$a_2 = \frac{P}{s_{n_1}^2\left(1 + s_{n_1}\alpha\right)}$$

$$a_3 = \frac{2P\sqrt{s_{n_1} + s_{n_1}^2\alpha}}{\sqrt{R}s_{n_1}^2\left(1 + s_{n_1}\alpha\right)\left(1 + 2s_{n_1}\alpha\right)\mathrm{Cosh}b_1}$$

$$a_4 = \frac{\sqrt{R}}{2}\frac{\left(1 + 2s_{n_2}\alpha\right)}{\sqrt{s_{n_2} + s_{n_1}^2\alpha}}$$

$$a_5 = \frac{P}{s_{n_2}^2\left(1 + s_{n_2}\alpha\right)}$$

$$a_6 = \frac{2P\sqrt{s_{n_2} + s_{n_2}^2\alpha}}{\sqrt{R}s_{n_2}^2\left(1 + s_{n_2}\alpha\right)\left(1 + 2s_{n_2}\alpha\right)\mathrm{Cosh}b_2}$$

$$a_7 = \frac{2\sqrt{s_{n_1} + s_{n_1}^2\alpha}}{\sqrt{R}\left(1 + 2s_{n_1}\alpha\right)s_{n_1}\mathrm{Cosh}b_1}$$

$$a_8 = \frac{2\sqrt{s_{n_2} + s_{n_2}^2\alpha}}{\sqrt{R}\left(1 + 2s_{n_2}\alpha\right)s_{n_2}\mathrm{Cosh}b_2}$$

$$a_9 = \frac{2\sqrt{s_{n_1} + s_{n_1}^2\alpha}}{\sqrt{R}\left(1 + 2s_{n_1}\alpha\right)s_{n_1}\mathrm{Cosh}b_1} - \frac{2P\sqrt{s_{n_1} + s_{n_1}^2\alpha}}{\sqrt{R}s_{n_1}^2\left(1 + s_{n_1}\alpha\right)\left(1 + 2s_{n_1}\alpha\right)\mathrm{Cosh}b_1}$$

$$a_{10} = \frac{2\sqrt{s_{n_2} + s_{n_2}^2\alpha}}{\sqrt{R}\left(1 + 2s_{n_2}\alpha\right)s_{n_2}\mathrm{Cosh}b_2} - \frac{2P\sqrt{s_{n_2} + s_{n_2}^2\alpha}}{\sqrt{R}s_{n_1}^2\left(1 + s_{n_2}\alpha\right)\left(1 + 2s_{n_2}\alpha\right)\mathrm{Cosh}b_2}$$

$$b_1 = \sqrt{R\left(s_{n_1} + s_{n_1}^2\alpha\right)}$$

$$b_2 = \sqrt{R\left(s_{n_2} + s_{n_2}^2\alpha\right)}$$

$$c_1 = \left(\frac{2\sqrt{s_{n_1} + s_{n_1}^2 \alpha}}{\sqrt{R}\left(1+2s_{n_1}\alpha\right)1+2s_{n_1}\alpha s_{n_1}\,\mathrm{Cosh}b_1} - \frac{2P\sqrt{s_{n_1} + s_{n_1}^2 \alpha}}{\sqrt{R}s_{n_1}^2\left(1+s_{n_1}\alpha\right)\left(1+2s_{n_1}\alpha\right)\mathrm{Cosh}b_1} \right)\sqrt{R\left(s_{n_1} + s_{n_1}^2 \alpha\right)}$$

$$c_2 = \left(\frac{2P\sqrt{s_{n_1} + s_{n_1}^2 \alpha}}{\sqrt{R}s_{n_1}^2\left(1+s_{n_1}\alpha\right)\left(1+2s_{n_1}\alpha\right)\mathrm{Cosh}b_1} \right)\sqrt{R\left(s_{n_1} + s_{n_1}^2 \alpha\right)}$$

$$c_3 = \left(\frac{2\sqrt{s_{n_2} + s_{n_2}^2 \alpha}}{\sqrt{R}\left(1+2s_{n_2}\alpha\right)s_{n_2}\,\mathrm{Cosh}b_2} - \frac{2P\sqrt{s_{n_2} + s_{n_2}^2 \alpha}}{\sqrt{R}s_{n_1}^2\left(1+s_{n_2}\alpha\right)\left(1+2s_{n_2}\alpha\right)\mathrm{Cosh}b_2} \right)\sqrt{R\left(s_{n_2} + s_{n_2}^2 \alpha\right)}$$

$$c_4 = \left[\frac{2P\sqrt{s_{n_2} + s_{n_2}^2 \alpha}}{\sqrt{R}s_{n_1}^2\left(1+s_{n_2}\alpha\right)\left(1+2s_{n_2}\alpha\right)\mathrm{Cosh}b_2} \right].\sqrt{R\left(s_{n_2} + s_{n_2}^2 \alpha\right)}$$

$$d_1 = \left(\frac{2\sqrt{s_{n_1} + s_{n_1}^2 \alpha}}{\sqrt{R}\left(1+2s_{n_1}\alpha\right)s_{n_1}\,\mathrm{Cosh}b_1} - \frac{2P\sqrt{s_{n_1} + s_{n_1}^2 \alpha}}{\sqrt{R}s_{n_1}^2\left(1+s_{n_1}\alpha\right)\left(1+2s_{n_1}\alpha\right)\mathrm{Cosh}b_1} \right)\frac{1}{\sqrt{R\left(s_{n_1} + s_{n_1}^2 \alpha\right)}}$$

$$d_2 = \left[\frac{2P\sqrt{s_{n_1} + s_{n_1}^2 \alpha}}{\sqrt{R}s_{n_1}^2\left(1+s_{n_1}\alpha\right)\left(1+2s_{n_1}\alpha\right)\mathrm{Cosh}b_1} \right]\frac{1}{\sqrt{R\left(s_{n_1} + s_{n_1}^2 \alpha\right)}}$$

$$d_3 = \left(\frac{2\sqrt{s_{n_2} + s_{n_2}^2 \alpha}}{\sqrt{R}\left(1+2s_{n_2}\alpha\right)s_{n_2}\,\mathrm{Cosh}b_2} - \frac{2P\sqrt{s_{n_2} + s_{n_2}^2 \alpha}}{\sqrt{R}s_{n_1}^2\left(1+s_{n_2}\alpha\right)\left(1+2s_{n_2}\alpha\right)\mathrm{Cosh}b_2} \right)\frac{1}{\sqrt{R\left(s_{n_2} + s_{n_2}^2 \alpha\right)}}$$

$$d_4 = \left[\frac{2P\sqrt{s_{n_2} + s_{n_2}^2 \alpha}}{\sqrt{R}s_{n_1}^2\left(1+s_{n_2}\alpha\right)\left(1+2s_{n_2}\alpha\right)\mathrm{Cosh}b_2} \right]\frac{1}{\sqrt{R\left(s_{n_2} + s_{n_2}^2 \alpha\right)}}$$

18

Bifurcations and Sequences of Elements in Non-Smooth Systems Cycles

Ivan Arango[1], Fabio Pineda[1], Oscar Ruiz[2]
[1]Mechatronics and Machine Design Group, Universidad EAFIT, Medellin, Colombia
[2]Laboratorio de CAD/CAM/CAE, Universidad EAFIT, Medellin, Colombia

ABSTRACT

This article describes the implementation of a novel method for detection and continuation of bifurcations in non-smooth complex dynamic systems. The method is an alternative to existing ones for the follow-up of associated phenomena, precisely in the circumstances in which the traditional ones have limitations (simultaneous impact, Filippov and first derivative discontinuities and multiple discontinuous boundaries). The topology of cycles in non-smooth systems is determined by a group of ordered segments and points of different regions and their boundaries. In this article, we compare the limit cycles of non-smooth systems against the sequences of elements, in order to find patterns. To achieve this goal, a method was used, which characterizes and records the elements comprising the cycles in the order that they appear during the integration process. The characterization discriminates: a) types of points and segments; b) direction of sliding segments; and c) regions or discontinuity boundaries to which each element belongs. When a change takes place in the value of a parameter of a system, our comparison method is an alternative to determine topological changes and hence bifurcations and associated phenomena. This comparison has been tested in systems with discontinuities of three types: 1) impact; 2) Filippov and 3) first derivative discontinuities. By coding well-known cycles as sequences of elements, an initial comparison database was built. Our comparison method offers a convenient approach for large systems with more than two regions and more than two sliding segments.

Keywords: Bifurcation Sequences; Non-Smooth Systems; Limit Cycles; Dynamic Systems

1. Introduction

Physical systems can often operate in different modes, and as the time of the transition from one mode to another mode is small, the transition is considered as instantaneous [1]. Events such as impact, dry friction, backlash, hysteresis, saturation and commutation carry a discontinuity or sudden change. Therefore, they can be modeled declaring at least two modes. Each mode is represented by differential equation or mixes of differential and difference equations. The mathematical modeling of these systems switches between different modes and they are classified as piecewise-smooth or non-smooth system.

Piecewise-smooth systems may be classified according to the degree of discontinuity that the orbits and vector fields present [1]. An updated classification by [2] discusses systems with three degrees of smoothness. In the zero level, one has jumps in the state variables. They are typically systems with impact, where the phenomenon is modeled assuming no deformation and a negligible impact time [3]. In the first degree of smoothness, we have systems described by differential equations with discontinuous right hand terms (Filippov systems) [4]. In these cases the vector field is discontinuous in the switching Boundary, as usual in mechanical systems with dry friction [5]. The second degree of smoothness, includes systems with continuous vector fields but discontinuities in the first derivative of the vector field. As an example for second degree, we might consider a mechanical system with a single mass, spring, damping element and limiting elastic support [6]. In general, a discontinuity in the i-th derivative implies that the system is classified as being $i + 1$ degree of smoothness.

Non-standard bifurcations in non-smooth systems have been intensively studied [6-9]. But, there are only mathematical tools to analyze phenomena in 2D or 3D systems with two vector fields and one discontinuity boundary [10,11]. The names assigned to the bifurcations vary according to the researcher. For example, [2] is used *Grazing*, *Switching*, *Crossing* and *Multisliding*. For the same bifurcations, in [12] is used *Touching*, *Bucking*, *Crossing* and *Adding*. Other sliding bifurcation types, recently reported in [8], have been characterized in systems with two *DBs*. Those bifurcations have been called *Exchanging*, *Sticking Disappearance* and *Non-smooth Fold*.

Article Outline. This article is organized as follows. Section 2 explains the notation and symbols used. Section 3 summarizes the solutions for the types of non-smooth systems. Section 4 describes the well-known bifurcations as sequences of elements. Section 5 analyzes the procedure of identification and comparison of the elements of the cycles versus the elements of an integration. Section 6 concludes the article.

2. Notation and Symbology for Points in the *DB*

The study of Non-smooth systems includes more information than a smooth system. The proposed method is based on the information of each element of the cycle. Therefore, we had to introduce a notation to see all the information of the points, segments and orbits. The information should be fully contained inside the textual or graphical symbols assigned to each element. Some distinguished symbols follow.

x: State variable vector, with $x=(x_1,x_2,\cdots,x_n)$.

Z_i: i-th smooth region of the space state.

α: Parameter of the physical system $(\alpha\in\mathbb{R})$.

$F_i(x,\alpha)$: Vector field on region Z_i.

DB: Discontinuity Boundary.

Σ_{ij}: Discontinuity Boundary between regions Z_i and Z_j. $\Sigma_{ij}=\bar{Z}_i\cap\bar{Z}_j=\{x\in\mathbb{R}^n : H_{ij}(x,\alpha)=0\}$.

$H_{ij}(x,\alpha)$: Smooth scalar function defining the *DB* between regions i and j. $H_{ij}(x,\alpha):\mathbb{R}^{n+1}\rightarrow\mathbb{R}^n$.

$H_{ijx}(x,\alpha)$: Gradient of $H_{ij}(x,\alpha)$.

$$H_{ijx}(x,\alpha)=\left(\frac{\partial H_{ij}(x,\alpha)}{\partial x_1},\cdots,\frac{\partial H_{ij}(x,\alpha)}{\partial x_n}\right).$$

Ω_{Ii}^-: i-th component of x before impact.

Ω_{Ii}^+: i-th component of x after impact.

γ: Impact restitution coefficient $\gamma=\left|\dot{\Omega}_I^-/\dot{\Omega}_I^+\right|$.

x_i: Point at the end of i-th integration step.

$G_{ij}(x,\alpha)$: Vector field that acts on the *DB* between regions i and j, for sliding.

Cycle equations include indicators, separators and elements (for cycles: points or segments). Cycles are identified with a letter C accompanied by a subscript number (e.g. C_4: 4-th cycle). If the cycle contains sliding segments they appear as S superscript preceding the C letter (e.g. ${}^SC_5$: cycle 5 has sliding segments). In the equations, the symbol Φ is used to represent a composed segment, determined by a sequence of points of a common type (e.g. Φ_5: a composed segment in region 5). The points are identified with the letter Ω with super-indices (− or +) indicating whether the point is an initial (−) or endpoint (+) of a sliding segment *S*. The symbol/notes a separator between consecutive elements.

The indicator $\grave{O}$ shows that the elements of the equation in an evolution are continuously repeated (e.g. $\Phi_i/\grave{O}$: segment Φ_i in region i is continuously repeated). Equations that describe the elements of Bifurcations (cycles) are identified by the symbol β. Sliding bifurcations are identified with a super-script S that precedes the β symbol and an alphabetic sub-script that indicates the bifurcation type (e.g. ${}^S\beta_c$ is a sliding crossing bifurcation).

3. Background of the Non-Smooth Solution

Typically, Non-smooth systems are modeled as piecewise-smooth systems (PWS) where the state space contains four kinds of spaces: *Smooth Zones*, *undefined Zones* associated to regions behind of impact boundaries, *Discontinuity Boundaries* with dynamics represented by convex combinations of the solution of the ODEs of each vector field and *Impact boundaries* with dynamic represented by algebraic equations. Equation (1) shows the state-space representation of the simplest non-smooth system with the three types of dynamics.

$$\dot{x}=\begin{cases}F_i(x,\alpha) & \text{if } x\in Z_i=\{x\in\mathbb{R}^n : H(x,\alpha)>0\}\\ F_j(x,\alpha) & \text{if } x\in Z_j=\{x\in\mathbb{R}^n : H(x,\alpha)<0\}\\ G(x,\alpha) & \text{if } x\in\Sigma_{ij}=\{x\in\mathbb{R}^{n-1} : H(x,\alpha)=0\}\\ I(x,\alpha) & \text{if } x\in\Sigma_{(i,j,k)}=\{x\in\mathbb{R}^{n-1} : H_I(x,\alpha)=0\}\end{cases}\tag{1}$$

In Equation (1), F_i and F_j are smooth vector fields; Z_i and Z_j are the corresponding regions and $\alpha\in\mathbb{R}^1$ is a parameter. The state space regions are determined by the smooth scalar function $H(x,\alpha)$ and the boundary of impact of Z_i or Z_j regions is determined by the scalar function $H_I(x,\alpha)$.

3.1. Zero Degree of Smoothness Systems

In electro-mechanical Non-smooth systems the impact phenomena is highly dynamical, then can be declared using an algebraic relation due to the impact time is negligible in relation with the time constant of mechanical systems. In this relation, γ is the restitution coefficient and $\dot{\Omega}_I^{(-)}$, $\dot{\Omega}_I^{(+)}$ are respectively the approximation and bounce speed.

$$I(x,\alpha)=\begin{cases}\Omega_I^{(+)}=\Omega_I^{(-)}\\ \dot{\Omega}_I^{(+)}=\gamma\dot{\Omega}_I^{(-)}\end{cases}\tag{2}$$

The first row of Equation (2) expresses that the position before and after the impact are identical. The second one expresses that the rebound velocity (+) equals the impact velocity (−) multiplied by the restitution coefficient γ.

3.2. First Degree of Smoothness Systems

Filippov systems, a set of first-order ordinary differential equations with a discontinuous right-hand side are a subclass of discontinuous dynamical systems. The trajectory of a sliding orbit remaining partially inside the discontinuity boundary may be calculated by the Filippov convex method as in [4]. Systems with multiple regions and *DBs* are treated in [13], where an extended equation for Filippov systems is described in order to deal with the intersection of several discontinuity surfaces.

In Filippov systems, between Z_i and Z_j in the discontinuity boundary, we assume that there is a region Σ_{ij}, which are a vector field of $\mathbb{R}^{n-1}$ dimension conformed by three types of points: crossing (Ω_C), sliding (Ω_S) and singular (Ω_{SO}), and each one with subtypes. The scalar function $\sigma(x)$ is used to determine the point type, according to the geometric condition of the vectors in the x point of analysis. Equation (3) describes the geometric conditions of an sliding point. Equation (4) helps to determine which is the nature of the point, according to the value of $\sigma(x)$ and the neighboring vector fields at x.

$$\sigma(x)=\{\langle H_x(x),F_i(x,\alpha)\rangle\langle H_x(x),F_j(x,\alpha)\rangle\} \quad (3)$$

$$x\in\Sigma_{i,j}:$$

$$\begin{cases}\Omega_C \Rightarrow \sigma(x)>0\\ \Omega_{SO} \Rightarrow (\sigma(x)=0)\wedge(\langle H_x(x),F_j(x)-F_i(x)\rangle=0)\\ \Omega_S \Rightarrow \sigma(x)<0\end{cases} \quad (4)$$

Crossing points (Ω_C), characterized by $\sigma(x)>0$, are points which the evolution of the trajectory will not remain in the DB. Instead, it crosses from the region in which has been previously evolving to the other.

Singular sliding points (Ω_{SO}), characterized by $\sigma(x)$ = 0, are points having the associated vectors with the normal component $\langle H_x(x),F_i\rangle$ equal to **0**. This is because the vectors are tangential to the *DB* or vanishes. At such points: a) F_i and F_j are tangent to the *DB*; b) either F_i or F_j vanishes while the other is tangent to the *DB*; or c) F_i and F_j vanish. To avoid the lack of definition of the Filippov solution for these points, in the examples, we adopt the methods presented in [14] which coincide with the topology of the normal forms VV, VI and II presented in [12].

Sliding points (Ω_S) are characterized by $\sigma(x)<0$. When a sliding motion is presented in the discontinuity boundary, the Filippov method gives as a solution a tangent vector to the *DB* which is a convex combination $G(x,\alpha)$, of the vector fields F_i and F_j at a point $x\in\Sigma_{i,j}$ (Equation (5)).

$$G(x,\alpha)=\lambda F_i(x,\alpha)+(1-\lambda)F_j(x,\alpha) \quad (5)$$

$$\lambda=\frac{\langle H_x(x),F_j(x,\alpha)\rangle}{\langle H_x(x),F_j(x,\alpha)-F_i(x,\alpha)\rangle} \quad (6)$$

λ is a scalar function defined through the projections of the vector fields in the direction of the normal vector $(H_x(x))$ to the discontinuity boundary. According to the direction of the normal components of the vectors, the sliding points are stable (or attractor) (Ω_{SS}), or unstable (or repulsive) (Ω_{SU}) (Equation (7)).

$$x\in\Sigma_{i,j}:\begin{cases}\Omega_{SS}\Rightarrow(\langle H_x(x),F_i\rangle>0)\wedge(\langle H_x(x),F_j\rangle<0)\\ \Omega_{SU}\Rightarrow(\langle H_x(x),F_i\rangle<0)\wedge(\langle H_x(x),F_j\rangle>0)\end{cases} \quad (7)$$

From Equation (4) the crossing set is open but the sliding set is closed, it is the union of the sliding segments, singular points and isolated or special sliding points. In this paper, the terms *special points* or *isolated points* refer to points whose neighbor points belong to a different class.

Special points define important dynamics in the sliding segments of 2d systems or areas in 3D systems. These points are: a) Equilibria points, in which both vectors F_i and F_j are attractive, transversal to the DB and are at the end of two sliding segments pointing each other. b) Quasi-equilibria points with both vectors F_i and F_j attractive transversal or anti collinear and which are at the start of two sliding segments pointing away each other. The contrary case have also quasi equilibria points: repulsive, transversal points which are at the end of two sliding segments pointing each other. c) Boundary equilibria points, in which one of the vector F_i or F_j vanishes. d) Tangent points, in which one of the vectors F_i or F_j is tangent to the *DB*. [15] is done a more strict classification giving the characterization of 42 types of points with the objective of differentiate topologies in order to detect bifurcations.

3.3. Second Degree of Smoothness Systems

The second degree of smoothness systems are represented as variable structure systems having different dynamics in each zone or region. The dynamics of the system does not allow sliding or stops on the boundary zone, all points are crossing and hence, there is not a particular dynamics defined in the limit zone, instead there is a change of the region equations set.

4. Sequences of Well Known Bifurcations

In this and the following sections, we will present the

cycles of the most referenced sliding bifurcations as sequences of elements. In each cycle are presented the constituent elements assuming that its presence was detected, in the same order, in the evolution of a dynamical system. In next equations the symbol Φ is used to represent segments composed by the same type of point. Arrows indicate the direction of the sliding segments related to the *DB*.

4.1. Grazing Bifurcation

The Grazing Bifurcation $\left({}^s\beta_G\right)$ occurs in the following sequence of changes. First, there is an orbit of a limit cycle C_1 evolving in only one of the regions i or j, without hitting the boundary, as shown in **Figure 1(a)**.

$$C_1 = \Phi_i/\dot{O} \tag{8}$$

Then, when the parameter α changes, for example, from α_1 to α_2, the cycle grows or moves toward the discontinuity and has a tangent contact with the last point of a sliding segment $\Omega_s^{(+)}$. The structure presented corresponds to a ${}^sC_2$ type cycle.

$${}^sC_2 = \Phi_i/\Omega_s^{(+)}/\dot{O} \tag{9}$$

Subsequently, as the parameter is moved further, the limit cycle changes again as is depicted in **Figure 1(c)**. The structure presented corresponds to a ${}^sC_3$ type cycle.

$${}^sC_3 = \Phi_i/\Phi_s^{\rightarrow}/\Omega_s^{(+)}/\dot{O} \tag{10}$$

The orbit of the limit cycle ${}^sC_3$ has now two different pieces: one without touching the discontinuity boundary and the other one, corresponding to a sliding segment $\Phi_s^{\rightarrow}$ that starts in any intermediate point of the discontinuity and ends at a tangent point $\Omega_s^{(+)}$. The equation describing the sequence of cycles is:

$${}^s\beta_G = (C_1)\left[{}^sC_2\right]\left({}^sC_3\right) \tag{11}$$

Impact systems also present grazing bifurcations. An orbit that is evolving in a region, due to a change in a parameter, makes contact with a boundary in only one point. This point has approximation speed equal to zero. Consequently, the bouncing speed is also zero. If the physical parameter continues changing, the approximation and rebound points separate. The corresponding cycles are:

$$\begin{aligned} C_1 &= \Phi_i/\dot{O} \\ {}^iC_2 &= \Phi_i/\Omega_I^{(+-)}/\dot{O} \\ {}^iC_3 &= \Phi_i/\Omega_I^{(+)}/\Omega_I^{(-)}\dot{O} \end{aligned} \tag{12}$$

4.2. Switching Bifurcation

The sequence of changes for a Switching Bifurcation $\left({}^s\beta_S\right)$ is as follows: the sliding piece of a limit cycle of type ${}^SC_3$ grows until it reaches the first point $\Omega_S^{(-)}$ of the sliding segment. See **Figure 1(d)**. The type of structure presented, corresponds to a cycle ${}^sC_4$. In general, the second cycle always characterizes the bifurcation type and it is only presented for one value of the parameter or a very narrow range in the numerical calculation terms.

$${}^sC_4 = \Phi_i/\Omega_S^{(-)}/\Phi_s^{\rightarrow}/\Omega_s^{(+)}\dot{O} \tag{13}$$

With a further change in the parameter, the orbit has now three segments: two of them, Φ_i and Φ_j are in two different regions separated by the discontinuity boundary, and the third piece is on the sliding region moving to the right. See **Figure 1(e)**. The structure presented corresponds to a ${}^sC_5$ type cycle.

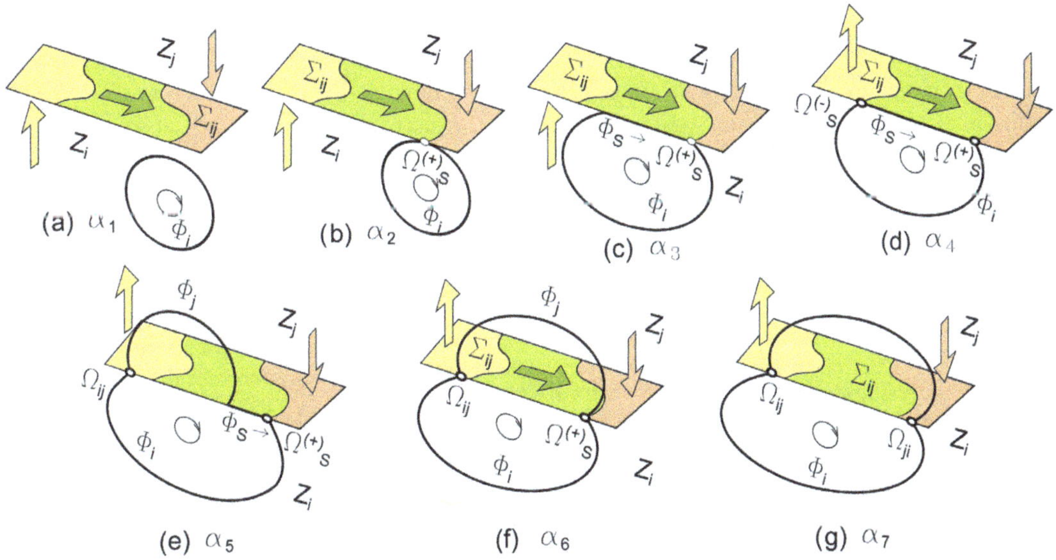

Figure 1. Grazing (a)-(c), Switching (c)-(e) and Crossing (e)-(g) bifurcations.

$$^{s}C_5 = \Phi_i / \Omega_{C_{i,j}} / \Phi_j / \Phi_s^{\rightarrow} / \Omega_s^{(+)} / \dot{O} \quad (14)$$

The equation describing the sequence of cycles is:

$$^{s}\beta_S = (^{s}C_3)[^{s}C_4](^{s}C_5) \quad (15)$$

4.3. Crossing Bifurcation

A Crossing Bifurcation $(^{s}\beta_C)$ occurs when the sliding piece of a cycle $^{S}C_5$ gets smaller and smaller. At a parameter value α_6, the piece of trajectory Φ_j hits the sliding region just at the last point of the sliding segment $\Omega_s^{(+)}$. See **Figure 1(f)**. The structure presented corresponds to a $^{s}C_6$ type cycle.

$$^{s}C_6 = \Phi_i / \Omega_{C_{i,j}} / \Phi_j / \Omega_s^{(+)} / \dot{O} \quad (16)$$

As the parameter further changes at some value α_7, the limit cycle has now two pieces without sliding. The structure presented corresponds to a $^{s}C_7$ type cycle.

$$C_7 = \Phi_i / \Omega_{C_{i,j}} / \Phi_j / \Omega_{C_{j,i}} / \dot{O} \quad (17)$$

The equation describing the sequence of cycles is:

$$^{s}\beta_C = (^{s}C_5)[^{s}C_6](^{s}C_7) \quad (18)$$

4.4. Adding or Multisliding Bifurcation

The sequence of changes for the Adding or Multisliding bifurcation is related to the addition or destruction of a second sliding segment in the discontinuity boundary as is described in [12]. Other sliding bifurcations recently reported are those including more than two discontinuity boundaries that are moving due to variations of a parameter. Those ones were introduced in [8] using an example.

5. The Implementation of the Sequences as a Method of Comparison

Next we will describe the tool which were developed to get the results obtained in the previous section. Additional to the numerical integrator, there are some databases, procedures and methods running in parallel. They perform the evaluation of information collected previously, and the information acquired in real time, when the system is evolving. These tools are:

5.1. Collection of Points

The collection of the values of the points is done in a vector, called vector of states. The new point includes the values of the states, the amount of time since the integration started and the data of the vector fields involved in the dynamics. As shown in **Figure 2(a)**, after each iteration of the numerical integration, one point is added to the vector of states and the graphic of the space states.

5.2. Database of Point Characteristics

Each point, additional to the characterization given by the states is classified by the region or *DB* it belongs. The orientation of the two vector fields for points in the *DB* determines types as anticollinear, transversal, tangent, also the attractiveness or repulsiveness and the direction relative to the *DB*. The magnitude of the vectors might tend to zero. The Equation (4) determines if is a crossing or sliding point. Finally the Equation (1), that represents its dynamics indicates if is an impact point. All points and their characteristics are listed in a 2 × 2 array called matrix of points, where the first column is the list of points and each row are the list of attributes that each point should to fulfill [15]. Other points presenting themselves in the evolution belonging only to one region, are the nodes and focus, stable and unstable.

5.3. Recognition of Points

From the states of the points and vector fields involved, secondary information is estimated. For a point in the *DB* it is evaluated if it is impacting or normal. Then it is evaluated if the point is crossing or sliding. If a point is crossing, it is evaluated to which vector field the evolution will move. The evolution of sliding points has direction tangent to the *DB*, spanning 42 possible subtypes [15]. Summarizing, each point should match all attributes listed in a row of the point matrix. The detected points are stored in vector of elements (**Figure 3**).

- While the vector of elements is being filled out other functions are debugging the information. Each point in a cell of the vector of elements is compared with the point that was met immediately before. Data of points having equal identity are removed from the vector. Instead, the repetition of points turns the first point in the repetition into a piece of curve of the same type. This procedure is carried out with the objective of avoiding a situation in which the vector is filled or saturated with the same data.
- While picking elements for the matrix, events with wrong result can be found and should be corrected. For example, it is impossible to accept the sequence Φ_i / Φ_j because implies a change of region Z_i to Z_j. In the change, a crossing point must be found, and an admissible sequence would be $\Phi_i / \Omega ij / \Phi_j$. Thus, a function to correct the sequences of elements is necessary. In [16] are listed 51 rules to correct errors.

5.4. Database of Cycle Elements

Each cycle as presented in the previous section, has a set of elements which could be points or segments of points.

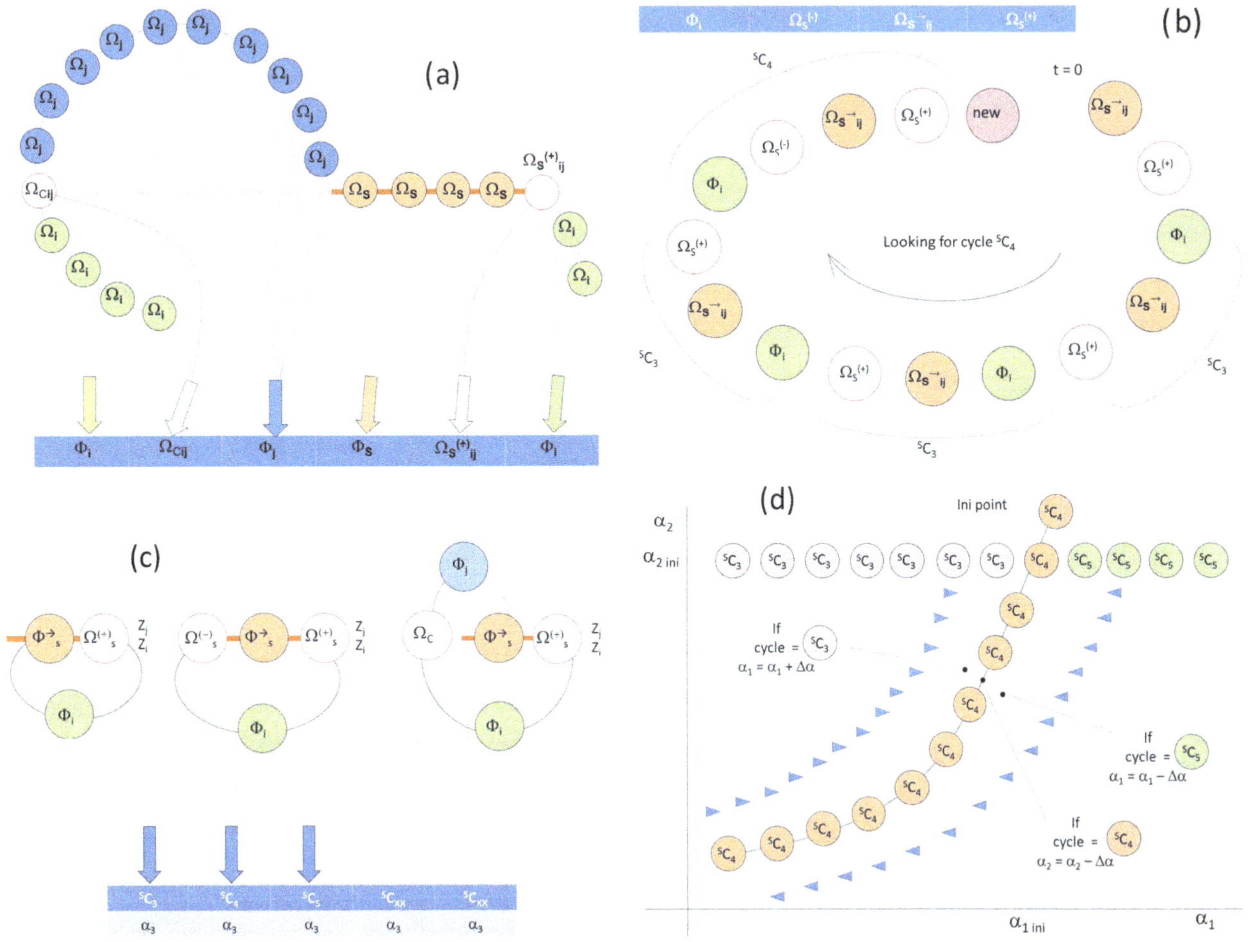

Figure 2. Implementation of Cycle Bifurcation. (a) Process of filling the vector with elements appearing in the numeric integration; (b) Searching process for a specific cycle; (c) Cycle tracking process for bifurcations detection; (d) Cycle continuation process.

The order of the elements also determines the cycle. In order to have a wider data base all papers in the literature should be analyzed and the cycles presented must be converted in sequences of elements. The information is stored in a bidimensional array, called matrix of cycles, in which each row are the identities of the elements of a cycle.

5.5. Comparison of Cycles

In this step the comparison between the matrix of cycles and the vector of elements is performed. We wish to know whether inside the vector of elements there exists a sub-vector of consecutive and ordered elements that matches with some row of the matrix of cycles. The sequence in the appearance of cycles (and other dynamics) in this step is recorded in a vector called vector of cycles. The result in the vector of cycles, for a given set of parameters, admits the presence of a) equilibrium points, b) limit cycles, and c) chaotic behavior. For time-varying parameters, the system evolution might be a sequence of n cycle types, whose order is dictated by the system nature (**Figures 2(b)** and **(c)**).

To prevent that a repetition of a cycle be mistaken as a single cycle, a function running in parallel with the integrator performs the evaluation and the correction. When a sub-sequence of the vector of elements, beginning in the position nb_1, is equal to the sub-sequence beginning in the position $nb_2 = nb_1 + l_j$ and l_j is the number of elements of the cycle, it is concluded that a cycle is repeating. A cycle is completed when a sequence of elements is continuously repeated and the time Γ to repeat becomes constant. Let us assume, as illustration, a sequence with a grazing cycle $\Phi_i / \Omega_s^{(+)}$. After some time 3Γ, the matrix of elements would contain a cycle with the sequence $\Phi_i / \Omega_s^{(+)} / \Phi_i / \Omega_s^{(+)} / \Phi_i / \Omega_s^{(+)}$, which is not correct.

If the search is for a specific cycle, the procedure is slightly different. In this case, the number of elements in the cycle under consideration is a date and then it is reserved the same amount of cells to store the elements during the integration process. When a new element

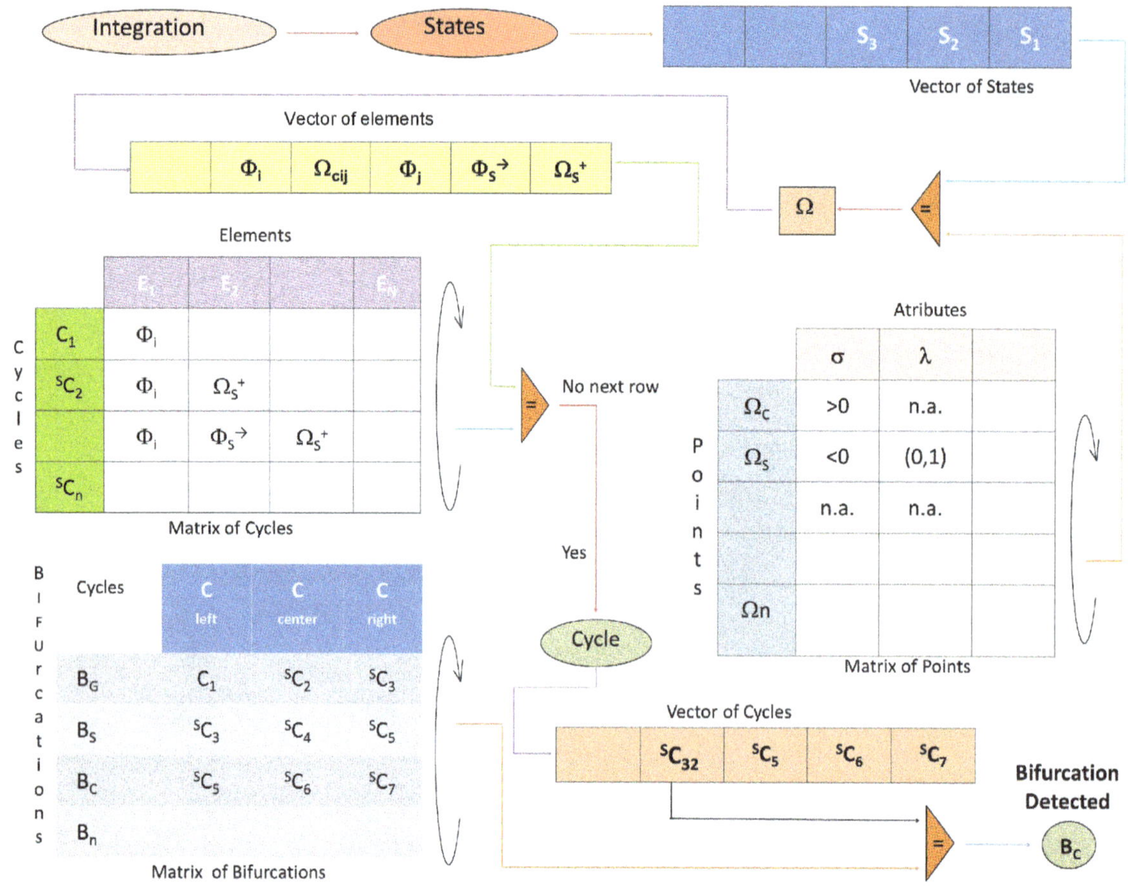

Figure 3. General method of comparison of sequences of cycles and bifurcations.

appears, a comparison is carried out until all the elements of the stored cycle are identical to the elements that are picked up from the integration (**Figure 2(b)**).

5.6. Change in Parameter and Storing of Cycles

When a cycle is already stored in vector of cycles and it is continuously repeating, a programmed disturbance is introduced in a physical parameter, to continue searching the bifurcations. The previous processes are repeated, and recorded in vector of cycles.

5.7. Database of Cycles Sequence

Each bifurcation is constituted by three ordered cycles, the first and third are presented for a wide range of the parameter but the second is only presented for a value of the parameter. The information of the bifurcations is then stored in a bidimensional array, called matrix of bifurcations, in which each row are the identities of the three cycles of the bifurcation.

5.8. Comparison of Cycles Sequence

The objective of the comparison is to identify if inside the vector of cycles there is a sub-vector of three consecutive and ordered cycles which matches a row of the bifurcation matrix (**Figure 2(c)**). Here we are looking for a specific sequence that corresponds to a known bifurcation. To achieve this, a double comparison must be performed: the first part is the comparison of elements that forms cycles, and the other part is referred to the comparison of the behavior of cycles in a specific sequence, until a full match is detected. When the phenomenon is poorly understood, the comparison could be used to identify sequences of cycles which occur when a parameter is modified within a range. For this purpose, the integrator uses the vector of cycles to store information regarding the cycles which have been found during the time that the method has been active. Each time the integrator detects a repeated sequence of elements, stores the information of the cycle, and changes the parameter value in order to continue with the next identification.

5.9. Continuation

To continue a bifurcation the parameters are adjusted corresponding to the central cycle of a previously detected bifurcation. Next, two additional parameters are slightly changed as per the rules of continuation. The first parameter is disturbed and the second changes accordingly, to keep the dynamics of the central cycle. This controlled disturbance of the two parameters is repeated, such that it determines a trajectory in a continuation-plot. The change of parameters could be done using methods like predictor-corrector described in [17] or [18]. In this cases, the predictive function is the cycle that generates the bifurcation, and the previous and posterior cycles to the bifurcation are used for correction.

Figure 2(d) shows an example of how is used the method of comparison. The first step is a sensibility analysis that indicates to which cycle, the system evolves when the parameters are increased or decreased. For example, the bifurcation ${}^{S}C_2$ has a sequence of cycles $({}^{S}C_3)[{}^{S}C_4]({}^{S}C_5)$. Assume that a direct proportional sensibility exists for parameter α_1. This implies that a small increment in the parameter value tends to change the cycle into ${}^{S}C_5$ and a small decrement tends to change the cycle into ${}^{S}C_3$. Changing α_1, the cycle ${}^{S}C_4$ is obtained. Then, the second parameter α_2 is decreased (in this case the initial point has a high value). After the change in parameter α_2, the cycle ${}^{S}C_4$ changes to ${}^{S}C_3$ or to ${}^{S}C_5$. In the first case, the continuation algorithm increases α_1 until the cycle type ${}^{S}C_4$ is found again. In the second case, the algorithm acts conversely. The process is continuously iterated until the prescribed final value of parameter α_2 is reached.

Two objectives of an application for automatic bifurcation detection are: 1) to perform the detection task without a close supervision; and 2) to track bifurcations through continuation. The procedures developed here can be used to achieve these goals.

6. Conclusions

This article presents an alternative method for detecting bifurcations of limit cycles in non-smooth systems. We focused on complex systems, which defy boundary-value methods. The comparison method, reported in this article, is not intended to focus in the same achievements of other methods. Instead, it addresses open issues left by them, such as multiple sliding segments and discontinuity boundaries (*DB*). The comparison method differs from other approaches in the identification and manipulation of the system information. While the methods in [10,11] consider a system as one entity to be solved by a group of equations, the comparison method uses previously collected information in a data base of points, cycles and bifurcations. This information allows comparisons and decision making. To enable the method for non-smooth systems, the cases when the evolution crosses the *DBs* of systems having simultaneously the three degrees of smoothness (impact, Filippov and first derivative discontinuities) was analyzed. To achieve the goal was used the method that characterizes and records the elements comprising the cycles in the order they appear in the integration process. The cycles were characterized as sequences of elements (points and segments). It must be noticed that the sequence of cycles has the topological changes (e.g. bifurcations) implicit. Some of the types of data considered as topological characteristic and collected during the evolution are: a) number of elements of the cycle; b) order in which the cycle elements are generated; c) position of the sliding elements in the sequence of cycle generation; d) way (e.g. extreme or interior) in which the cycle reaches and leaves the sliding segment; e) discontinuity boundary to which the element belongs; f) direction (CW, CCW) in which the cycle evolves. In this article we also report a textual notation to describe the elements of the cycles. The comparison method is also able to handle continuation of sliding bifurcations.

The method of comparison could be implemented using tools of the sequence theory, suffix-trees and string-matching, which offer procedures to drive a large number of elements and allow us to discriminate subsets with low computing time investment. The procedure of comparison fulfills the two tasks required by an application for automatic bifurcations detection: perform the detection task without a closed supervision and track bifurcations through continuation.

REFERENCES

[1] M. Amorin, S. Divenyi, L. Franca and H. I. Weber, "Numerical and Experimental Investigations of the Nonlinear Dynamics and Chaos in Non-Smooth Systems," *Journal of Sound and Vibration*, Vol. 301, No. 1-2, 2007, pp. 59-73.

[2] I. Arango, "Singular Point Tracking: A Method for the Analysis of Sliding Bifurcations in Non-smooth Systems," PhD Thesis, Universidad Nacional de Colombia, 2011.

[3] I. Arango and J. A. Taborda, "Integration-Free Analysis of Non-smooth Local Dynamics in Planar Filippov Systems," *International Journal of Bifurcation and Chaos*, Vol. 19, No. 3, 2009, pp. 947-975.

[4] M. Bernardo, C. Budd, A. R. Champneys and P. Kowalczyk, "Piecewise-Smooth Dynamical Systems: Theory and Applications," Springer, Berlin, 2008.

[5] P. Casini and F. Vestroni, "Nonstandard Bifurcations in Oscillators with Multiple Discontinuity Boundaries," *Nonlinear Dynamics*, Vol. 35, No. 1, 2004, pp. 41-59.

[6] A. Colombo, M. Di Bernardo, E. Fossas and M. R. Jeffrey, "Teixeira Singularities in 3D Switched Feedback Control Systems," *Systems & Control Letters*, Vol. 59, No. 10, 2010, pp. 615-622.

[7] F. Dercole and Y. Kuznetsov, "SlideCont: An Auto97 Driver for Bifurcation Analysis of Filippov Systems," *ACM Transactions on Mathematical Software* (*TOMS*), Vol. 31, No. 1, 2005, pp. 95-119.

[8] L. Dieci and L. Lopez, "Sliding Motion in Filippov Differential Systems: Theoretical Results and a Computational Approach," *SIAM Journal on Numerical Analysis*, Vol. 47, No. 3, 2009, pp. 2023-2051.

[9] A. Filippov and F. Arscott, "Differential Equations with Discontinuous Righthand Sides: Control Systems, Volume 18 of Mathematics and Its Applications: Soviet Series," Springer, Berlin, 1988.

[10] M. Guardia, T. M. Seara and M. A. Teixeira, "Generic Bifurcations of Low Codimension of Planar Filippov Systems," *Journal of Differential Equations*, Vol. 250, No. 4, 2011, pp. 1967-2023.

[11] Y. Kuznetsov, S. Rinaldi and A. Gragnani, "One-Parameter Bifurcations in Planar Filippov Systems," *International Journal of Bifurcation and Chaos*, Vol. 13, No. 8, 2003, pp. 2157-2188.

[12] R. I. Leine, "Bifurcations in Discontinuous Mechanical Systems of Filippov-Type," PhD Thesis, Technische Universiteit Eindhoven, 2000.

[13] W. Marszalek and Z. Trzaska, "Singular Hopf Bifurcations in DAE Models of Power Systems," *Energy and Power Engineering*, Vol. 3, No. 1, 2011, pp. 1-8.

[14] I. Merillas, "Modeling and Numerical Study of Nonsmooth Dynamical Systems: Applications to Mechanical and Power," PhD Thesis, Technical University of Catalonia, 2006.

[15] I. Arango and J. A. Taborda, "Integration-Free Analysis of Non-Smooth Local Dynamics in Planar Filippov System," *International Journal of Bifurcation and Chaos*, Vol. 19, No. 3, 2009, pp. 947-975.

[16] A. Nordmark, "Existence of Periodic Orbits in Grazing Bifurcations of Impacting Mechanical Oscillators," *Nonlinearity*, Vol. 14, No. 6, 2001, pp. 1517-1542.

[17] T. S. Parker and L. Chua, "Practical Numerical Algorithms for Chaotic Systems," Springer Limited, London, 2011.

[18] P. Thota and H. Dankowicz, "TC-HAT (TC): A Novel Toolbox for the Continuation of Periodic Trajectories in Hybrid Dynamical Systems," *SIAM Journal on Applied Dynamical Systems*, Vol. 7, No. 4, 2008, pp. 1283-1322.

Frequency-domain Elastic wave Simulation Based on the Nonoverlapping Domain Decomposition Method

Wensheng Zhang, Yinyun Dai
Institute of Computational Mathematics and Scientific/Engineering Computing, LSEC Academy of Mathematics and Systems Science, Chinese Academy of Sciences, Beijing, P.R.China

ABSTRACT

A new wave simulation technique for the elastic wave equation in the frequency domain based on a no overlapping domain decomposition algorithm is investigated. The boundary conditions and the finite difference discrimination of the elastic wave equation are derived. The algorithm of no overlapping domain decomposition method is given. The method solves the elastic wave equation by iteratively solving sub problems defined on smaller sub domains. Numerical computations both for homogeneous and inhomogeneous media show the effectiveness of the proposed method. This method can be used in the full-waveform inversion.

Keywords: Finite Difference; Nonoverlapping DDM; Elastic wave Equation; Wave Simulation; Preconditioner; Absorbing Boundary Conditions

1. Introduction

The numerical solution of wave equations is a very important problem. The applications can be found in electromagnetic, acoustic wave propagation, seismic inversion. The wave equations can be solved in either the time or the frequency domain. When applying the Fourier transformation with respect to time to the acoustic wave equation, the Helmholtz equation is obtained. The Helmholtz equation has important applications in underwater and seismic imaging [1, 2]. In the geophysical frequency-domain inversion one needs to do forward modeling which means solving the Helmholtz equation. During the inversion process, the synthetic model is continuously updated until a convergence is reached. Thus finding an efficient numerical method to solve the Helmholtz equation is important. For a finite element or finite difference discrimination of the 2D Helmholtz, it is necessary to use large number of grid points to handle large domains and high wave numbers. Moreover, the requirement of ten points per wavelength at least leads to very large complex linear systems. In seismic full-waveform inversion, the number of grid points grows very rapidly in order to maintain good accuracy. The error is proportional to $k^{p+1}h^p$, where k is the wave number, p is the order of discrimination and h is the grid size [3,4]. The linear system becomes extremely large and highly indefinite. This makes the problem even harder to solve. Direct methods easily suffer from inacceptable computational work. The iterative methods for the Helmholtz equation have been an active research field since the 1980s. Since then many attempts have been spent to develop a powerful iterative method for solving it, see, for instance, [2, 5-6].

The domain decomposition method (DDM) is an effective technique for solving large-scale problems [7-10]. The idea is to split the computational domain Ω into several smaller m sub domains Ω_i, $i=1,\cdots,m$ and solve a sequence of similar sub problems on these subdomains. The boundary conditions are adjusted iteratively the transmission conditions between adjacent subdomains. The number and size of sub domains can now be chosen so as to enable direct methods to solve the sub problems. Several classes of DDM exist, e.g. multiplicative and additive Schwartz methods [7-10].

Absorbing boundary conditions (ABCs) are required in numerical simulations of unbounded problems. They are enforced at edges of a computational domain to absorb outgoing waves and thereby model an unbounded region. A number of ABCs have been introduced for use in finite-difference simulation of elastic wave propagation. There exist several ABC methods, for instance, the sponge layer method [11], the paraxial approximation method [12] and the perfectly matched layer method [13]. The approach based on a paraxial approximation of the elastic wave equations was introduced by Clayton and Engquist [12]. It is often adopted in frequency-domain simulation. The most recent advance in ABCs theory is the so-called perfectly matched layer (PML) technique proposed by Berenger [13] which offers an almost complete elimination of boundary reflections. The PML method has been widely used in wave simulation.

The DDM solving procedure for the Helmholtz has been investigated; for example, see [14-16]. In this paper, we focus on the elastic wave simulation in the frequency domain rather than the acoustic wave based on the no overlapping domain decomposition method. Here we use the no overlapping DDM algorithm to solve the elastic wave equation in the frequency domain.

2. Theoretical Methods

2.1. Finite Difference Discretization

The 2-D, second-order, frequency-domain elastic wave equation in a homogeneous, isotropic, and source-free media consists of the two coupled equations

$$\rho\omega^2 u+(\lambda+2\mu)\frac{\partial^2 u}{\partial x^2}+\mu\frac{\partial^2 u}{\partial z^2}+(\lambda+\mu)\frac{\partial^2 v}{\partial x\partial z}=0, \quad (1)$$

$$\rho\omega^2 v+(\lambda+2\mu)\frac{\partial^2 v}{\partial z^2}+\mu\frac{\partial^2 v}{\partial z^2}+(\lambda+\mu)\frac{\partial^2 u}{\partial x\partial z}=0, \quad (2)$$

where $\omega=2\pi f$ is the angular frequency, ρ is the density, and λ and μ are Lame parameters. The wavefield variables u and v are, respectively, the horizontal and vertical components of the Fourier transformed displacements. In a more general formulation, one also included a source term in the right-hand side of (1) or (2).

We wish to find the solution to system (1)-(2) for a finite model domain, Ω, but without reflected energy from the domain boundary. For simplicity we use a simple absorbing boundary condition firstly proposed by Clayton and Engquist [11]

$$\frac{\partial}{\partial n}\begin{pmatrix}u\\ v\end{pmatrix}+i\omega S\begin{pmatrix}u\\ v\end{pmatrix}=0,\ S=\begin{pmatrix}1/v_p^2 & 0\\ 0 & 1/v_s^2\end{pmatrix}, \quad (3)$$

where n is the direction normal to the absorbing model boundary, and

$$v_p=\sqrt{(\lambda+2\mu)/\rho},\ v_s=\sqrt{\mu/\rho}, \quad (4)$$

are the velocities of compressive wave and shear wave, respectively. The condition (3) is only perfectly absorbing when the out-going wave impacts the boundary at normal incidence. We discredited (1)-(3) using an implicit finite-difference scheme based on the 2nd order accurate centered-difference operators. We assume the computational domain Ω is rectangular. A 2D regular mesh with constant grid spacing is used for deriving computational schemes. The difference schemes for (1)-(2) are the following:

$$\rho\omega^2 u_{i,j}+(\lambda+2\mu)\frac{u_{i+1,j}-2u_{i,j}+u_{i-1,j}}{\Delta x^2}+\mu\frac{u_{i,j+1}-2u_{i,j}+u_{i,j-1}}{\Delta z^2}$$
$$+(\lambda+\mu)\frac{v_{i+1,j+1}+v_{i-1,j-1}-v_{i+1,j-1}-v_{i-1,j+1}}{4\Delta x\Delta z}=0, \quad (5)$$

$$\rho\omega^2 v_{i,j}+(\lambda+2\mu)\frac{v_{i+1,j}-2v_{i,j}+v_{i-1,j}}{\Delta z^2}$$
$$+\mu\frac{v_{i,j+1}-2v_{i,j}+v_{i,j-1}}{\Delta x^2} \quad (6)$$
$$+(\lambda+\mu)\frac{u_{i+1,j+1}+u_{i-1,j-1}-u_{i+1,j-1}-u_{i-1,j+1}}{4\Delta x\Delta z}=0,$$

The difference schemes for the absorbing boundary condition (3) can be written as

$$\begin{pmatrix}u_{i,0}\\ v_{i,0}\end{pmatrix}=\begin{pmatrix}u_{i,2}\\ v_{i,2}\end{pmatrix}+2i\Delta x\omega\begin{pmatrix}u_{i,1}\\ v_{i,1}\end{pmatrix},\quad i=2,\cdots,N-1, \quad (7)$$

$$\begin{pmatrix}u_{i,N+1}\\ v_{i,N+1}\end{pmatrix}=\begin{pmatrix}u_{i,N-1}\\ v_{i,N-1}\end{pmatrix}-2i\Delta z\omega\begin{pmatrix}u_{i,N}\\ v_{i,N}\end{pmatrix},\quad i=2,\cdots,N-1, \quad (8)$$

$$\begin{pmatrix}u_{0,j}\\ v_{0,j}\end{pmatrix}=\begin{pmatrix}u_{2,j}\\ v_{2,j}\end{pmatrix}-2i\Delta x\omega\begin{pmatrix}u_{i,1}\\ v_{i,1}\end{pmatrix},\quad j=2,\cdots,M-1, \quad (9)$$

$$\begin{pmatrix}u_{N+1,j}\\ v_{N+1,j}\end{pmatrix}=\begin{pmatrix}u_{N-1,j}\\ v_{N-1,j}\end{pmatrix}-2i\Delta x\omega\begin{pmatrix}u_{N,j}\\ v_{N,j}\end{pmatrix}, \quad (10)$$

where Δx and Δz are the spatial steps in x and z directions respectively. The difference schemes (7)-(10) are used at the top, bottom, left and right boundary respectively. Combining all the difference schemes above, we have the following concise form

$$xA\begin{pmatrix}u\\ v\end{pmatrix}=\begin{pmatrix}A_{11} & A_{12}\\ A_{21} & A_{22}\end{pmatrix}\begin{pmatrix}u\\ v\end{pmatrix}=\begin{pmatrix}b_1\\ b_2\end{pmatrix}, \quad (11)$$

where A is the known matrix, A_{ij} can be considered the sub matrixes of A, b_1 and b_2 are the known vectors. They can be obtained based on the difference schemes above. For example, the elements of A_{11} can be determined by the following expressions:

$$A_{11}(i+(j-1)n_x,i+(j-1)n_z)=2(\lambda+2\mu)+2\mu r^2-\rho\Delta x^2\omega^2, \quad (12)$$

$$A_{11}(i+(j-1)n_x,i+1+(j-1)n_z)=-(\lambda+2\mu), \quad (13)$$

$$A_{11}(i+(j-1)n_x,i-1+(j-1)n_z)=-(\lambda+2\mu), \quad (14)$$

$$A_{11}(i+(j-1)n_x,i+jn_z)=-\mu r^2, \quad (15)$$

$$A_{11}(i+(j-1)n_x,i+(j-1)n_z)=-\mu r^2, \quad (16)$$

where $i=2,\cdots,n_x-1$; $j=2,\cdots,n_z-1$; r is the grid ratio

$r = \Delta x / \Delta z$; $n_x = N$ and $n_z = M$ are the discretization points number in the x and z directions respectively. We omit the detail expressions of other A_{ij} for saving space. Solving (11) by a direct method is impractical when the system is huge. For this reason, various iterative methods, e.g. the Krylov subspace methods, are the interesting alternative. However, Krylov subspace methods are not competitive without a good preconditioned [17]. Here we adopt the preconditioned Bi-conjugate gradient stabilized (Bi-CGSTAB) to solve (11). The preconditioned is the so-called "shifted Laplace" preconditioner [18].

2.2. Nonoverlapping Domain Decomposition

In no overlapping domain decomposition methods, the computational domain Ω is decomposed into K subdomains Ω_k, $k = 1, \cdots, K$, which are no overlapped. The basic idea of DDM is to find a solution in Ω by solving the sub domain problems in Ω_k and then ex- changing solutions in the interface between two neighbor- ing domains. Let us consider a trivial case where Ω is split into two sub domains Ω_1 and Ω_2, see **Figure 1(a)**. We denote by u_1, v_1 and u_2, v_2 the restrictions of (1) on Ω_1 and Ω_2 respectively. They satisfy the follow- ing interface conditions on $\partial\Omega_1$ and $\partial\Omega_2$:

$$\begin{pmatrix} u_1 \\ v_1 \end{pmatrix} = \begin{pmatrix} u_2 \\ v_2 \end{pmatrix}, \quad \frac{\partial}{\partial n_1}\begin{pmatrix} u_1 \\ v_1 \end{pmatrix} = \frac{\partial}{\partial n_2}\begin{pmatrix} u_2 \\ v_2 \end{pmatrix}, \tag{17}$$

where n_i is the exterior normal to Ω_i . However, the conditions (17) may cause ill-posed problem. We propose to use the following equivalent boundary conditions:

$$\frac{\partial}{\partial n_1}\begin{pmatrix} u_1 \\ v_1 \end{pmatrix} + i\omega S\begin{pmatrix} u_1 \\ v_1 \end{pmatrix} = -\frac{\partial}{\partial n_2}\begin{pmatrix} u_2 \\ v_2 \end{pmatrix} + i\omega S\begin{pmatrix} u_2 \\ v_2 \end{pmatrix}, \tag{18}$$

and

$$\frac{\partial}{\partial n_2}\begin{pmatrix} u_2 \\ v_2 \end{pmatrix} + i\omega S\begin{pmatrix} u_2 \\ v_2 \end{pmatrix} = -\frac{\partial}{\partial n_1}\begin{pmatrix} u_1 \\ v_1 \end{pmatrix} + i\omega S\begin{pmatrix} u_1 \\ v_1 \end{pmatrix}, \tag{19}$$

The sub problem on Ω_1 or Ω_2 together with condition (18) or (19) is now well posed. We describe the general domain decomposition algorithm in the following. The idea is to adjust iteratively the boundary conditions at the interface between sub domains to obtain the transmission conditions of the type (3) and (17). Initialize u_k^0, v_k^0 for all $k = 1, \cdots, K$, then iterating for $n > 0$, where n is the iterative number:

1) Solve the following equations for $(x, z) \in \Omega_k$:

$$\begin{aligned} &\rho\omega^2 u_k^{n+1} + (\lambda + 2\mu)\frac{\partial^2 u_k^{n+1}}{\partial x^2} + \mu\frac{\partial^2 u_k^{n+1}}{\partial z^2} \\ &+(\lambda+\mu)\frac{\partial^2 v_k^{n+1}}{\partial x \partial z} = 0 \end{aligned} \tag{20}$$

$$\begin{aligned} &\rho\omega^2 v_k^{n+1} + (\lambda + 2\mu)\frac{\partial^2 v_k^{n+1}}{\partial z^2} + \mu\frac{\partial^2 v_k^{n+1}}{\partial z^2} \\ &+(\lambda+\mu)\frac{\partial^2 u_k^{n+1}}{\partial x \partial z} = 0 \end{aligned} \tag{21}$$

2) Solve the following interface equations for $(x, z) \in \partial\Omega \cap \partial\Omega_k$:

$$\frac{\partial}{\partial n}\begin{pmatrix} u_k^{n+1} \\ v_k^{n+1} \end{pmatrix} + i\omega S\begin{pmatrix} u_k^{n+1} \\ v_k^{n+1} \end{pmatrix} = 0, \tag{22}$$

and for $(x, z) \in \partial\Omega_j \cap \partial\Omega_k$:

$$\frac{\partial}{\partial n_k}\begin{pmatrix} u_k^{n+1} \\ v_k^{n+1} \end{pmatrix} + i\omega S\begin{pmatrix} u_k^{n+1} \\ v_k^{n+1} \end{pmatrix} = -\frac{\partial}{\partial n_j}\begin{pmatrix} u_j^{n+1} \\ v_j^{n+1} \end{pmatrix} + i\omega S\begin{pmatrix} u_j^{n+1} \\ v_j^{n+1} \end{pmatrix}, \tag{23}$$

where Ω_j and Ω_k denote the j th and kth subdomains respectively.

3) Extrapolate the u_k^{n+1} and v_k^{n+1} in the sub domains Ω_k to the whole domain Ω :

$$\begin{pmatrix} \overline{u}_k^{n+1} \\ \overline{v}_k^{n+1} \end{pmatrix}_{\Omega_k} = \begin{pmatrix} u_k^{n+1} \\ v_k^{n+1} \end{pmatrix}, \begin{pmatrix} \overline{u}_k^{n+1} \\ \overline{v}_k^{n+1} \end{pmatrix}_{\Omega/\Omega_k} = \begin{pmatrix} u_k^{n} \\ v_k^{n} \end{pmatrix}, \tag{24}$$

then set

$$\begin{pmatrix} \overline{u}_k^{n+1} \\ \overline{v}_k^{n+1} \end{pmatrix}_{\Omega} = \frac{1}{K}\sum_{k=1}^{K}\begin{pmatrix} \overline{u}_k^{n+1} \\ \overline{v}_k^{n+1} \end{pmatrix}, \tag{25}$$

4) Set $n = n + 1$, return to step (1) to start the next iteration until the solutions convergence.

3. Numerical Computations

Numerical tests are performed on a homogeneous model and a three-layered model. The source function is set to be a point source. First we consider a homogeneous model with $v_p = 3000$ m/s and $v_s = 2000$ m/s. The frequency is 100 Hz. The convergent solutions are shown in **Figures 2** and **3**. In **Figure 2**, the strategy of two no overlapping sub domains is adopted. **Figure 2(a)** is the horizontal displacement and **Figure 2(b)** is vertical component. In **Figure 3**, the strategy of three nonoverlapping subdomains is used. **Figure 3(a)** is the horizontal

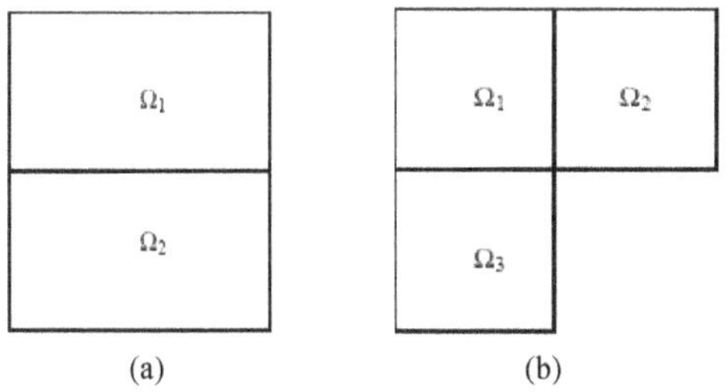

Figure 1. Sketch map of domain decomposition. The computational domain is decomposed into no overlapping (a) two subdomains and (b) three subdomains.

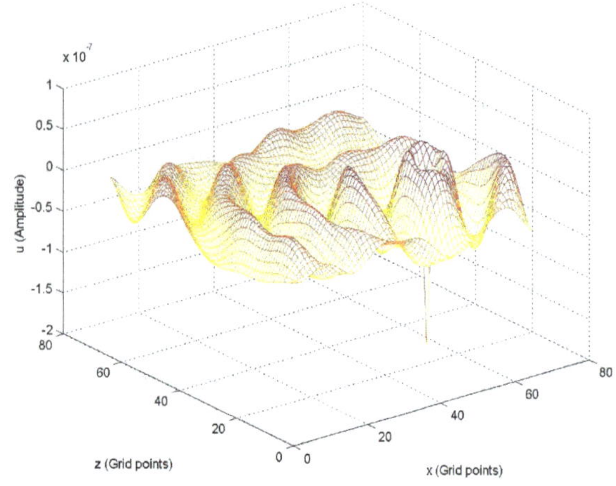

(a)

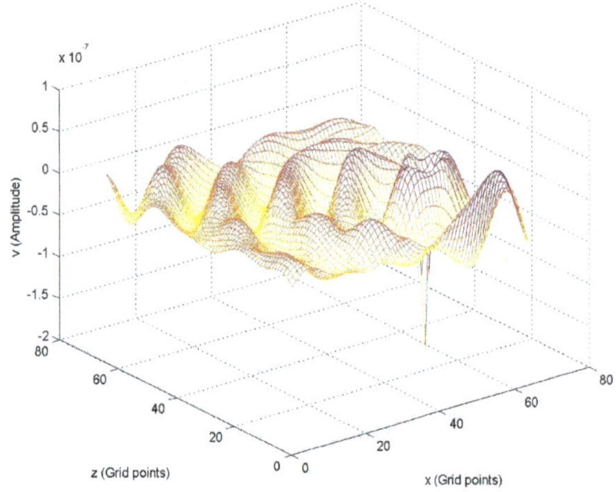

(b)

Figure 2. Horizontal (a) and vertical (b) displacements for a homogeneous model. The solutions are obtained based on the nonoverlapping DDM with two subdomains. The frequency in computations is 100Hz.

displacement and **Figure 3(b)** is the vertical component. The waveform is correct through verification. The solutions on the common interface are almost same and we conclude that the solutions convergence.

Next we consider an inhomogeneous model with three-layered media shown in **Figure 4**. From top to bottom in **Figure 4**, the velocities of compressive wave are 2000 m/s, 3000 m/s and 4500 m/s respectively, and the velocities of shear wave are 1500 m/s, 2000 m/s and 2500 m/s respectively. The solutions are shown in **Figure 5**. **Figure 5(a)** is the horizontal displacement and **Figure 5(b)** is the vertical displacement. The frequency is still 100 Hz. In computations the strategy of two no overlapping subdomains is adopted. We also present the contours of waveform. The contours corresponding to **Figures 5(a)** and **(b)** are shown in **Figures 6(a)** and **(b)** respectively. **Figure 7** are the solutions with a high frequency 200Hz. **Figure 7(a)** is the horizontal displacement and **Figure 7(b)** is the vertical displacement. Comparing **Figure 7** with **Figure 6**, we see that the waves at high frequency have more oscillation which coincides with the law of physics.

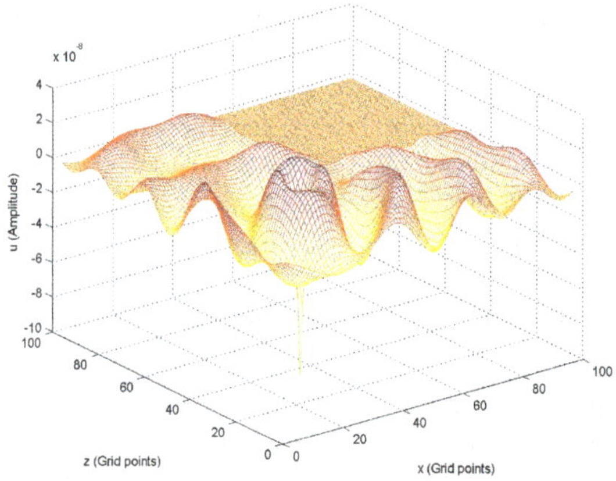

(a)

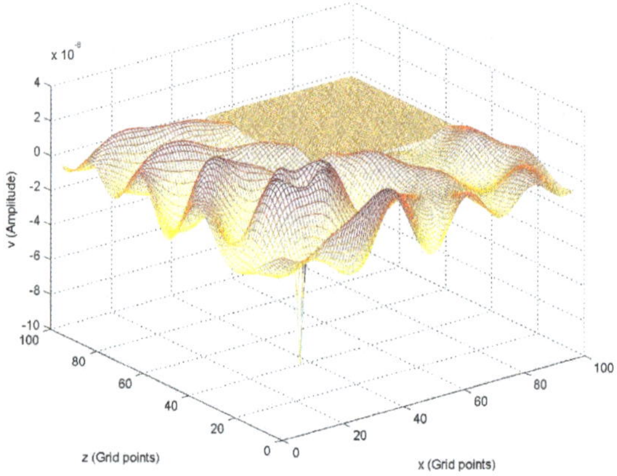

(b)

Figure 3. Horizontal (a) and vertical (b) displacements for a homogeneous model. The solutions are obtained base on the nonoverlapping DDM with three subdomains. The frequency in computations is 100Hz.

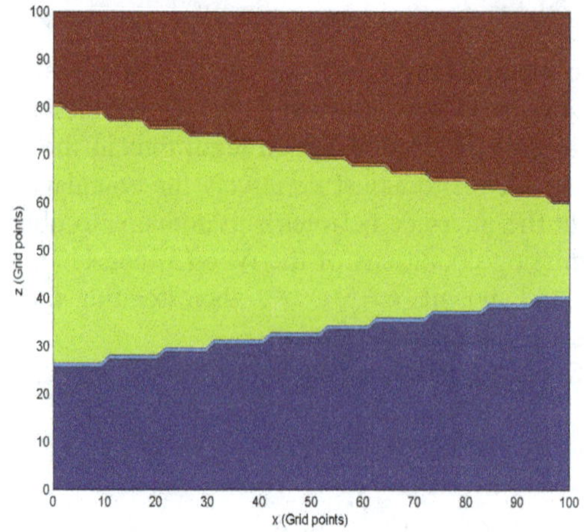

Figure 4. A three-layered velocity model.

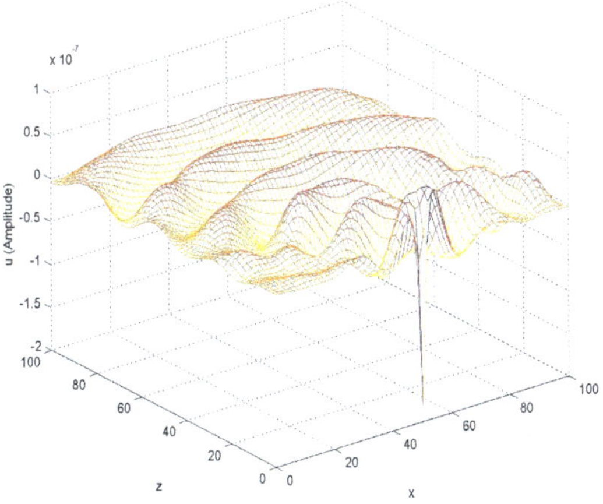

(a)

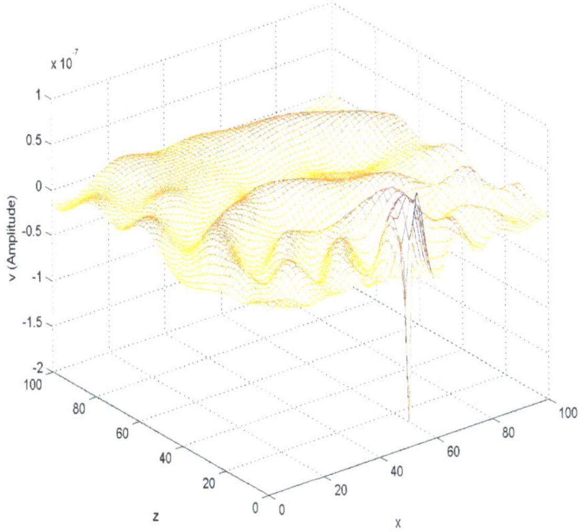

(b)

Figure 5. Horizontal (a) and vertical (b) displacements for a three-layered model. The solutions are obtained based on the nonoverlapping DDM with two subdomains. The frequency in computations is 100Hz.

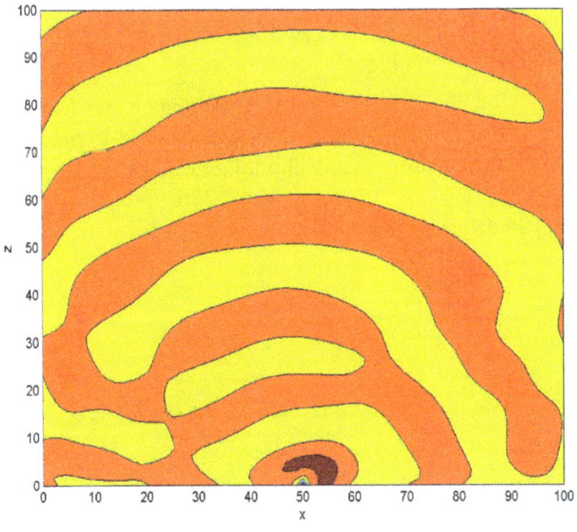

(a)

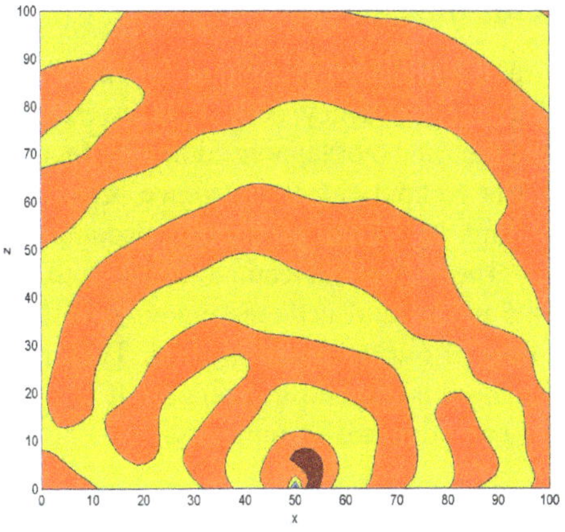

(b)

Figure 6. Waveform contours of horizontal (a) and vertical (b) displacements for a three-layered model. The solutions are obtained based on the nonoverlapping DDM with two subdomains. The frequency in computations is 100Hz.

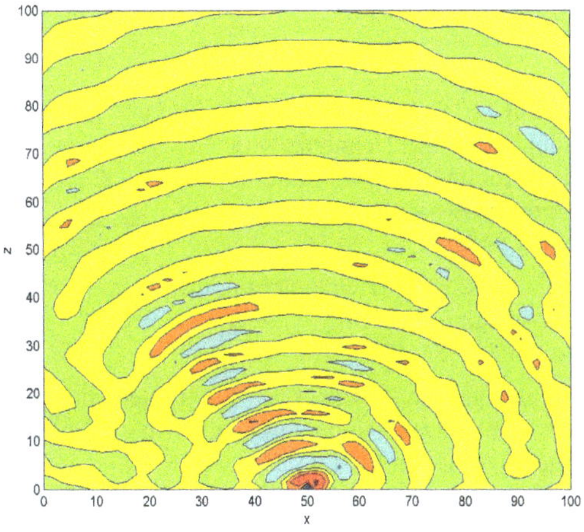

(a)

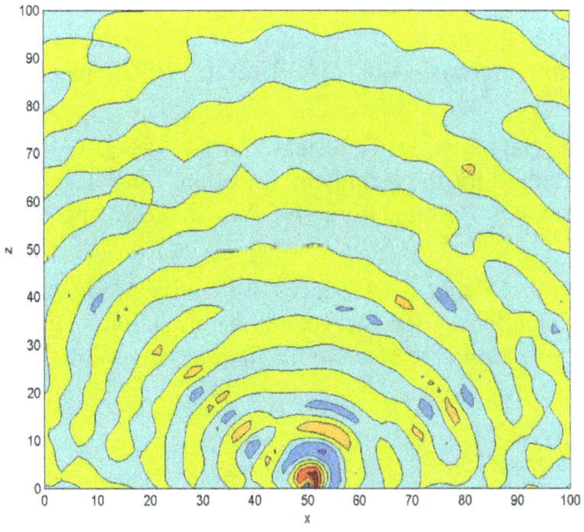

(b)

Figure 7. Waveform contours of horizontal (a) and vertical displacements for a three-layered model. The solutions are obtained based on the nonoverlapping DDM with two subdomains. The frequency in computations is 200Hz.

4. Conclusions

A new wave simulation technique for the elastic wave equation in the frequency domain is investigated. It is based on the no overlapping domain decomposition method. The computational difference schemes and the corresponding algorithm of domain decomposition are presented. The numerical computations both for a homogeneous model and a three-layered model show the effectiveness of our proposed method. This method can be used in the full-waveform inversion. It can sometimes reduce the computational complexity.

5. Acknowledgements

This research is supported by the State Key project with grant number 2010CB731505 and the Foundation of National Center for Mathematics and Interdisciplinary Sciences, CAS. The computations are implemented in the State Key Lab. of Sci. and Eng. Computing (LSEC).

REFERENCES

[1] A. Bayliss, C. I. Goldstein and E. Turkel, "The Numerical Solution of the Helmholtz Equation for Wave Propagation Problmes in Underwater Acoustics," *Computers and Mathematics with Applications*, Vol. 11, No. 7-8, 1985, pp. 655-665.

[2] R. E. Plexxis, "A Helmholtz Iterative Solver for 3D Seismic-imaging Problems," *Geophysics*, Vol. 72, No. 5 2007, pp. SM185-SM197.

[3] F. Ihlenburg and I. Babuska, "Finite Element Solution of the Helmholtz Equation with High Wave Number. Part I: The H-version of the FEM," *Computers and Mathematics with Applications,* Vol. 30, No. 9, 1995, pp. 9-37.

[4] F. Ihlenburg and I. Babuska, "Finite Element Solution of the Helmholtz Equation with High Wave Number. Part II: The Hp-version of the FEM," *SIAM Journal on Numerical Analysis*, Vol. 34, No. 1, 1997, pp. 315-358.

[5] A. Bayliss, C. I. Goldstein and E. Turkel, "An Iterative Method for the Helmholtz Equation," *Journal of Computational Physics*, Vol. 49, No. 3, 1983, pp. 443-457.

[6] Y. A. Erlangga, "Advances in Iterative Methods and Pre-Conditioners for the Helmholtz Equation," *Archives of Computational Methods in Engineering*, Vol. 15, No. 1 2008, pp. 37-66.

[7] T. F. Chan and T. P. Mathew, "Domain Decomposition Algorithms," *Acta Numerica*, Vol. 3, 1994, pp. 61-143.

[8] A. Tosseli and O. Widlund, "Domain Decomposition Methods-Algorithms and Theory," Springer, Berlin, 2005.

[9] J. Xu, "Iterative Methods by Space Decomposition and Subspace Correction," *SIAM Review*, Vol. 34, No. 4, 1992, pp. 581-613.

[10] J. Xu and J. Zou, "Some Nonoverlapping Domain Decomposition Methods," *SIAM Review*, Vol. 40, No. 4, 1998, pp. 857-914.

[11] C. Cerjan, D. Kosloff, R. Kosloff and M. Reshef, "A Non-reflecting Boundary Condition for Discrete Acoustic and Elastic Wave Equations," *Geophysics*, Vol. 50, No. 4, 1985, pp. 705-708.

[12] R. Clayton and B. Engquist, "Absorbing Boundary Conditions for Acoustic and Elastic Wave Equations," *Bulletin of the Seismological Society of America,* Vol. 67, No. 6, 1977, pp. 1529-1540.

[13] J. P. Berenger, "A Perfectly Matched Layer for Absorbing of Electromagnetic Waves," *Journal of Computational Physics*, Vol. 114, No. 2, 1994, pp. 185-200.

[14] S. Kim, "Domain Decomposition Iterative Procedures for Solving Scalar Waves in the Frequency Domain", *Numerische Mathematik*, Vol. 79, No. 2, 1998, pp. 231-259.

[15] S. Larsson, "A Domain Decomposition Method for the Helmholtz Equation in a Multilayer Domain," *SIAM Journal on Scientific Computing*, Vol. 20, No. 5, 1999, pp. 1713-1731.

[16] F. Magoulès, F. X. Roux and S. Salmon, "Optimal Discrete Transmission Conditions for a Nonoverlapping Domain Decomposition Method for the Helmholtz Equation," *SIAM Journal on Scientific Compting*, Vol. 25, No. 5, 2004, pp. 1497-1515.

[17] Y. A. Erlangga, C. Vuik and C. W. Oosterlee, "On a Class of Preconditioners for Solving the Helmholtz Equation," *Applied Numerical Mathematics*, Vol. 50, No. 3-4, 2003, pp. 409-425.

[18] T. Airaksinen, E. Heikkola, A. Pennanen and J. Toivanen, "An Algebraic Multigrid based Shifted-Laplacian P[e-conditioner for the Helmholtz Equation," *Journal of Computational Physics*, Vol. 226, No. 1, 2007, pp. 1196-1210.

On the Rate of Convergence of Some New Modified Iterative Schemes

Renu Chugh, Sanjay Kumar*
Department of Mathematics, M.D.U, Rohtak, India

ABSTRACT

In this article, following Bizare and Amriteimoori [1] and B. Parsad and R. Sahni [2], we modify Ishikawa, Agarwal *et al.*, Noor, SP iterative schemes and compare the rate of convergence of Ishikawa, Agarwal *et al.*, Noor, SP and new modified Ishikawa, Agarwal *et al.*, Noor, SP iterative schemes not only for particular fixed value of $\alpha_n, \beta_n, \gamma_n$ but also for varying the value of $\alpha_n, \beta_n, \gamma_n$. With the help of two numerical examples, we compare the converging step.

Keywords: Metric Space; New Modified Ishikawa; New Modified Agarwal *et al.*; New Modified SP; New Modified Noor

1. Introduction

Let X be a complete metric space and T be self map, then $F_T = \{x \in X, Tx = x\}$ is called set of fixed points of T. Now in literature, there are several iteration processes to find the fixed point of any equation. In complete metric space, Picard iteration process is defined as

$$x_{n+1} = T(x_n), n = 0,1,\cdots$$

which is used to approximate fixed points of mappings satisfying the condition

$$d(Tx,Ty) \le ad(x,,y), \text{ where } 0 \le a < 1$$

called Banach contraction condition.

In Ishikawa iteration process [3], $\{x_n\}_{n=0}^{\infty}$ is defined as

$$x_{n+1} = (1-\alpha_n)x_n + \alpha_n Ty_n$$
$$y_n = (1-\beta_n)x_n + \beta_n Tx_n$$

where α_n, β_n are real sequences in [0,1].

Now for Agarwal *et al.* iteration [4], $\{x_n\}_{n=0}^{\infty}$ is defined as

$$x_{n+1} = (1-\alpha_n)Tx_n + \alpha_n Ty_n$$
$$y_n = (1-\beta_n)x_n + \beta_n Tx_n$$

where α_n, β_n are real sequences in [0,1].

M. A. Noor defines [5] $\{x_n\}_{n=0}^{\infty}$ as

$$x_{n+1} = (1-\alpha_n)x_n + \alpha_n Ty_n$$
$$y_n = (1-\beta_n)x_n + \beta_n Tz_n$$
$$z_n = (1-\gamma_n)x_n + \gamma_n Tx_n$$

where $\alpha_n, \beta_n, \gamma_n$ are real sequences in [0,1].

For SP iteration [6] $\{x_n\}_{n=0}^{\infty}$ is defined as

$$x_{n+1} = (1-\alpha_n)y_n + \alpha_n Ty_n$$
$$y_n = (1-\beta_n)z_n + \beta_n Tz_n$$
$$z_n = (1-\gamma_n)x_n + \gamma_n Tx_n$$

where $\alpha_n, \beta_n, \gamma_n$ are real sequences in [0,1].

2. Preliminaries

In this paper following Bizare and Amriteimoori [1] and B. Parsad and R. Sahni [2] we prove the basic results in sequel. In [1] Bizare and Amriteimoori improved the picard iteration under following conditions:

1) Initial approximation is chosen in the interval $[a, b]$, where function is defined.
2) Function has continuous derivative on (a, b).
3) $|T'(x)| < 1$ for all $x \in [a,b]$
4) $a \le T(x) \le b$ for all $x \in [a,b]$

*Corresponding author.

Definition 2.1 [1]. Let $\{x_n\}$ converges to α. If there exists an integer constant q and a real +ve constant C such that

$$\lim_{n\to\infty}\left|\frac{x_{n+1}-\alpha}{(x_n-\alpha)^q}\right|=C$$

q is called order and C is called constant of convergence

Theorem 2.2([1,7]). Let $f\in C^q[a,b]$, if $f^k(x)=0$ for $k=1,2,\cdots,q-1$ and $f^q(x)\neq 0$ then sequence $\{x_n\}$ is of order q.

To improve the order of convergence of fixed iterative schemes, such that $f'(\alpha),f''(\alpha),\cdots,f^k(\alpha)=0$. We determines $\lambda_i\,(i=1,2,\cdots,k)$ from the following equation

$$\begin{aligned}&x+\lambda_1x+\lambda_2x^2+\cdots+\lambda_kx^k\\&=f(x)+\lambda_1x+\lambda_2x^2+\cdots+\lambda_kx^k\end{aligned}$$

which becomes

$x=\dfrac{f(x)+\lambda_1x+\lambda_2x^2+\cdots+\lambda_kx^k}{1+\lambda_1+\lambda_2x+\cdots+\lambda_kx^{k-1}}=f_\lambda(x)$ this is fixed point equation form .Now the assumption that $f'_\lambda(\alpha)=f''_\lambda(\alpha)=\cdots=f_\lambda^{k-1}(\alpha)=0$ yields to a system of linear equations which after solving [1] converted into upper triangular matrix which have nonzero diagonal entries. It means determinant is nonzero. So we determine $\lambda_i\,(i=1,2,\cdots,k)$ uniquely.

Now the new Picard iteration becomes

$$x_{n+1}=f_\lambda(x_n)\quad n=1,2,\cdots. \tag{1}$$

where

$$f_\lambda(x)=\frac{f(x)+\lambda_1x+\lambda_2x^2+\cdots+\lambda_kx^k}{1+\lambda_1+\lambda_2x+\cdots+\lambda_kx^{k-1}} \tag{2}$$

Following Bhagwati Parsad and Ritu Shani [2] the new modified Ishikawa, Agarwal *et al.*, Noor, SP iterations are defined as:

New modified Ishikawa iteration scheme

$$\begin{aligned}x_{n+1}&=(1-\alpha_n)x_n+\alpha_nf_\lambda(y_n)\\y_n&=(1-\beta_n))x_n+\beta_nf_\lambda(x_n)\end{aligned} \tag{3}$$

New modified Agarwal et al. iteration scheme

$$\begin{aligned}x_{n+1}&=(1-\alpha_n)f_\lambda(x_n)+\alpha_nf_\lambda(y_n)\\y_n&=(1-\beta_n)x_n+\beta_nf_\lambda(x_n)\end{aligned} \tag{4}$$

New modified Noor iteration scheme

$$\begin{aligned}x_{n+1}&=(1-\alpha_n)x_n+\alpha_nf_\lambda(y_n)\\y_n&=(1-\beta_n)x_n+\beta_nf_\lambda(z_n)\\z_n&=(1-\gamma_n)x_n+\gamma_nf_\lambda(x_n)\end{aligned} \tag{5}$$

New modified SP iteration scheme

$$\begin{aligned}x_{n+1}&=(1-\alpha_n)y_n+\alpha_nf_\lambda(y_n)\\y_n&=(1-\beta_n)z_n+\beta_nf_\lambda(z_n)\\z_n&=(1-\gamma_n)x_n+\gamma_nf_\lambda(x_n)\end{aligned} \tag{6}$$

where $\{\alpha_n\},\{\beta_n\}$ and $\{\gamma_n\}$ are real sequences in [0, 1].

In this article, we compare the rate of convergence of new modified iterative schemes and simple iterative schemes with the help of the following examples

$$p_1(x)=e^{(1-x)^2}-1-x \tag{7}$$

$$p_2(x)=\frac{3}{8}+\frac{35}{8}x^4-\frac{15}{4}x^2 \tag{8}$$

To find the fixed point we write $p_1(x)$ and $p_2(x)$ as

$$e^{(1-x)^2}-1-x=0\quad\text{and}\quad \frac{3}{8}+\frac{35}{8}x^4-\frac{15}{4}x^2=0$$

both equations has unique root in the interval (0,1) so we convert this in the fixed point form

$$x=e^{(1-x)^2}-1=f(x)\quad\text{and}\quad x=\frac{\sqrt{90+1050x^4}}{30}=f(x)$$

and take $\alpha=0.5$ and $\alpha=0.35$ respectively. Now we solve it by

$$f_\lambda(x)=\frac{f(x)+\lambda_1x+\lambda_2x^2+\cdots+\lambda_kx^k}{1+\lambda_1+\lambda_2x+\cdots+\lambda_kx^{k-1}}$$

For respective value of α, $\lambda_1,\lambda_2,\lambda_3,\cdots,\lambda_k$ can be determined uniquely from system of linear equations as in [5] for $\alpha=0.5$ and $f(x)=e^{(1-x)^2}-1$ we have

$$\begin{pmatrix}1&\alpha&\alpha^2&\alpha^3&\alpha^4\\0&1&2\alpha&3\alpha^2&4\alpha^3\\0&0&2&6\alpha&12\alpha^2\\0&0&0&6&24\alpha\\0&0&0&0&24\end{pmatrix}\begin{pmatrix}\lambda_1\\\lambda_2\\\lambda_3\\\lambda_4\\\lambda_5\end{pmatrix}=\begin{pmatrix}f'_\lambda(\alpha)\\f_\lambda^{(2)}(\alpha)\\f_\lambda^{(3)}(\alpha)\\f_\lambda^{(4)}(\alpha)\\f_\lambda^{(5)}(\alpha)\end{pmatrix}=\begin{pmatrix}1.28403\\-3.85208\\8.98818\\-32.1006\\104.006\end{pmatrix} \tag{9}$$

After solving the system of linear equations we have $\lambda_1=6.0858,\lambda_2=-17757,\lambda_3=22.2698,$ $\lambda_4=-14.0173,\lambda_5=4.3336$

for second polynomial equation where α=0.35 and

$$f(x)=\frac{\sqrt{90+1050x^4}}{30}$$

System of linear equations become

$$\begin{pmatrix} 1 & \alpha & \alpha^2 & \alpha^3 & \alpha^4 \\ 0 & 1 & 2\alpha & 3\alpha^2 & 4\alpha^3 \\ 0 & 0 & 2 & 6\alpha & 12\alpha^2 \\ 0 & 0 & 0 & 6 & 24\alpha \\ 0 & 0 & 0 & 0 & 24 \end{pmatrix} \begin{pmatrix} \lambda_1 \\ \lambda_2 \\ \lambda_3 \\ \lambda_4 \\ \lambda_5 \end{pmatrix} = \begin{pmatrix} -f'_\lambda(\alpha) \\ -f_\lambda^{(2)}(\alpha) \\ -f_\lambda^{(3)}(\alpha) \\ -f_\lambda^{(4)}(\alpha) \\ -f_\lambda^{(5)}(\alpha) \end{pmatrix} = \begin{pmatrix} -0.291842 \\ -2.25304 \\ -8.53984 \\ 32.6662 \\ 422.235 \end{pmatrix} \qquad (10)$$

Hence we get

$$\lambda_1 = 0.0129005, \lambda_2 = -0.330016, \lambda_3 = 3.01516,$$
$$\lambda_4 = 17.5910, \lambda_5 = -19.186$$

3. Experiments

Now using the value of $\lambda_1, \lambda_2, \lambda_3, \cdots, \lambda_k$ and iterative schemes we have following **Tables 1-24**, and **Figures 1-16**.

We take $\alpha_n = a, \beta_n = b, \gamma_n = c$

Table 1. Simple Ishikawa for $p_1(x)$.

	$a = 0.2, b = 0.9999$		$a = 0.2, b = 0.9$		$a = 0.3, b = 0.6$		$a = 0.2, b = 0.2$	
N	x_{n+1}	x_{n+1}	x_{n+1}	Tx_n	x_{n+1}	Tx_n	x_{n+1}	Tx_n
0	0.0804248	0.0100502	0.162309	0.0100502	0.446501	0.0100502	0.736069	0.0100502
1	0.56607	1.32942	0.519967	1.01723	0.41845	0.358473	0.622944	0.0721434
2	0.291442	0.207189	0.346672	0.259144	0.413572	0.402422	0.548051	0.152774
7	0.443316	0.483717	0.417293	0.426179	0.412392	0.412388	0.427266	0.375276
8	0.388159	0.363285	0.409449	0.404313	**0.412391**	**0.412391**	0.421969	0.388222
9	0.431032	0.454046	0.414152	0.417296	**0.412391**	**0.412391**	0.418561	0.39671
---	---	---	---	---	---	---	---	---
27	0.412589	0.412814	0.412391	0.412392			0.412393	0.412385
28	0.412238	0.412063	**0.412391**	**0.412391**			0.412393	0.412387
29	0.41251	0.412646	**0.412391**	**0.412391**			0.412392	0.412389
---	-----	----	---	----	----	----	----	
31	0.412463	0.412545					0.412392	0.41239
32	0.412335	0.412272					**0.412391**	**0.412391**
33	0.412435	0.412484					**0.412391**	**0.412391**
	--------	--------						
55	0.412391	0.412392						
56	**0.412391**	**0.412391**						
57	**0.412391**	**0.412391**						

Table 2. Modified Ishikawa for $p_1(x)$.

	$a = 0.1$ to 0.9, $b = 0.99$		$a = 0.1$ to 0.9, $b = 0.9$		$a = 0.1$ to 0.9, $b = 0.6$		$a = 0.1$ to 0.9, $b = 0.1$	
N	x_{n+1}	Tx_n	x_{n+1}	Tx_n	x_{n+1}	Tx_n	x_{n+1}	Tx_n
0	0.424618	0.472629	0.481638	0.472629	0.654425	0.472629	0.942404	0.472629
1	0.412376	0.412344	0.416257	0.408094	0.488042	0.370271	0.886798	0.420223
2	**0.412391**	**0.412391**	0.412795	0.412411	0.440202	0.407234	0.834615	0.384598
3	**0.412391**	**0.412391**	0.412434	0.412395	0.423255	0.411826	0.786785	0.364432
4	**0.412391**	**0.412391**	0.412396	0.412392	0.41673	0.412365	0.743706	0.356248
5			0.412392	0.412391	0.414139	0.412411	0.705366	0.356001
6			**0.412391**	**0.412391**	0.413097	0.412404	0.671492	0.360233
7			**0.412391**	**0.412391**	0.412677	0.412397	0.641687	0.366495
------					------------	------------		
14					0.412392	0.412391	0.512968	0.400961
15					**0.412391**	**0.412391**	0.502184	0.403247
16					**0.412391**	**0.412391**	0.492625	0.405092
100							0.412403	0.412391
134							0.412392	0.412391
135							**0.412391**	**0.412391**
136							**0.412391**	**0.412391**

Table 3. Simple Agarwal *et al.* for $p_1(x)$.

	$a = 0.2$, $b = 0.9999$		$a = 0.2$, $b = 0.9$		$a = 0.3$, $b = 0.6$		$a = 0.5$, $b = 0.5$	
N	x_{n+1}	Tx_n	x_{n+1}	Tx_n	x_{n+1}	Tx_n	x_{n+1}	Tx_n
0	0.0803358	0.0100502	0.0733136	0.0100502	0.090521	0.0100502	0.177931	0.0100502
1	0.566185	1.3298	0.644644	1.3602	0.727194	1.2868	0.583438	0.965599
2	0.291318	0.20707	0.222863	0.134597	0.177672	0.0772623	0.323279	0.189489
25	0.412721	0.413097	0.418074	0.423399	0.420618	0.428339	0.412391	0.412392
26	0.412135	0.411843	0.407474	0.403035	0.405294	0.398897	**0.412391**	**0.412391**
27	0.412591	0.412817	0.416645	0.42061	0.418546	0.424292	**0.412391**	**0.412391**
55	**0.412391**	**0.412392**	0.412465	0.412533	0.412498	0.412596		
56	**0.412391**	**0.412391**	0.412327	0.412269	0.412299	0.412214		
95			0.412391	0.412392	0.412391	0.412392		
96			**0.412391**	**0.412391**	0.412391	0.412391		
97					0.412391	0.412392		
98					**0.412391**	**0.412391**		

Table 4. Modified Agarwal *et al.* for $p_1(x)$.

	$a=0.9, b=0.99$		$a=0.1, b=0.9$		$a=0.3, b=0.6$		$a=0.9, b=0.1$	
N	x_{n+1}	Tx_n	x_{n+1}	Tx_n	x_{n+1}	Tx_n	x_{n+1}	Tx_n
0	0.408091	0.472629	0.428901	0.472629	0.408937	0.472629	0.465458	0.472629
1	0.412381	0.412331	0.412289	0.412252	0.412356	0.412346	0.410188	0.409941
2	**0.412391**	**0.412391**	0.41239	0.41239	0.412391	0.412391	0.412368	0.412365
3	**0.412391**	**0.412391**	**0.412391**	**0.412391**	**0.412391**	**0.412391**	**0.412391**	**0.412391**
4	**0.412391**	**0.412391**	**0.412391**	**0.412391**	**0.412391**	**0.412391**	**0.412391**	**0.412391**

Table 5. Simple SP for $p_1(x)$.

	$a=0.1, b=0.1, c=0.999$		$a=0.1, b=0.1, c=0.9$		$a=0.5, b=0.5, c=0.5$		$a=0.1, b=0.1, c=0.1$	
N	x_{n+1}	Tx_n	x_{n+1}	Tx_n	x_{n+1}	Tx_n	x_{n+1}	Tx_n
0	0.0736477	0.0100502	0.139578	0.0100502	0.416514	0.0100502	0.667545	0.0100502
1	0.701496	1.35874	0.616499	1.09662	0.412245	0.405588	0.531266	0.116866
2	0.182726	0.0931953	0.271814	0.158438	0.412396	0.412634	0.463428	0.245717
3	0.617407	0.950211	0.519974	0.699365	0.412391	0.412382	0.433328	0.333637
4	0.242048	0.157633	0.334615	0.259135	**0.412391**	**0.412391**	0.420799	0.378667
5	0.566158	0.776228	0.471698	0.556962	**0.412391**	**0.412391**	0.415737	0.398604
---	-----		---	---			-----	------
16	0.371841	0.341091	0.409891	0.406887			**0.412391**	**0.412391**
17	0.448911	0.483766	0.41427	0.416556			**0.412391**	**0.412391**
---	----	----	---	-----	---	---	---	---
50	0.41149	0.41072	0.412391	0.412391				
51	0.413197	0.413889	**0.412391**	**0.412391**				
---	-----	------	-----	-----	----	---	---	---
126	**0.412391**	**0.412391**						

Table 6. Modified SP for $p_1(x)$.

	$a=0.9, b=0.9, c=0.9999$		$a=0.9, b=0.9, c=0.9$		$a=0.2, b=0.9, c=0.1$		$a=0.1, b=0.1, c=0.1$	
N	x_{n+1}	Tx_n	x_{n+1}	Tx_n	x_{n+1}	Tx_n	x_{n+1}	Tx_n
0	0.412399	0.472629	0.412496	0.472629	0.436509	0.472629	0.844335	0.472629
1	**0.412391**	**0.412391**	0.412391	0.412392	0.413946	0.411994	0.712973	0.367256
2	**0.412391**	**0.412391**	**0.412391**	**0.412391**	0.412512	0.412403	0.618933	0.359118
3	**0.412391**	**0.412391**	**0.412391**	**0.412391**	0.412401	0.412392	0.55555	0.37889
4	**0.412391**	**0.412391**	**0.412391**	**0.412391**	0.412392	0.412391	0.512865	0.394474
5					**0.412391**	**0.412391**	0.483669	0.403265
6					**0.412391**	**0.412391**	0.463373	0.407829
-----	-----------	---------	-----------	--------	-----------	---------		
----							----------	----------
43							0.412392	0.412391
44							0.412392	0.412391
45							**0.412391**	**0.412391**
46							**0.412391**	**0.412391**

Table 7. Simple Noor for $p_1(x)$.

	$a=0.3, b=0.3, c=0.5$		$a=0.3, b=0.5, c=0.7$		$a=0.5, b=0.7, c=0.9$		$a=0.5, b=0.7, c=0.9$	
N	x_{n+1}	Tx_n	x_{n+1}	Tx_n	x_{n+1}	Tx_n	x_{n+1}	Tx_n
0	0.506337	0.0100502	0.449615	0.0100502	0.331926	0.0100502	0.381042	0.0100502
1	0.417997	0.275966	0.410239	0.353812	0.506867	0.562555	0.442088	0.466838
2	0.412427	0.40316	0.41258	0.415975	0.34732	0.275298	0.389512	0.365152
3	0.412391	0.412331	0.412375	0.412077	0.485348	0.531107	0.433466	0.451642
5	**0.412391**	**0.412391**	0.412391	0.412389	0.470593	0.508772	0.395663	0.378452
6	**0.412391**	**0.412391**	**0.412391**	**0.412391**	0.367544	0.323489	0.427488	0.440836
7			**0.412391**	**0.412391**	0.459703	0.491825	0.400159	0.38787
75							0.412391	0.412392
76							**0.412391**	**0.412391**
155					0.412391	0.412392		
156					**0.412391**	**0.412391**		
157					**0.412391**	**0.412391**		

Table 8. Modified Noor for $p_1(x)$.

	$a=0.9, b=0.9, c=0.999$		$a=0.9, b=0.9, c=0.9$		$a=0.7, b=0.5, c=0.5$		$a=0.1, b=0.1, c=0.1$	
N	x_{n+1}	Tx_n	x_{n+1}	Tx_n	x_{n+1}	Tx_n	x_{n+1}	Tx_n
0	0.410381	0.472629	0.469342	0.472629	0.681606	0.472629	0.942022	0.472629
1	0.412389	0.412368	0.418102	0.409542	0.539107	0.364431	0.886261	0.419927
2	**0.412391**	**0.412391**	0.412967	0.412409	0.473981	0.398109	0.834065	0.384326
3	**0.412391**	**0.412391**	0.412449	0.412396	0.442827	0.409028	0.78628	0.364285
6			**0.412391**	**0.412391**	0.416194	0.4124	0.67112	0.360296
7			**0.412391**	**0.412391**	0.4143	0.412411	0.641343	0.366574
20					**0.412391**	**0.412391**	0.463812	0.409495
21					**0.412391**	**0.412391**		
134							0.412392	0.412391
135							**0.412391**	**0.412391**

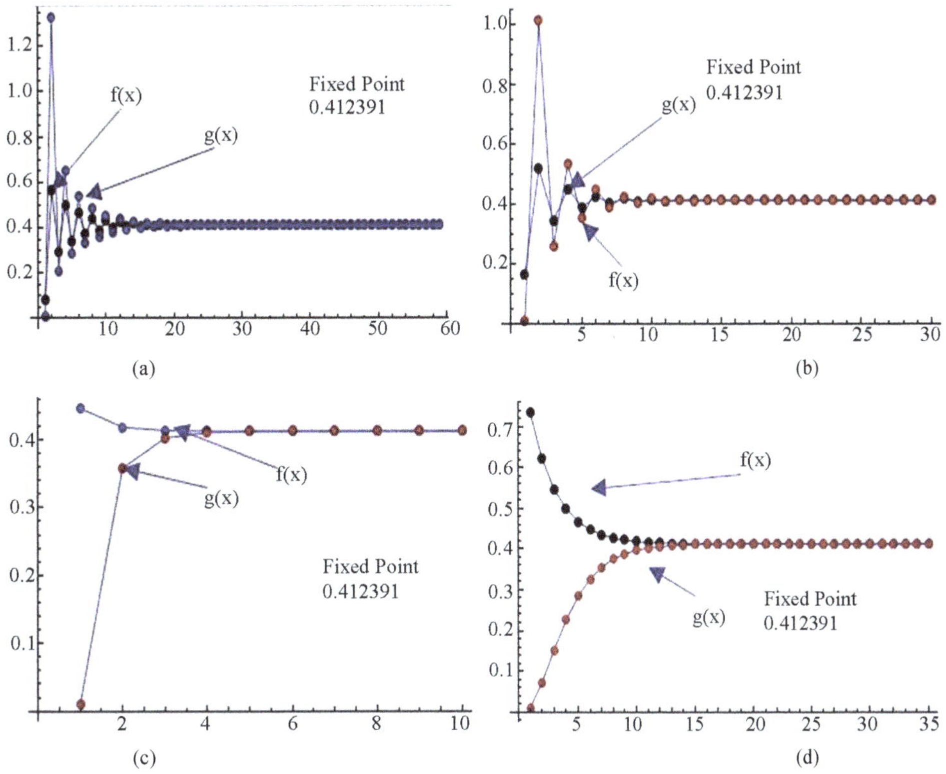

Figure 1. Graphical observations of simple Ishikawa iteration for $p_1(x)$. Here (a)-(d) show the graph for Table 1. The merging point with value 0.412391 is fixed point.

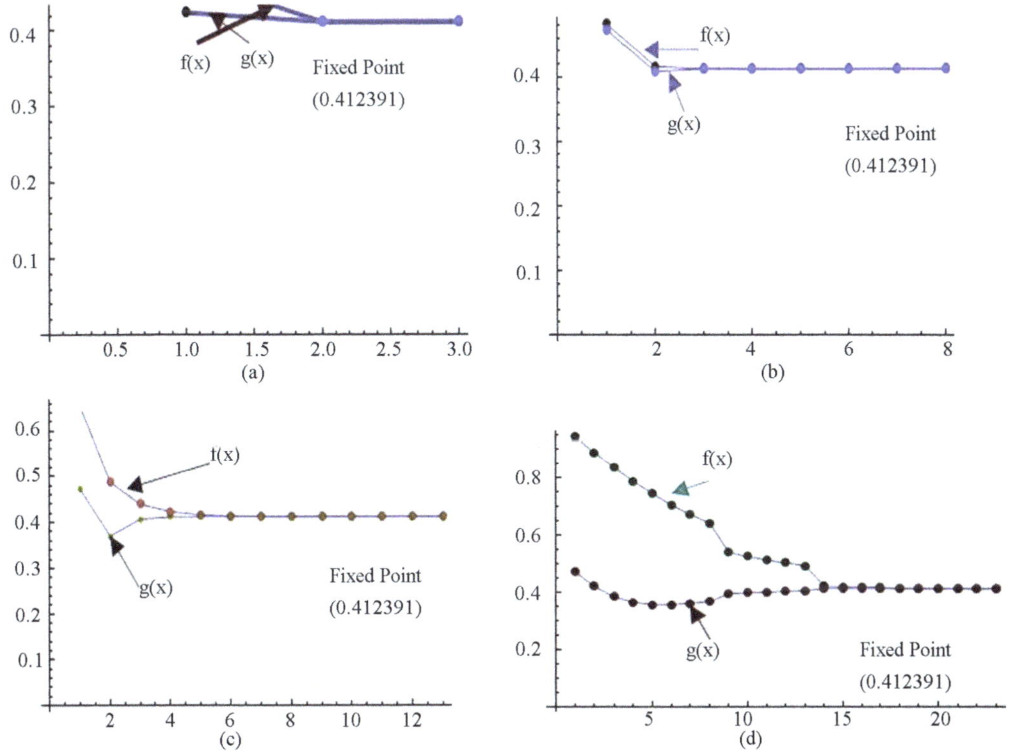

Figure 2. Graphical observations of new modified Ishikawa iteration for $p_1(x)$. Here (a)-(d) show the graph for Table 2. The merging point with value 0.412391 is fixed point.

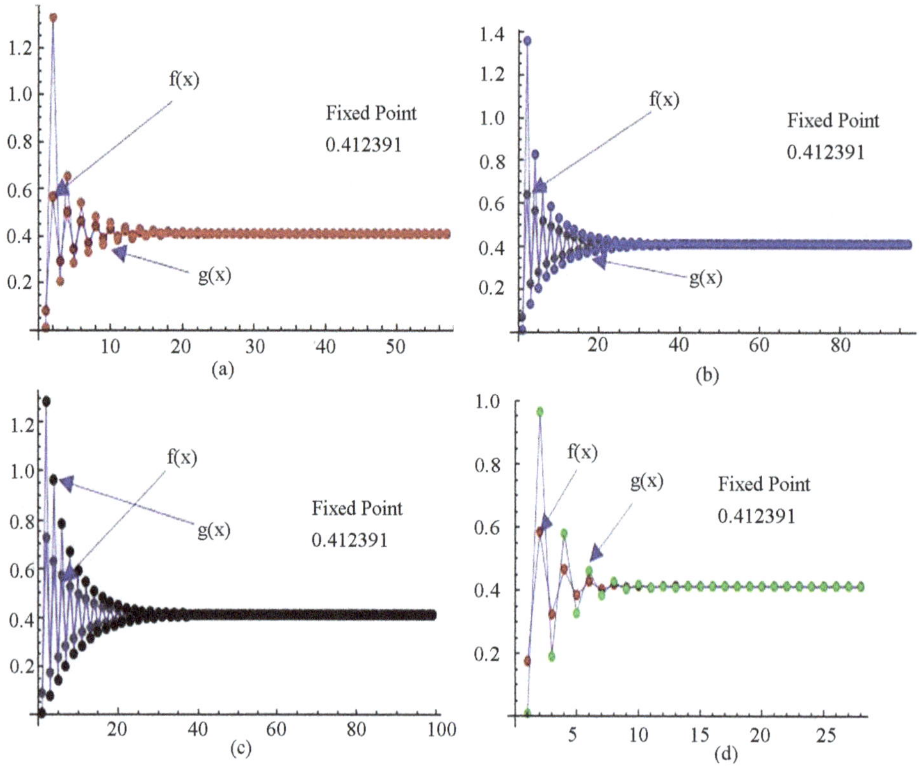

Figure 3. Graphical observations of simple Agarwal *et al.* iteration for $p_1(x)$. Here (a)-(d) show the graph for Table 3. The merging point with value 0.412391 is fixed point.

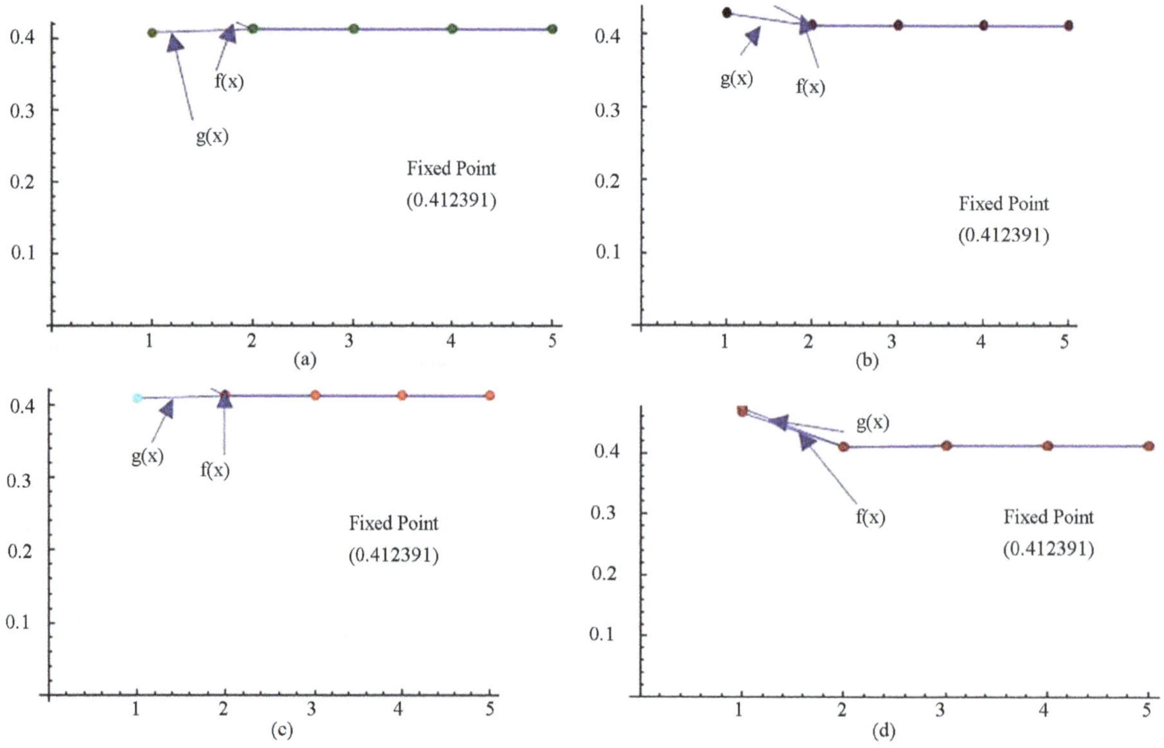

Figure 4. Graphical observations of new modified Agarwal *et al.* iteration for $p_1(x)$. Here (a)-(d) show the graph for Table 4. The merging point with value 0.412391 is fixed point.

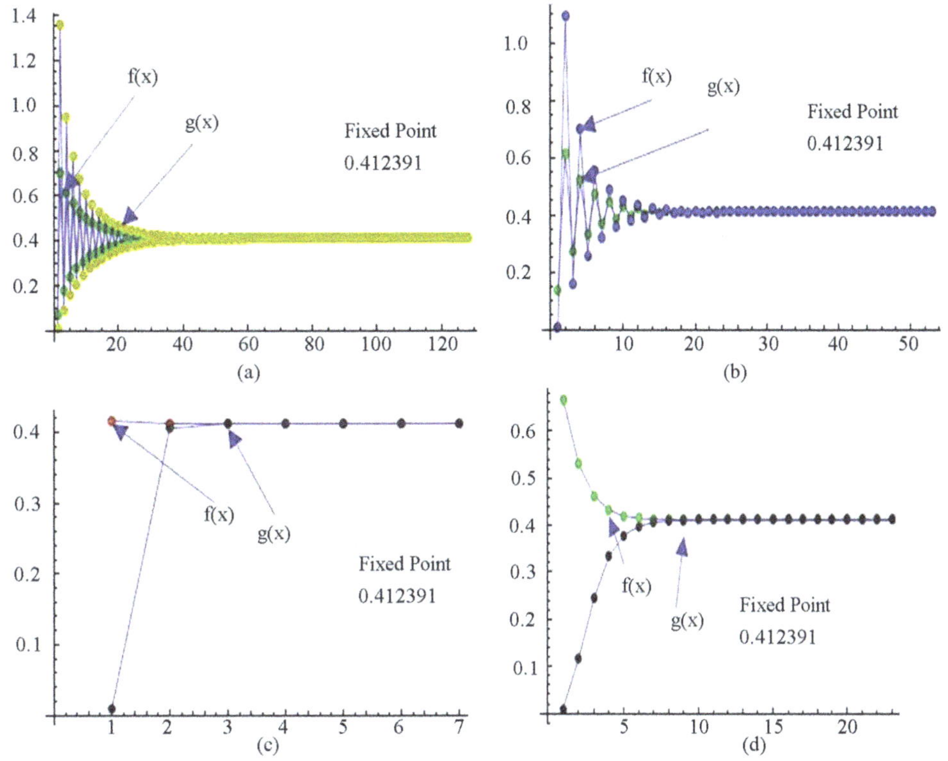

Figure 5. Graphical observations of simple SP iteration for $p_1(x)$. Here Here (a)-(d) show the graph for Table 5. The merging point with value 0.412391 is fixed point.

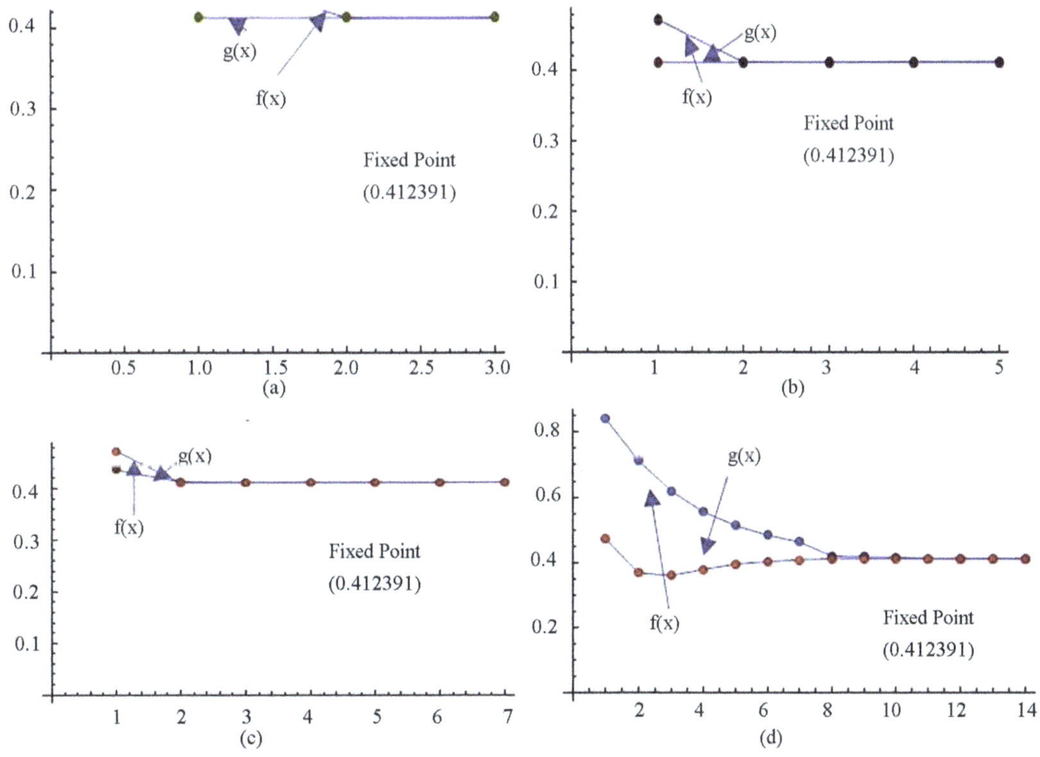

Figure 6. Graphical observations of new modified SP iteration for $p_1(x)$. Here (a)-(d) show the graph for Table 6. The merging point with value 0.412391 is fixed point.

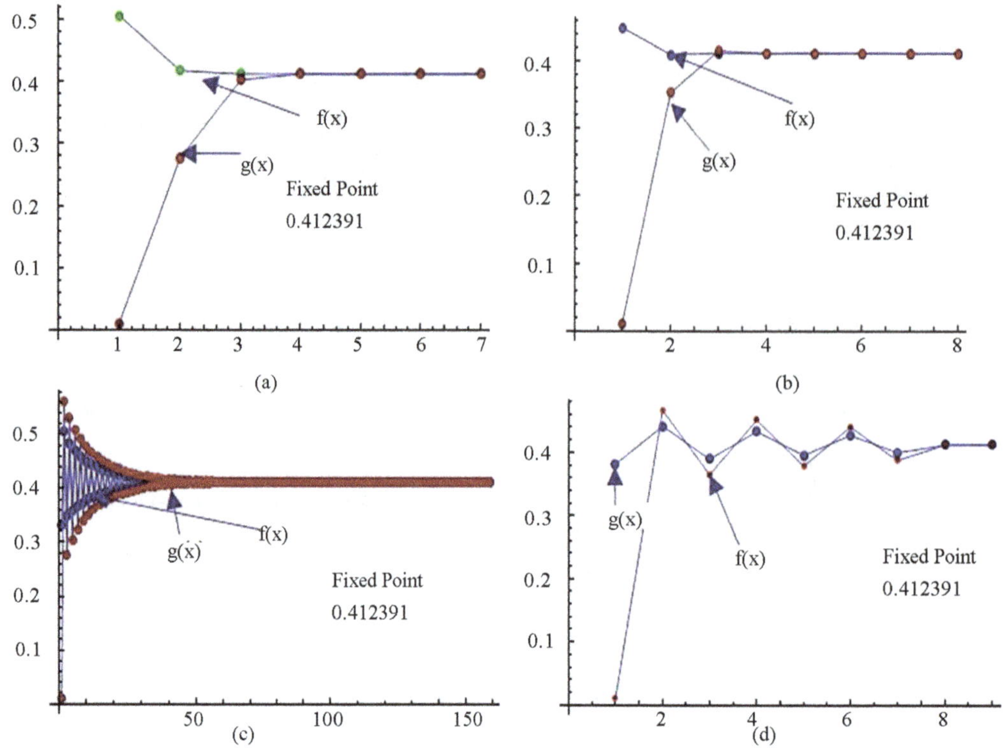

Figure 7. Graphical observations of simple Noor iteration for $p_1(x)$. Here (a)-(d) show the graph for Table 7. The merging point with value 0 .412391 is fixed point.

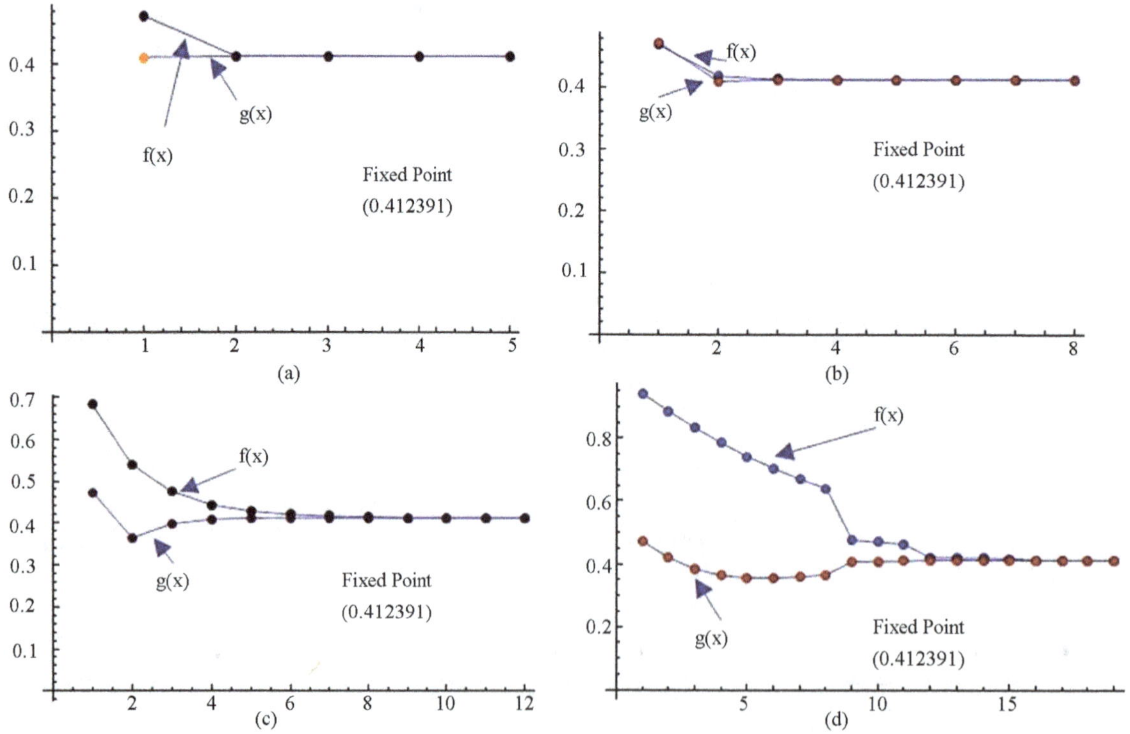

Figure 8. Graphical observations of new modified Noor iteration for $p_1(x)$. Here (a)-(d) show the graph for Table 8. The merging point with value 0.412391 is fixed point.

Table 9. Simple Ishikawa for $p_2(x)$.

	$a=0.9, b=0.999$		$a=0.1, b=0.9$		$a=0.3, b=0.5$		$a=0.1, b=0.1$	
N	x_{n+1}	Tx_n	x_{n+1}	Tx_n	x_{n+1}	Tx_n	x_{n+1}	Tx_n
0	0.340948	0.342793	0.343325	0.342793	0.346084	0.342793	0.349258	0.342793
1	0.340071	0.340243	0.341078	0.340895	0.343689	0.341668	0.348571	0.342578
2	0.339989	0.340005	0.340338	0.340278	0.342231	0.340996	0.347934	0.342379
3	0.339982	0.339983	0.340097	0.340078	0.341345	0.340593	0.347344	0.342195
6	**0.339981**	**0.339981**	0.339985	0.339984	0.340284	0.340116	0.345821	0.341726
7	**0.339981**	**0.339981**	0.339982	0.339982	0.340164	0.340063	0.345386	0.341593
9			**0.339981**	**0.339981**	0.340048	0.340011	0.344611	0.341357
10			**0.339981**	**0.339981**	0.340022	0.339999	0.344266	0.341253
18					0.339982	0.339981	0.342284	0.340659
20					**0.339981**	**0.339981**	0.341952	0.340561
21					**0.339981**	**0.339981**	0.341805	0.340517
127							0.339982	0.339981
128							**0.339981**	**0.339981**
129							**0.339981**	**0.339981**

Table 10. Modified Ishikawa for $p_2(x)$.

	$a=0.9, b=0.9999$		$a=0.1, b=0.9$		$a=0.3, b=0.6$		$a=0.1, b=0.1$	
N	x_{n+1}	Tx_n	x_{n+1}	Tx_n	x_{n+1}	Tx_n	x_{n+1}	Tx_n
0	0.33998	0.339823	0.340869	0.339823	0.343944	0.339823	0.348985	0.339823
1	**0.339981**	**0.339981**	0.340069	0.33998	0.341559	0.339957	0.348075	0.339854
2	**0.339981**	**0.339981**	0.33999	0.339981	0.340611	0.339977	0.347257	0.339879
4			**0.339981**	**0.339981**	0.340082	0.339981	0.345864	0.339915
5			**0.339981**	**0.339981**	0.340021	0.339981	0.345271	0.339928
----					----------	-----------		
9					0.339982	0.339981	0.343443	0.339959
10					**0.339981**	**0.339981**	0.343096	0.339963
11					**0.339981**	**0.339981**	0.342783	0.339966
							---------	--------
93							0.339982	0.339981
94							**0.339981**	**0.339981**
95							**0.339981**	**0.339981**

Table 11. Simple Agarwal *et al.* for $p_2(x)$.

	$a=0.9, b=0.99999,$		$a=0.1, b=0.9$		$a=0.3, b=0.5$		$a=0.1, b=0.1$	
N	x_{n+1}	Tx_n	x_{n+1}	Tx_n	x_{n+1}	Tx_n	x_{n+1}	Tx_n
0	0.340949	0.342793	0.342605	0.342793	0.342481	0.342793	0.342772	0.342793
1	0.340071	0.340243	0.340649	0.340696	0.340587	0.340662	0.340737	0.340742
2	0.339989	0.340005	0.34015	0.340162	0.340127	0.340145	0.340184	0.340185
5	**0.339981**	**0.339981**	0.339984	0.339984	0.339983	0.339983	0.339985	0.339985
6	**0.339981**	**0.339981**	0.339982	0.339982	0.339982	0.339982	0.339982	0.339982
7	**0.339981**	**0.339981**	**0.339981**	**0.339981**	**0.339981**	**0.339981**	**0.339981**	**0.339981**
8			**0.339981**	**0.339981**	**0.339981**	**0.339981**	**0.339981**	**0.339981**

Table 12. Modified Aggarwal *et al.* for $p_2(x)$.

	$a = 0.99999, b = 0.999$		$a = 0.1, b = 0.9$		$a = 0.1, b = 0.1$		$a = 0.1, b = 0.5$	
N	x_{n+1}	Tx_n	x_{n+1}	Tx_n	x_{n+1}	Tx_n	x_{n+1}	Tx_n
0	0.339981	0.339823	0.339851	0.339823	0.339827	0.339823	0.339839	0.339823
1	**0.339981**	**0.339981**	**0.339981**	**0.339981**	**0.339981**	**0.339981**	**0.339981**	**0.339981**
2	**0.339981**	**0.339981**	**0.339981**	**0.339981**	**0.339981**	**0.339981**	**0.339981**	**0.339981**

Table 13. Simple SP for $p_2(x)$.

	$a = 0.9, b = 0.9, c = 0.999$		$a = 0.1, b = 0.1, c = 0.9$		$a = 0.3, b = 0.5, c = 0.7$		$a = 0.1, b = 0.1, c = 0.1$	
N	x_{n+1}	Tx_n	x_{n+1}	Tx_n	x_{n+1}	Tx_n	x_{n+1}	Tx_n
0	0.340313	0.342793	0.343012	0.342793	0.342455	0.342793	0.347988	0.342793
1	0.339992	0.340071	0.340881	0.340808	0.340583	0.340655	0.346375	0.342211
2	0.339981	0.339984	0.340247	0.340225	0.340127	0.340144	0.345085	0.34175
3	**0.339981**	**0.339981**	0.340059	0.340053	0.340016	0.34002	0.344053	0.341386
4	**0.339981**	**0.339981**	0.340004	0.340002	0.33999	0.339991	0.343229	0.341097
--	--	----------	---------	----	------------	------------	--------	--------
8			**0.339981**	**0.339981**	**0.339981**	**0.339981**	0.341292	0.340428
9			**0.339981**	**0.339981**	**0.339981**	**0.339981**	0.341026	0.340337
--	--	----------	---------	----	------------	------------	--------	--------
44							**0.339981**	**0.339981**
45							**0.339981**	**0.339981**

Table 14. Modified SP for $p_2(x)$.

	$a = 0.9, b = 0.9, c = 0.999$		$a = 0.1, b = 0.1, c = 0.9$		$a = 0.3, b = 0.5, c = 0.7$		$a = 0.1, b = 0.1, c = 0.1$	
N	x_{n+1}	Tx_n	x_{n+1}	Tx_n	x_{n+1}	Tx_n	x_{n+1}	Tx_n
0	0.340066	0.339823	0.34807	0.339823	0.343427	0.339823	0.34807	0.339823
1	0.339982	0.339981	0.346515	0.339879	0.34118	0.339963	0.346515	0.339879
2	**0.339981**	**0.339981**	0.345263	0.339915	0.3404	0.339979	0.345263	0.339915
3	**0.339981**	**0.339981**	0.344252	0.339938	0.340128	0.339981	0.344252	0.339938
--	--	---------	---------	----	------------	------------	--------	--------
8			0.341463	0.339976	0.339982	0.339981	0.341463	0.339976
9			0.341181	0.339978	**0.339981**	**0.339981**	0.341181	0.339978
10			0.340953	0.339979	**0.339981**	**0.339981**	0.340953	0.339979
--	--	----------	---------	----	------------	------------	--------	--------
46			0.339982	0.339981			0.339982	0.339981
47			**0.339981**	**0.339981**			**0.339981**	**0.339981**
48			**0.339981**	**0.339981**			**0.339981**	**0.339981**

Table 15. Simple Noor for $p_2(x)$.

	$a = 0.9, b = 0.9, c = 0.999$		$a = 0.1, b = 0.1, c = 0.9$		$a = 0.3, b = 0.5, c = 0.7$		$a = 0.1, b = 0.1, c = 0.1$	
N	x_{n+1}	Tx_n	x_{n+1}	Tx_n	x_{n+1}	Tx_n	x_{n+1}	Tx_n
0	0.340491	0.342793	0.34332	0.342793	0.344161	0.342793	0.349258	0.342793
1	0.340006	0.340119	0.341074	0.340894	0.341717	0.341127	0.34857	0.342577
2	0.339982	0.339988	0.340337	0.340277	0.3407	0.340452	0.347932	0.342378
3	**0.339981**	**0.339981**	0.340097	0.340077	0.340279	0.340176	0.347342	0.342195
4	**0.339981**	**0.339981**	0.340019	0.340012	0.340104	0.340061	0.346794	0.342025
5	**0.339981**	**0.339981**	0.339993	0.339991	0.340032	0.340014	0.346288	0.341869
6			0.339985	0.339984	0.340002	0.339995	0.345818	0.341725
7			0.339982	0.339982	0.33999	0.339987	0.345383	0.341592
8			**0.339981**	**0.339981**	0.339985	0.339983	0.344981	0.34147
9			**0.339981**	**0.339981**	0.339983	0.339982	0.344608	0.341357
10					0.339982	0.339981	0.344263	0.341252
11					**0.339981**	**0.339981**	0.343943	0.341156
12					**0.339981**	**0.339981**	0.343647	0.341067
13					**0.339981**	**0.339981**	0.343373	0.340984
--	--	----------	---------	----	------------	-----------	--------	---------
127							0.339982	0.339981
128							**0.339981**	**0.339981**
129							**0.339981**	**0.339981**

Table 16. Modified Noor for $p_2(x)$.

	$a = 0.9, b = 0.9, c = 0.999$		$a = 0.1, b = 0.1, c = 0.9$		$a = 0.3, b = 0.5, c = 0.7$		$a = 0.1, b = 0.1, c = 0.1$	
N	x_{n+1}	Tx_n	x_{n+1}	Tx_n	x_{n+1}	Tx_n	x_{n+1}	Tx_n
0	0.340982	0.339823	0.348985	0.339823	0.344953	0.339823	0.348985	0.339823
1	0.340081	0.33998	0.348075	0.339854	0.342458	0.339943	0.348075	0.339854
2	0.339991	0.339981	0.347257	0.339879	0.341217	0.339972	0.347257	0.339879
3	0.339982	0.339981	0.346523	0.339899	0.340599	0.339979	0.346523	0.339899
4	**0.339981**	**0.339981**	0.345864	0.339915	0.34029	0.33998	0.345864	0.339915
5	**0.339981**	**0.339981**	0.345271	0.339928	0.340135	0.339981	0.345271	0.339928
13			0.342249	0.339972	0.339982	0.339981	0.342249	0.339972
14			0.342022	0.339973	**0.339981**	**0.339981**	0.342022	0.339973
15			0.341817	0.339975	**0.339981**	**0.339981**	0.341817	0.339975
93			0.339982	0.339981			0.339982	0.339981
94			**0.339981**	**0.339981**			**0.339981**	**0.339981**

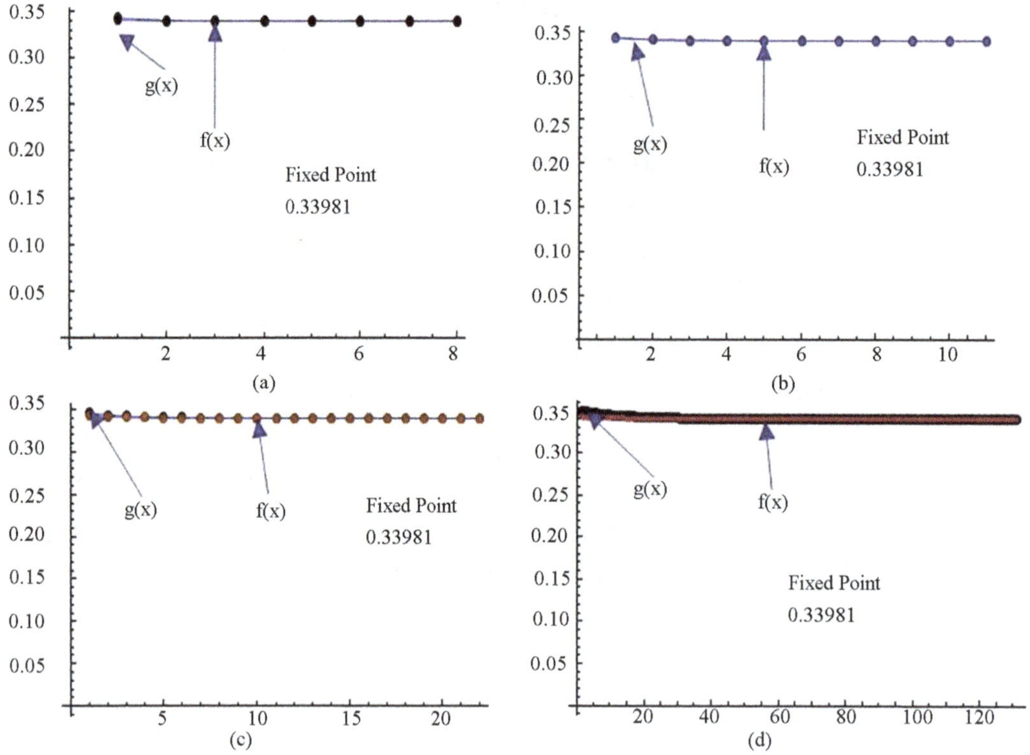

Figure 9. Graphical observations for simple Ishikawa iteration for $p_2(x)$. Here (a)-(d) show the graph for Table 9. The merging point with value 0.33981 is fixed point.

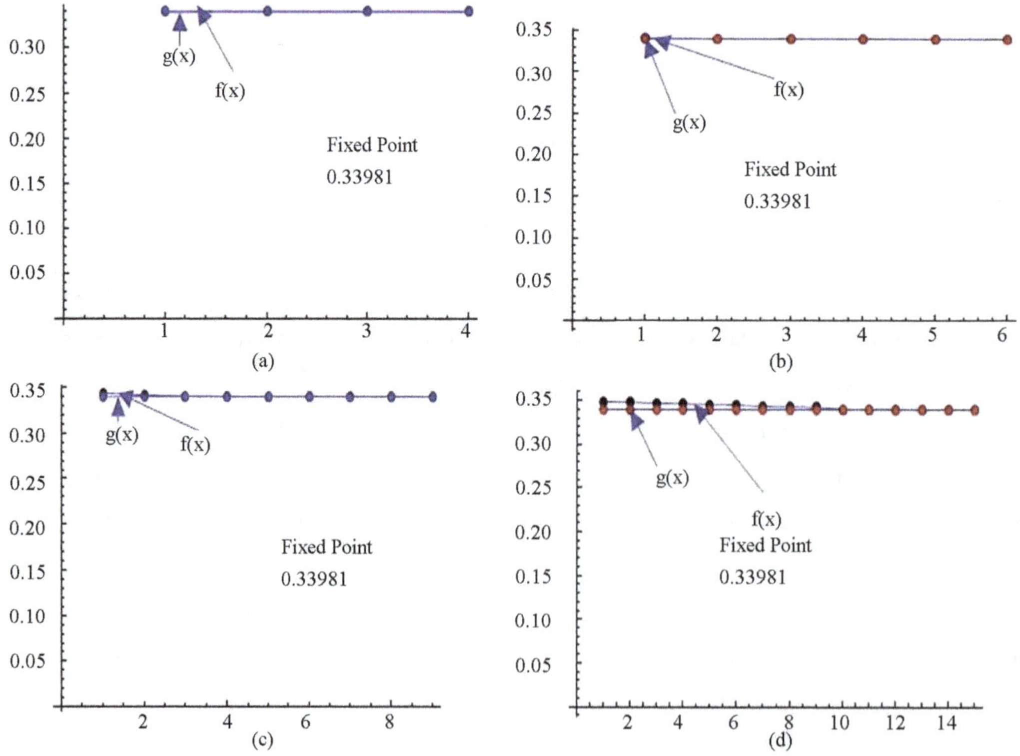

Figure 10. Graphical observations for new modified Ishikawa iteration for $p_2(x)$. Here (a)-(d) show the graph for Table 10. The merging point with value 0.33981 is fixed point.

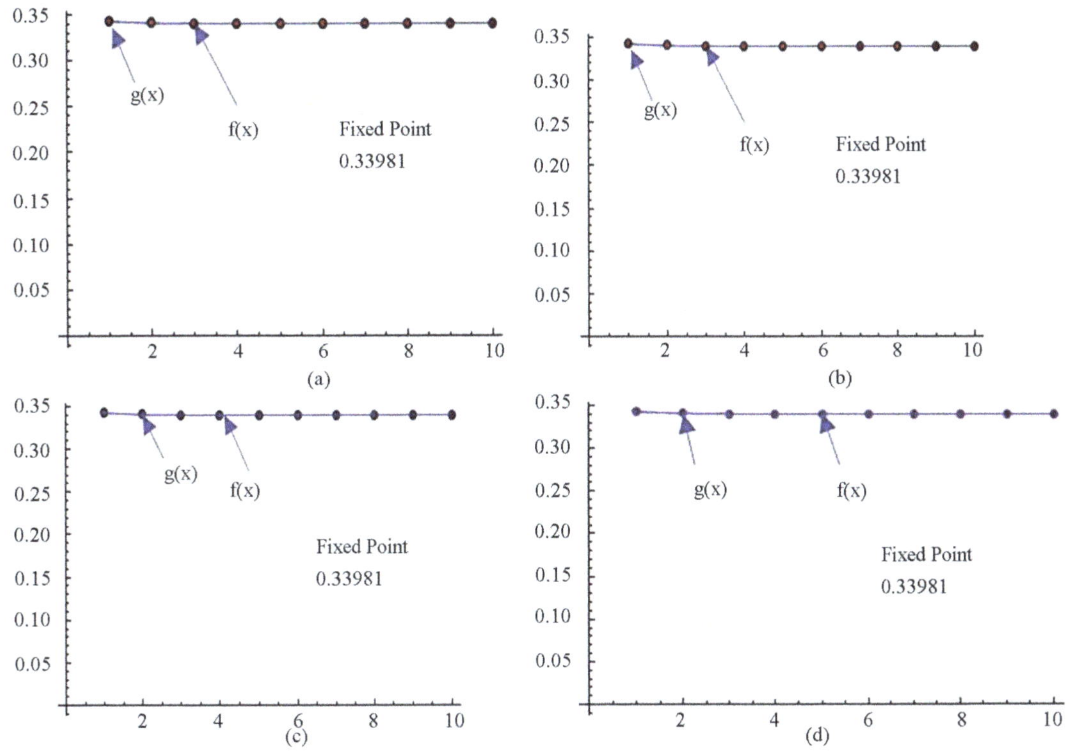

Figure 11. Graphical observations for simple Agarwal iteration for $p_2(x)$. Here (a)-(b) show the graph for Table 11. The merging point with value 0.33981 is fixed point.

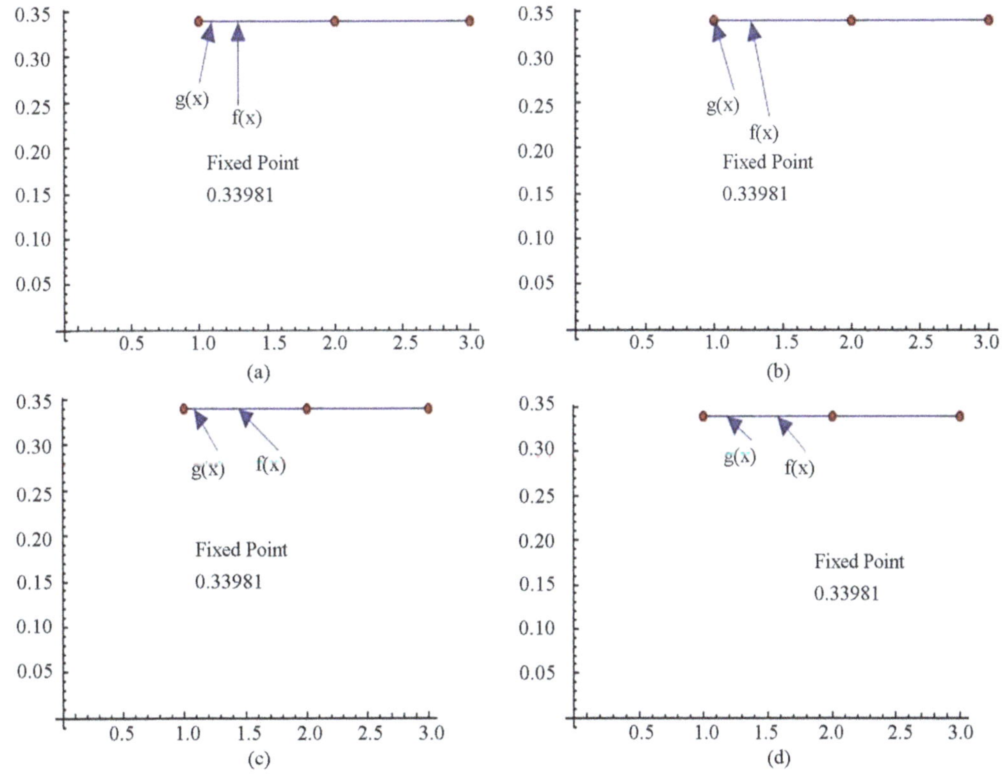

Figure 12. Graphical observations for simple Agarwal iteration for $p_2(x)$. Here (a)-(b) show the graph for Table 12. The merging point with value 0.33981 is fixed point.

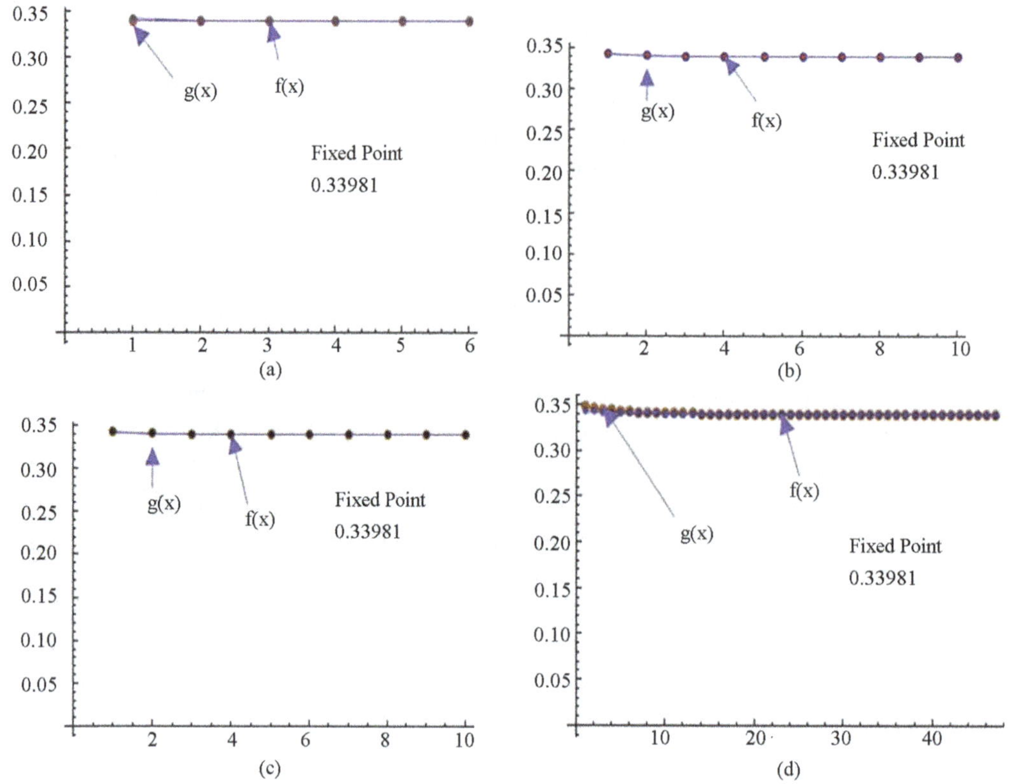

Figure 13. Graphical observations for simple SP iteration for $p_2(x)$. Here (a)-(b) show the graph for Table 13. The merging point with value 0.33981 is fixed point.

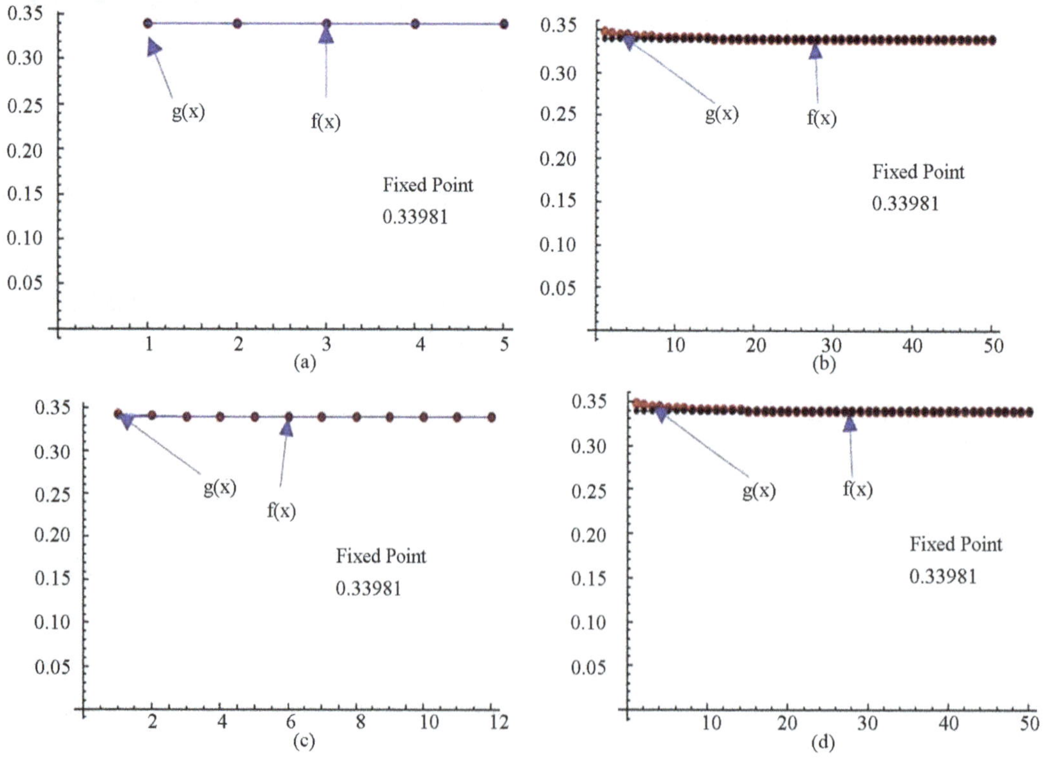

Figure 14. Graphical observations for new modified SP iteration for $p_2(x)$. Here (a)-(b) show the graph for Table 14. The merging point with value 0.33981 is fixed point.

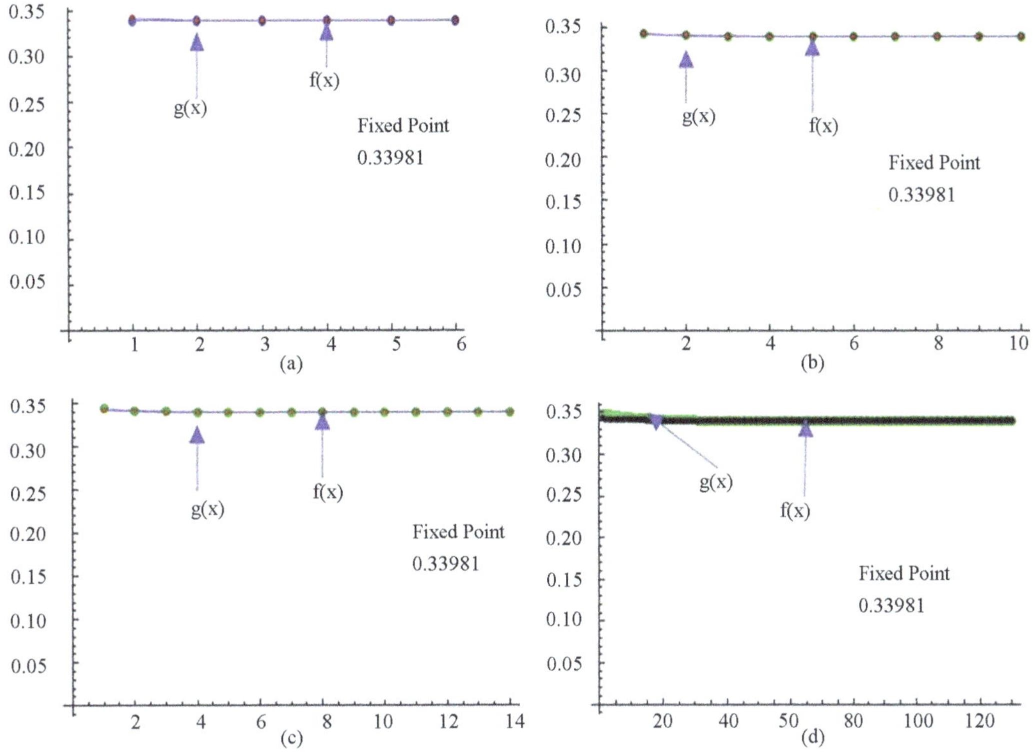

Figure 15. Graphical observations for simple Noor iteration for $p_2(x)$. Here (a)-(b) show the graph for Table 15. The merging point with value 0.33981 is fixed point.

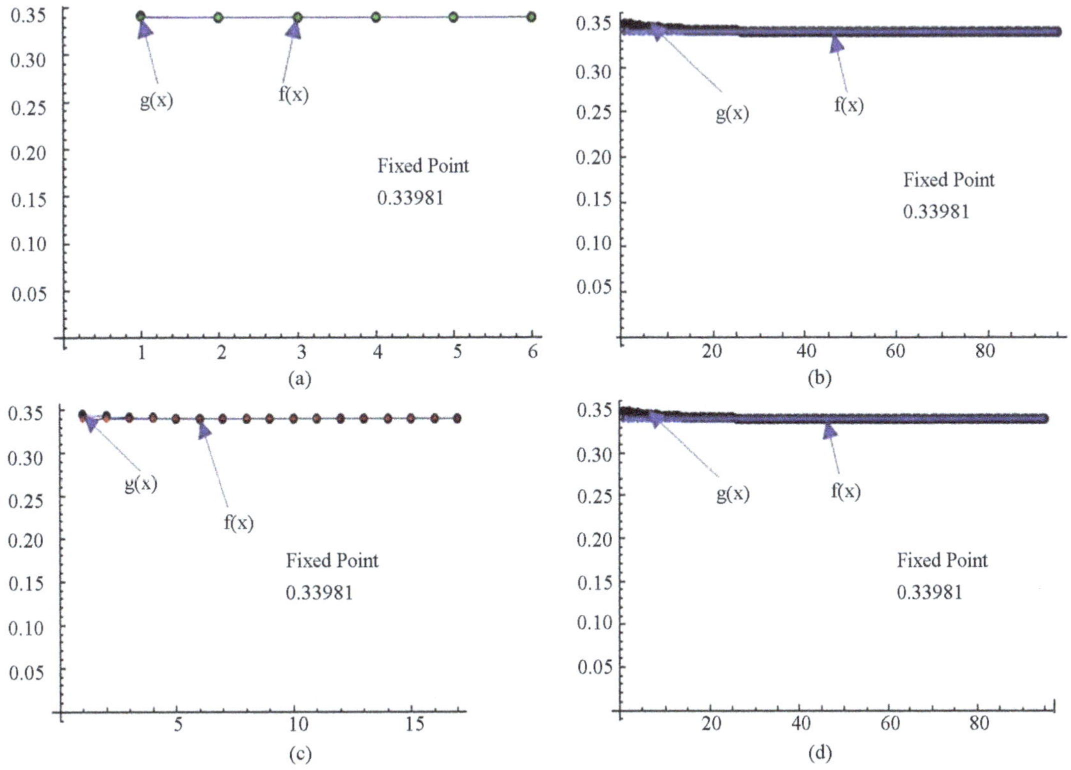

Figure 16. Graphical observations for new modified Noor iteration for $p_2(x)$. Here(a)-(b) show the graph for Table 16. The merging point with value 0.33981 is fixed point.

4. Observations

Table 17. Simple-Ishikawa for $p_1(x)$.

a/b	0.001	0.01	0.1	0.2	0.3	0.4	0.5	0.6	0.7	0.8	0.9	0.99	0.999	0.9999
0.001	5513	546	49	16	11	6	13	24	86	not	not	not	not	Not
0.01	5590	554	50	22	12	7	12	24	72	not	not	not	not	not
0.1	6503	645	59	26	15	9	8	12	24	53	3476	not	not	Not
0.2	7994	794	74	34	20	13	8	6	11	17	28	50	54	54
0.3	10455	1040	99	46	29	20	14	10	7	7	9	13	13	13
0.4	15285	1523	147	70	45	32	24	18	14	12	10	9	9	9
0.5	29149	2909	285	140	91	66	52	42	34	28	21	23	22	22
0.6	916911	91684	9161	4577	3048	2284	1826	1520	1301	1137	1009	916	907	906
0.7	not	not	not	not	not	not	not	not	not	not	not	not	not	not
0.8	not	not	not	not	not	not	not	not	not	not	not	not	not	not
0.9	not	not	not	not	not	not	not	not	not	not	not	not	not	not
0.99	not	not	not	not	not	not	not	not	not	not	not	not	not	not
0.999	not	not	not	not	not	not	not	not	not	not	not	not	not	not

Table 18. Modified-Ishikawa for $p_1(x)$.

a/b	0.001	0.01	0.1	0.2	0.3	0.4	0.5	0.6	0.7	0.8	0.9	0.99	0.999	0.9999
0.001	14030	1397	134	63	40	28	21	16	12	9	6	4	4	4
0.01	14032	1397	134	63	40	28	21	16	12	9	6	4	4	4
0.1	14056	1400	134	63	40	28	21	16	12	9	6	5	4	4
0.2	14086	1403	134	64	40	28	21	16	12	9	5	5	4	4
0.3	14119	1406	134	64	40	28	21	16	12	9	5	5	4	4
0.4	14152	1409	135	64	40	28	21	16	12	9	6	4	4	4
0.5	14181	1412	135	64	40	28	21	16	12	9	6	4	4	4
0.6	14204	1414	135	64	40	28	21	16	12	9	6	4	4	4
0.7	14220	1416	136	64	40	28	21	16	12	9	7	4	4	4
0.8	14224	1416	136	64	40	28	21	16	12	9	7	3	3	3
0.9	14216	1416	135	64	40	28	21	16	12	9	7	4	3	3
0.99	14197	1414	135	64	40	28	21	16	12	9	7	4	3	3
0.999	14195	1414	135	64	40	28	21	16	12	9	7	4	3	3

Table 19. Simple-Ishikawa for $p_1(x)$.

a/b	0.001	0.01	0.1	0.2	0.3	0.4	0.5	0.6	0.7	0.8	0.9	0.99	0.999	0.9999
0.001	13701	1366	132	65	41	29	23	18	15	12	10	8	8	8
0.01	13667	1363	132	64	41	29	22	18	14	12	10	8	8	8
0.1	13332	1329	129	62	40	29	22	17	14	11	9	8	8	8

Continued

0.2	12998	1296	125	60	39	28	21	17	13	11	9	8	7	7
0.3	12671	1263	122	59	38	27	20	16	13	11	9	7	7	7
0.4	12360	1232	119	57	36	26	20	16	12	10	8	7	7	7
0.5	12066	1203	116	56	36	25	19	15	12	10	8	7	6	6
0.6	11785	1174	113	54	35	25	19	15	12	9	8	6	6	6
0.7	11516	1148	110	53	34	24	18	14	11	9	7	6	6	6
0.8	11260	1122	108	52	33	23	18	14	11	8	7	6	6	5
0.9	11015	1096	106	50	32	23	17	13	10	8	7	5	5	5
0.99	10803	1076	104	49	31	22	17	13	10	8	6	5	5	5
0.999	10783	1074	103	49	31	22	17	13	10	8	6	5	5	5

Table 20. Modified-Ishikawa for $p_2(x)$.

a/b	0.001	0.01	0.1	0.2	0.3	0.4	0.5	0.6	0.7	0.8	0.9	0.99	0.999	0.9999
0.001	9977	994	95	45	28	20	15	11	9	7	5	3	2	2
0.01	9977	994	95	45	28	20	15	11	9	7	5	2	2	2
0.1	9980	994	95	45	28	20	15	11	9	7	5	2	2	2
0.2	9983	994	95	45	28	20	15	11	9	7	5	2	2	2
0.3	9985	994	95	45	28	20	15	11	9	7	5	2	2	2
0.4	9987	995	95	45	29	20	15	11	9	7	5	3	2	2
0.5	9988	995	95	45	29	20	15	11	9	7	5	3	2	2
0.6	9990	995	95	45	29	20	15	11	9	7	5	3	2	2
0.7	9991	995	95	45	29	20	15	11	9	7	5	3	2	2
0.8	9992	995	95	45	29	20	15	11	9	7	5	3	2	2
0.9	9992	995	95	45	29	20	15	11	9	7	5	3	2	2
0.99	9992	995	95	45	29	20	15	11	9	7	5	3	2	2
0.999	9992	995	95	45	29	20	15	11	9	7	5	3	2	2

Table 21. Simple-Agarwal for $p_1(x)$.

a/b	0.001	0.01	0.1	0.2	0.3	0.4	0.5	0.6	0.7	0.8	0.9	0.99	0.999	0.9999
0.001	not	not	not	not	not	not	Not	not	not	not	not	not	not	not
0.01	not	not	not	not	not	not	Not	not	not	not	not	not	not	not
0.1	not	not	not	not	not	not	Not	not	not	not	not	not	not	not
0.2	not	not	not	not	not	not	Not	not	not	291	97	59	57	57
0.3	not	not	not	not	not	not	5475	101	47	29	19	14	14	14
0.4	not	not	not	not	not	307	59	29	19	11	8	9	10	10
0.5	not	not	not	not	not	59	27	15	9	9	15	23	24	24
0.6	not	not	not	not	101	29	15	7	10	17	37	288	741	885
0.7	not	not	not	not	47	18	9	8	20	62	not	not	not	not
0.8	not	not	not	287	29	11	8	19	68	not	not	not	not	not
0.9	not	not	not	95	17	7	15	53	not	not	not	not	not	not
0.99	not	not	not	55	14	10	33	not	not	not	not	not	not	not
0.999	not	not	not	53	14	10	39	not	not	not	not	not	not	not

Table 22. Modified-Agarwal for $p_1(x)$.

a/b	0.001	0.01	0.1	0.2	0.3	0.4	0.5	0.6	0.7	0.8	0.9	0.99	0.999	0.9999
0.001	4	4	4	4	4	4	4	4	4	4	4	4	4	4
0.01	4	4	4	4	4	4	4	4	4	4	4	4	4	4
0.1	4	4	4	4	4	4	4	4	4	4	4	4	4	4
0.2	4	4	4	4	4	4	4	4	4	4	4	4	4	4
0.3	4	4	4	4	4	4	4	4	4	4	4	4	4	4
0.4	4	4	4	4	4	4	4	4	4	4	4	4	4	4
0.5	4	4	4	4	4	4	4	4	4	4	4	4	4	4
0.6	4	4	4	4	4	4	4	4	4	4	4	4	4	4
0.7	4	4	4	4	4	4	4	4	4	4	4	4	4	4
0.8	4	4	4	4	4	4	4	4	4	3	3	3	3	3
0.9	4	4	4	4	4	4	4	3	3	3	3	3	3	3
0.99	4	4	4	4	4	4	4	3	3	3	3	3	3	3
0.999	4	4	4	4	4	4	4	3	3	3	3	3	3	3

Table 23. Simple-Agarwal for $p_2(x)$.

a/b	0.001	0.01	0.1	0.2	0.3	0.4	0.5	0.6	0.7	0.8	0.9	0.99	0.999	0.9999
0.001	8	8	8	8	8	8	8	8	8	8	8	8	8	8
0.01	8	8	8	8	8	8	8	8	8	8	8	8	8	8
0.1	8	8	8	8	8	8	8	8	8	8	8	8	8	8
0.2	8	8	8	8	8	8	8	8	8	8	8	8	7	7
0.3	8	8	8	8	8	8	8	8	7	7	7	7	7	7
0.4	8	8	8	8	8	8	7	7	7	7	7	7	7	7
0.5	8	8	8	8	8	7	7	7	7	7	7	6	6	6
0.6	8	8	8	8	8	7	7	7	7	7	6	6	6	6
0.7	8	8	8	8	7	7	7	7	6	6	6	6	6	6
0.8	8	8	8	8	7	7	7	7	6	6	6	6	5	5
0.9	8	8	8	8	7	7	7	6	6	6	5	5	5	5
0.99	8	8	8	7	7	7	6	6	6	6	5	5	5	5
0.999	8	8	8	7	7	7	6	6	6	5	5	5	5	5

Table 24. Modified-Agarwal for $p_2(x)$.

a/b	0.001	0.01	0.1	0.2	0.3	0.4	0.5	0.6	0.7	0.8	0.9	0.99	0.999	0.9999
0.001	2	2	2	2	2	2	2	2	2	2	2	2	2	2
0.01	2	2	2	2	2	2	2	2	2	2	2	2	2	2
0.1	2	2	2	2	2	2	2	2	2	2	2	2	2	2
0.2	2	2	2	2	2	2	2	2	2	2	2	2	2	2
0.3	2	2	2	2	2	2	2	2	2	2	2	2	2	2
0.4	2	2	2	2	2	2	2	2	2	2	2	2	2	2
0.5	2	2	2	2	2	2	2	2	2	2	2	2	2	2
0.6	2	2	2	2	2	2	2	2	2	2	2	2	2	2

Continued

0.7	2	2	2	2	2	2	2	2	2	2	2	2	2	2
0.8	2	2	2	2	2	2	2	2	2	2	2	2	2	2
0.9	2	2	2	2	2	2	2	2	2	2	2	2	2	2
0.99	2	2	2	2	2	2	2	2	2	2	2	2	2	2
0.999	2	2	2	2	2	2	2	2	2	2	2	2	2	2

We have noted the converging step of different iterations in tabular form and compare the conversing step for different value of a, b, c. Now by comparative analysis we noted that

1) For $p_1(x)$, simple Ishikawa do not converge for $0 < a \le 0.1$, $0.8 < b \le 1$ and $0.6 < a < 1$, $0 < b < 1$ but new modified Ishikawa converges for all values of a and b converges faster than Ishikawa iteration for corresponding values of a, b. Also it converges at lesser step as a and b both approaches one but not so in case of simple Ishikawa as observe from **Tables 17** and **18**. Similarly if we compare the both iterations for $p_2(x)$ as observed from **Tables 19** and **20** that as we increase values of a and b simultaneously than converging step decreases for both iterations but modified Ishikawa iteration converges at lesser step for $p_2(x)$.

2) As observed from **Tables 21** and **22** for $p_1(x)$ simple Agarwal *et al.* do not converge for all values of a and b it converges for

$\{a = 0.2, 0.8 \le b < 1\}$, $\{a = 0.3, 0.5 \le b < 1\}$, $\{a = 0.4, 0.5, 0.4 \le b < 1\}$, $\{a = 0.6, 0.3 \le b < 1\}$, $\{a = 0.7, 0.3 \le b \le 0.8\}$, $\{0.8 \le a < 1, 0.2 \le b \le 0.5\}$

but modified new Agarwal *et al.* iteration converges at lesser step for all values of a, b. For $p_2(x)$ both iterations converge for all values of a and b but modified iteration converges faster than simple iteration

3) The simple SP iteration converges at lesser step for $p_1(x)$ when $a = 1/2$, $b = 1/2$, $c = 1/2$ as we increases a and b the step of convergence increases. But do not converge if a and b approaches to one whereas modified new SP iteration converges for all values of a, b, c and at lesser step than simple SP. For $p_2(x)$ both iteration converges for all values of a, b and c but modified new SP converge faster than Simple SP iteration.

4) For simple Noor and modified new Noor iteration result is same as for SP and modified new SP iterations.

5. Conclusion

By the observation formed from the program and graph drawn in C^{++} and Mathematica for $p_1(x)$ and $p_2(x)$ polynomial, we conclude that the modified Ishikawa, Agarwal *et al.*, SP, Noor are faster than simple Ishikawa, Agarwal *et al.*, SP, Noor; but if we compare modified Ishikawa, Agarwal *et al.*, SP, Noor with decreasing order of rate of convergence of modified Agarwal *et al.*, SP, Noor, Ishikawa, modified new Agarwal *et al.* have consistent rate of convergence. The graphs drawn are based on data formed from C^{++} program and plot the data in mathematica to show the fixed point.

REFERENCES

[1] J. Biazar and A. Amirteimoori, "An Improvement to the Fixed Point Iterative Method," *Applied Mathematics and Computation*, Vol. 182, No. 1, 2006, pp. 567-571.

[2] B. Prasad and R. Sahni, "Convergence of Some General Iterative Schemes," *International Journal of Mathematical Analysis*, Vol. 5, No. 25, 2011, pp. 1237-1242.

[3] S. Ishikawa, "Fixed Points by a New Iteration Method," *Proceedings of the American Mathematical Society*, Vol. 44, No. 1, 1974, pp. 147-150.

[4] R. P. Agarwal, D. O'Regan and D. R. Sahu, "Iterative Construction of Fixed Points of Nearly Asymptotically Non-Expansive Mappings," *Journal of Nonlinear and Convex Analysis*, Vol. 8, No. 1, 2007, pp. 61-79.

[5] M. A. Noor, "New Approximation Schemes for General Variational Inequalities," *Journal of Mathematical Analysis and Applications*, Vol. 251, No. 1, 2000, pp. 217-229.

[6] W. Phuengrattana and S. Suantai, "On the Rate of Convergence of Mann, Ishikawa, Noor and SP Iterations for Continuous Functions on an Arbitrary Interval," *Journal of Computational and Applied Mathematics*, Vol. 235, No. 9, 2011, pp. 3006-3014.

[7] E. Babolian and J. Biazar, "On the Order of Convergence of Adomian Method," *Applied Mathematics and Computation*, Vol. 130, No. 2-3, 2002, pp. 383-387.

21

Predicting Rainfall Using the Principles of Fuzzy Set Theory and Reliability Analysis

Mahbub Hasan[1], Salam Md. Mahbubush Khan[2], Chandrasekhar Putcha[3], Ashraf Al-Hamdan[4], Chance M. Glenn[5]

[1]Department of Civil Engineering, Alabama Agricultural and Mechanical University, Normal, USA
[2]Department of Mathematics, Alabama Agricultural and Mechanical University, Normal, USA
[3]Department of Civil and Environmental Engineering, California State University, Fullerton, USA
[4]Department of Civil & Environmental Engineering, University of Alabama, Huntsville, USA
[5]College of Engineering, Technology and Physical Sciences, Alabama Agricultural and Mechanical University, Normal, USA

ABSTRACT

The paper presents occurrence of rainfall using principles of fuzzy set theory and principles of reliability analysis. Both the abstract and the rest of the paper are discussed from these two points of view. First, a fuzzy inference model for predicting rainfall using scan data from the USDA Soil Climate Analysis Network Station at Alabama Agricultural and Mechanical University (AAMU) campus for the year 2004 is presented. The model further reflects how an expert would perceive weather conditions and apply this knowledge before inferring a rainfall. Fuzzy variables were selected based on judging patterns in individual monthly graphs for 2003 and 2004 and the influence of different variables that caused rainfall. A decrease in temperature (TP) and an increase in wind speed (WS) when compared between the *i*th and $(i - 1)$th day were found to have a positive relation with a rainfall (RF) occurrence in most cases. Therefore, TP and WS were used in the antecedent part of the production rules to predict rainfall (RF). Results of the model showed better performance when threshold values for 1) Relative Humidity (RH) of *i*th day; 2) Humidity Increase (HI) between the *i*th and $(i - 1)$th day; and 3) Product (P) of decrease in temperature (TP) and an increase in wind speed (WS) were introduced. The percentage of error was 12.35 when compared the calculated amount of rainfall with actual amount of rainfall. This is followed by prediction of rainfall using principles of reliability analysis. This is done by comparing theoretical probabilities with experimental probabilities for the occurrence of two main events, namely, Relative Humidity (RH) and Humidity Increase (HI) being in between specified threshold values. The experimental values of probability are falling in between $\mu - \sigma$ and $\mu + \sigma$ for both RH and HI parameters, where μ is the mean value and σ is the standard deviation.

Keywords: Fuzzy Sets; Prediction; Reliability; Rainfall; Water Resources

1. Introduction

First fuzzy set concepts are discussed followed by principles of reliability analysis. This work is an extension of the work done by Hasan *et al.* [1]. In predicting weather conditions, factors in the antecedent and consequent parts that exhibit vagueness and ambiguity are being treated with logic and valid algorithms by Hasan *et al.* [2]. Use of fuzzy set theory has been proved by scientists to be applicable with uncertain, vague and qualitative expressions of the system. Application of fuzzy set theory in soil, crop, and water management is still in its infant stage due to the lack of awareness of the potentials of fuzzy set theory. Weather forecasting is one of the most important and demanding operational responsibilities carried out by meteorological services worldwide. It is a complicated procedure that includes numerous specialized technological fields. The task is complicated in the field of meteorology because all decisions are made within a visage of uncertainty associated with weather systems. Chaotic features associated with atmospheric phenomena have also attracted the attention of modern scientists. The drawback of statistical models is a foundation, in most cases, upon several tacit assumptions regarding the system mentioned by Wilks [3]. Carrano *et al.*, [4] compared non-linear regression modeling and fuzzy knowledge-based modeling, and explained that fuzzy models were most appropriate when subjective and qualitative data were utilized and the numbers of empirical observations were small. Brown-Brandl *et al.* [5] used four modeling techniques to predict respiration rate as an indicator of stress in livestock. Four modeling techniques consisted of two multiple regression and two fuzzy inference systems. Fuzzy inference models offered better results than the two multiple regression models (Brown-Brandl *et al.* [5]). Fuzzy inference models yielded a lower percentage of error when compared to the linear multiple regression model (Hasan *et al.*, [2]). Similar research by Wong *et al.* [6] compared the results of fuzzy rule based rainfall prediction with an established method which used radial basis function networks and orographic effect. They concluded that fuzzy rule based methods could provide similar results from the established method. However, the method has an advantage of allowing the analyst to understand and interact with the model using fuzzy rules. Lee *et al.* [7] considered two smaller areas where they assumed precipitation was proportional to elevation. Predictions of those two areas were made using a simple linear regression based on elevation information only. Comparison with the observed data revealed that the radial basis function (RBF) network produced better results than the linear regression models. Hence, considering the advantage of using the concept of fuzzy logic for predicting rainfall as stated by other researchers was justifiable. The advantage of fuzzy inference modeling can reflect expert knowledge and yield results with precision and accuracy. In fuzzy rule basics, knowledge acquisition is the main concern for building an expert system. Knowledge in the form of IF-THEN rules can be provided by experts or can be extracted from data. Each rule has an antecedent part and a consequent part. The antecedent part is the collection of conditions connected by AND, OR, NOT logic operators and the consequent part represents its action (Pant and Ashwagosh [8]). In a fuzzy inference engine, the truth-value for the premise of each rule is computed and applied to the conclusion part of each rule. This result is one fuzzy subset being assigned to each output variable for each rule. For composite rules, usually, min-max inference technique is used.

Defuzzification is used to convert fuzzy output sets to a crisp value. The widely used methods for defuzzification are center of gravity and mean of maxima.

Generating production rules for fuzzy inference modeling is cumbersome if they are not derived as they are being perceived by an expert. Production rules have the form:

$$\text{IF } X \text{ is } A1 \text{ AND } Y \text{ is } B1 \text{ THEN } Z \text{ is } C1 \quad (1)$$

$$\text{IF } X \text{ is } A2 \text{ AND } Y \text{ is } B2 \text{ THEN } Z \text{ is } C2 \quad (2)$$

Here X, Y represent two antecedent variables (the conditional part of the production rule, like TP and WS as explained above), and Z is the variable yielding the consequent part of the production rule. $A1$, $A2$, $B1$, $B2$, $C1$, $C2$ are the linguistic and vague expressions with ambiguities. Focusing this idea of production rule, an example for such production rule that can be employed in the present research is shown as:

$$\begin{array}{l}\text{IF WP is very high AND TP} \\ \text{is lower THEN RF is moderate}\end{array} \quad (3)$$

Equation (3) shows the qualitative form of explanation, such as very high, lower and moderate, which are all fuzzy in nature. These are explained linguistically without specific quantity or as a crisp value. The relationship of the variables between antecedent and consequent parts represents a production rule in Equation (3) based on valid logic. In the complex reality of the world, it is usually not easy to construct rules due to the limitations of manipulation and verbalization of experts, Abe and Ming-Shong [9]. This method is termed as the Fuzzy Adaptive System (FAS).

A brief discussion of principles of reliability analysis as related to prediction of rainfall is discussed in the paper.

A large set of data for rainfall have been collected for various years from several sources as various locations. These are—AAMU 2004, WATARS 2004, BRAGG 2004, AAMU 2005, WATARS 2005 and BRAGG 2005. It has been established that there are mainly two parameters—Relative Humidity (RH) and Humidity Increase (HI) after the occurrence of rainfall. Hence, these two variables are the main random variables (RV) in this study. Since the data set is large, it can be reasonably assumed, from central limit theorem, that both RH and HI follow normal distribution. Normal distribution is a 2-parameter distribution as N(μ, σ), where μ is mean value of the random variable and σ is standard deviation of the variable under consideration.

2. Definitions

2.1. Fuzzy Set

Fuzzy sets are collection of objects with the same properties, and in crisp sets the objects either belong to the set or do not. In practice, the characteristic value for an object belonging to the considered set is coded as 1 and if it is outside the set then the coding is 0. In crisp sets, there is no ambiguity or vagueness about each object belongs to the considered set. On the other hand, in daily life humans are always confronted with objects that may be similar to one other with quite different properties. Therefore uncertainty always arises concerning the assessment of membership values 0 or 1. Logically, of course, some of the similar objects may partially belong to the same set, therefore, an ambiguity emerges in the decision of belonging or not. In order to alleviate such situations [10] generalized the crisp set membership degree as having any value continuously between 0 and 1. Fuzzy sets are a generalization of conventional set theory. The basic idea of fuzzy sets is easy to grasp. An object with membership function 1 belongs to the set with no doubt and those with 0 membership functions again absolutely do not belong to the set, but objects with intermediate membership functions partially belong to the same set. The greater the membership function, the more the object belongs to the set [11].

The membership function of a fuzzy set is a generalization of the indicator function in classical sets. In fuzzy logic, it represents the degree of truth as an extension of valuation. Degrees of truth are often confused with probabilities, although they are conceptually distinct, because fuzzy truth represents membership in vaguely defined sets, not likelihood of some event or condition.

For the universe X and given the membership-degree function $\mu \rightarrow [0,1]$ the fuzzy set is defined as:

$$A = \{(x, \mu_A(x)) | x \in X\} \tag{4}$$

The following holds good for the functional values of the membership function $\mu_A(x)$

$$\mu_A(x) \geq 0, \ \forall x \in X, \tag{5}$$

$$\sup_{x \in X}[\mu_A(x)] = 1 \tag{6}$$

2.2. Fuzzy Levels

Range between the minimum (Min) and maximum (Max) value of any fuzzy variable is divided into suitable numbers which are denoted in ascending order starting from the minimum (Min) to maximum (Max) value of a fuzzy set. **Figure 1** shows the range and the fuzzy levels for a fuzzy set of objects, in a triangular functional diagram. Here the range has been divided into five fuzzy levels which are NL, NS, ZE, PS, and PL. A fuzzy inference model consists of 3 modules. **Figure 2** shows a schematic diagram of steps involved in fuzzy rule based system. Definitions and methods of calculations are presented below.

2.3. Fuzzification

As per Lee [12], fuzzification is a process which involves the following:

1) measures the values of input variables,

2) performs a scale mapping that transfers the range of values of input variables into a corresponding universe of discourse,

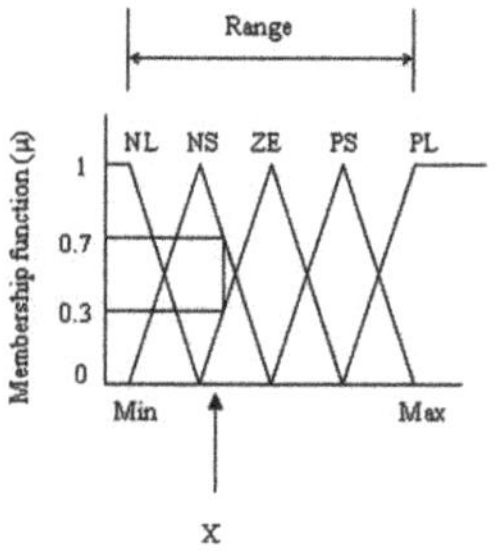

Abbreviation	Stands for	Meaning
NL	Negative large	Very low
NS	Negative small	Low
ZE	Zero	Normal
PS	Positive small	High
PL	Positive large	Very high

Figure 1. Triangular functional diagram and method for calculating membership function (μ) and corresponding fuzzy levels.

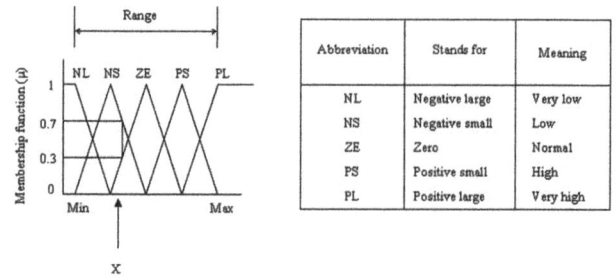

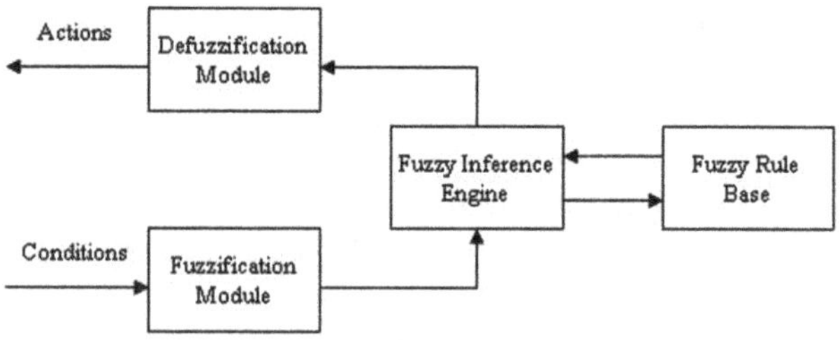

Figure 2. General scheme of a fuzzy sastem.

3) performs the function that converts input data into suitable linguistic values which may be viewed as labels of fuzzy sets.

Figure 1 shows a value of a fuzzy variable x intersecting the triangles with fuzzy levels of ZE and NS and their respective membership functions (μ) of 0.3 and 0.7. Hence, Fuzzification is the process that involves:

1) inputting the value of the fuzzy variable in the universe of discourse,

2) obtaining the intersecting points on the arms of the triangles to calculate the fuzzy levels, and

3) obtaining the corresponding membership functions (μ).

2.4. Min-Max Composition

From **Figure 1**, it is observed that one fuzzy variable (x) yields two membership functions (0.3 and 0.7) and their respective fuzzy levels are NS and ZE. Hence, if there are two fuzzy variables in the antecedent part, an increase in wind speed and a decrease in temperature when compared between the ith day and (i − 1)th day, hereafter denoted as *WS* and *TP*, respectively as in Equations (7) and (8) below, there will be four membership functions and four respective fuzzy levels obtained after fuzzification. The mathematical method followed by fuzzification is termed as "min-max composition". Considering *WS* and *TP* as the two fuzzy variable inputs and rainfall, hereafter denoted as *RF* as the output in each of the production rules:

IF *WS* is *strong* and *TP* is *lower*
THEN *RF* is *moderate* (7)

IF *WS* is *strong* and *TP* is *moderate*
THEN *RF* is *moderate* (8)

Fuzzifying any of these production rules will yield fuzzy levels and membership functions as shown in **Figure 3**. Here the values of *WS* and *TP* are the two fuzzy variables representing the antecedent part of a production rule yielding *RF* as its consequence that is shown inside **Figure 3**. Suppose a value of *WS* yields the membership function values of 0.2 and 0.8 belonging to the fuzzy levels of ZE and PS, respectively. Similarly, *TP*, another fuzzy variable in the antecedent part, yields membership functions values of 0.3 and 0.7 for the fuzzy levels of ZE and NS, respectively. Inferring fuzzy level for *RF* is NS which is shown in the production rule table of **Figure 3**. The following equation holds good:

IF *WS* is ZE and *TP* is NS THEN *RF* is NS (9)

Figure 3 shows that a value of membership function μ for *WS* equals 0.2 with its fuzzy level of ZE. This figure further shows another value of membership function μ for *TP* equals 0.7 with its fuzzy level of NS. A value of membership function for *RF* is taken to be 0.2 as it is the minimum value of μ between 0.2 and 0.7. A similar mathematical approach for the same fuzzy variables of ZE for 3 other production rules inside the table are presented for *RF* in **Figure 3**. Hence, the three minimum values 0.7, 0.2, and 0.3 for the same fuzzy levels of ZE are obtained. Finally, the maximum value 0.7 is taken out of the three minimum values of 0.7, 0.2, and 0.3 for the next step of the calculation process for defuzzification. Let us give an example to show the generalized form of the equation for min-max composition. Considering two equations for the four production rules presented in the table written here as follows:

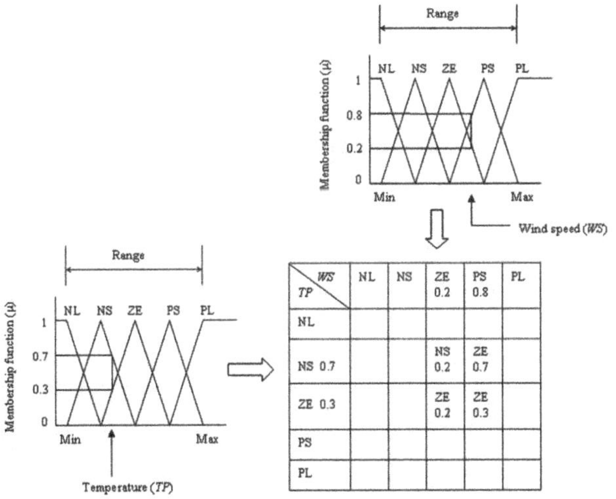

Figure 3. Triangular functional diagram and method for calculating membership functions (μ) and corresponding fuzzy levels.

IF *WS* is LW_1 and *TP* is LT_1 then *RF* is ZE (10)

IF *WS* is LW_2 and *TP* is LT_2 then *RF* is ZE (11)

Equations (10) and (11) have the same fuzzy levels of ZE for *RF*. Hence, the general form of the equation for calculating the membership function $\mu^{(ZE)}(RF)$ having the same fuzzy levels ZE for the consequent part can be shown as:

$$\mu^{(ZE)}(RF)=\bigcup_{i=1}^{3}\left[\mu^{(LW_{ii})}(WS)\cap\mu^{(TP_{ii})}(LT)\right] \quad (12)$$

Here, $\mu^{(ZE)}(RF)$ is the membership function for *RF* for fuzzy level ZE, *LW* is the fuzzy level for *WS*, *LT* is the fuzzy level for *TP*, and $\cap$ indicates selecting the minimum value of membership function out of $\mu^{(LWi)}(WS)$ and $\mu^{(TP_i)}(LT)$. $\cup$ indicates selecting the maximum value of the calculated minimum membership function values. *i* is the number of production rules having the same fuzzy levels (here it is ZE). Equation (12) is valid only when $i > 1$.

If the fuzzy levels of *RF* are not the same, then the membership functions of *RF* can be calculated by the following equation:

$$\mu^{(LV)}(RF)=\mu^{(LW_i)}(WS)\cap\mu^{(TP_i)}(LT) \quad (13)$$

Here, *LV*, the abbreviation for fuzzy level for *RF*, is different for various production rules. In these cases, only the minimum value of the membership functions between $\mu^{(LWi)}(WS)$ and $\mu^{(TP_i)}(LT)$ is considered.

2.5. Defuzzification

Defuzzification is the calculation method to yield the quantified value for the consequent part of a fuzzy statement described by production rule. Defuzzification performs the following functions:

1) a scale mapping which converts the range of values of output variables into corresponding universe of discourse, and

2) yields a non-fuzzy control action from an inferred fuzzy control action.

Figure 4 illustrates the mathematical procedure followed to calculate the center of gravity for the defuzzification method. The following are the possible cases:

Case 1. fuzzy levels of the inference part of production rules belong to NL and NS with their corresponding values of membership functions (μ),

Case 2. fuzzy levels of the inference part of production rules are in the region from NS to PS range with their corresponding values of membership functions (μ), and

Case 3. fuzzy levels of the inference part of production rules belong to PS and PL with their corresponding values of membership functions (μ),

Considering **Figure 4(a)** as a description of the mathematical procedure for calculating center of gravity for

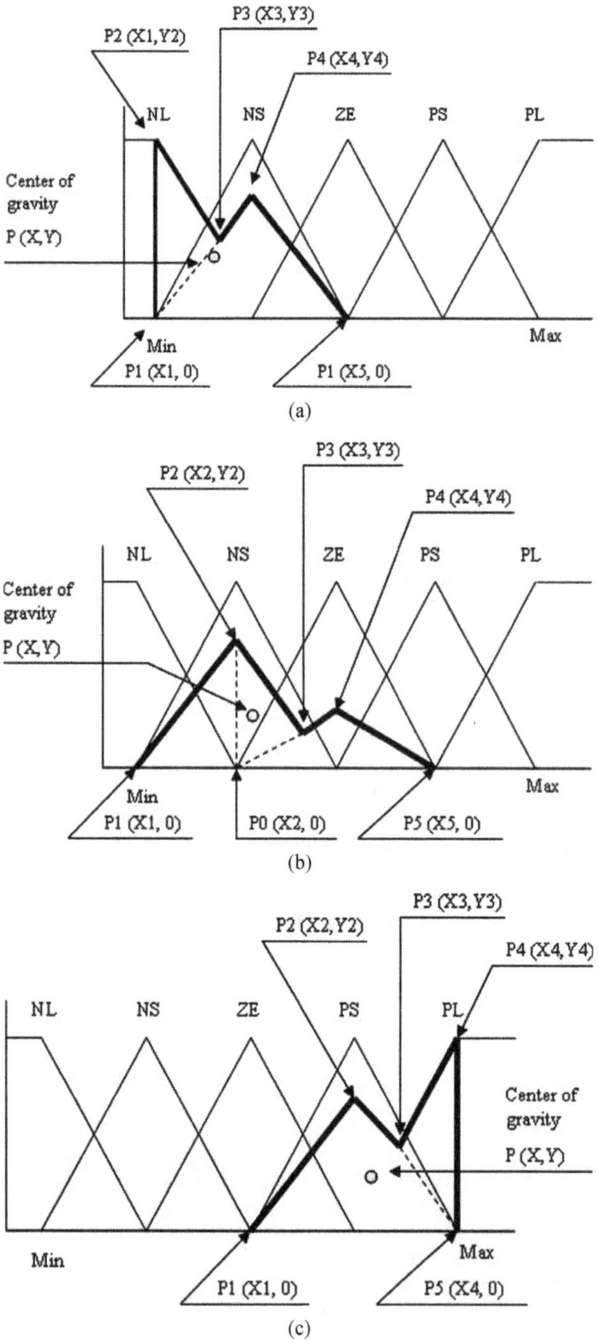

Figure 4. (a) Calculation method for defuzzification if the fuzzy levels for inference part of the rule table belong to NL and NS; (b) Calculation method for defuzzification if the fuzzy levels for inference part of the rule table are between NS to PS; (c) Calculation method for defuzzification if the fuzzy levels for inference part of the rule table belong to PS to PL.

Case 1, the point of intersections may be defined as $P1(X1,0)$, $P2(X1,Y2)$, $P3(X3,Y3)$, $P4(X4, Y4)$ and $P5(X5,0)$. Let the co-ordinate of center of gravity for the area bounded by the above five co-ordinates be $P(X,Y)$. There are two triangles, one which can be shown by the co-ordinates $P1(X1,0)$, $P2(X1, Y2)$ and $P3(X3,Y3)$; and the other triangle can be shown by the co-ordinates $P1(X1, 0)$, $P4(X4,Y4)$ and $P5(X5, 0)$.

Now, the average of X values in triangle formed by $P1(X1,0)$, $P2(X1,Y2)$, and $P3(X3,Y3)$ is

$$XC1 = (X1 + X1 + X3)/3.0 \quad (14)$$

and the Y value in the same triangle formed by $P1(X1, 0)$, $P2(X1,Y2)$, and $P3(X3,Y3)$ is

$$YC1 = (0 + Y2 + Y3)/3.0 \quad (15)$$

Similarly, X value in triangle formed by $P1(X1,0)$, $P4(X4,Y4)$ and $P5(X5,0)$ is

$$XC2 = (X1 + X4 + X5)/3.0 \quad (16)$$

and the Y value in the same triangle formed by $P1(X1, 0)$, $P4(X4,Y4)$ and $P5(X5,0)$ is

$$YC2 = (0 + Y4 + 0)/3.0 \quad (17)$$

Area formed by $P1(X1,0)$, $P2(X1,Y2)$, and $P3(X3, Y3)$ is

$$\text{Area 1} = \{(X1 - X1)*(Y3 - 0) - (X3 - X1)*(Y2 - 0)\}/2.0 \quad (18)$$

Similarly, area formed by $P1(X1,0)$, $P4(X4, Y4)$and $P5(X5,0)$ is

$$\text{Area 2} = \{(X4 - X1)*(0 - 0) - (X5 - X1)*(Y4 - 0)\}/2.0 \quad (19)$$

$$\text{Therefore, Total area} = \text{area 1} + \text{area 2} \quad (20)$$

and the area covered by $P1(X1,0)$, $P2(X1,Y2)$, and $P3(X3,Y3)$ is

$$\text{Fraction 1} = \text{area 1}/(\text{area 1} + \text{area 2}) = \text{area 1}/\text{total area} \quad (21)$$

and the area covered by $P1(X1,0)$, $P4(X4,Y4)$ and $P5(X5,0)$ is

$$\text{Fraction 2} = \text{area 2}/(\text{area 1} + \text{area 2}) = \text{area 2}/\text{total area} \quad (22)$$

Therefore, the co-ordinate for center of gravity is

$$X = (XC1 * \text{Fraction 1} + XC2 * \text{Fraction 2}) \quad (23)$$

$$\text{and } Y = (YC1 * \text{Fraction 1} + YC2 * \text{Fraction 2}) \quad (24)$$

Considering **Figure 4(b)** as describing the mathematical procedure for calculating the center of gravity for Case 2, the point of intersections may be defined as $P0(X2,Y0)$, $P1(X1,0)$, $P2(X2,Y2)$, $P3(X3,Y3)$, $P4(X4,Y4)$ and $P5(X5,0)$. Let the co-ordinate of the center of gravity of the area bounded by the above five co-ordinates represented by thick lines be $P(X,Y)$. Let us consider that total area under the thick lines is consisting of three small triangles which are as follows:

triangle 1 which is formed by the co-ordinates ($P0$, $P1$, and $P2$),

triangle 2 which is formed by the co-ordinates ($P0$, $P2$, and $P3$), and

triangle 3 which is formed by the co-ordinates ($P0$, $P4$, and $P5$).

The co-ordinates for triangle 1 are $P0(X2, 0)$, $P1(X1, 0)$, and $P2(X2,Y2)$; triangle 2 consisting of $P0(X2, 0)$, $P2(X2,Y2)$ and $P3(X3,Y3)$; and triangle 3 shown with the co-ordinates $P0(X2,0)$, $P4(X4,Y4)$ and $P5(X5,0)$.

Now, the average of X value in triangle formed by $P0(X2,0)$, $P1(X1,0)$, and $P2(X2,Y2)$ is

$$XC1 = (X2 + X1 + X2)/3.0 \tag{25}$$

and the Y value in the same triangle formed by $P0(X2, 0)$, $P1(X1,0)$, and $P2(X2,Y2)$ is

$$YC1 = (0 + 0 + Y2)/3.0 \tag{26}$$

Similarly, X value in triangle formed by $P0(X2,0)$, $P2(X2,Y2)$ and $P3(X3,Y3)$ is

$$XC2 = (X2 + X2 + X3)/3.0 \tag{27}$$

and the Y value in the same triangle formed by $P0(X2, 0)$, $P2(X2,Y2)$ and $P3(X3,Y3)$ is

$$YC2 = (0 + Y2 + Y3)/3.0 \tag{28}$$

Similarly, X value in triangle formed by $P0(X2,0)$, $P4(X4,Y4)$ and $P5(X5,0)$ is

$$XC2 = (X2 + X4 + X5)/3.0 \tag{29}$$

and the Y value in the same triangle formed by $P0(X2, 0)$, $P4(X4,Y4)$ and $P5(X5,0)$ is

$$YC2 = (0 + Y4 + 0)/3.0 \tag{30}$$

Area formed by $P0(X2,0)$, $P1(X1,0)$, and $P0(X2, Y2)$ is

$$\text{Area 1} = \{(X1 - X2)*(Y2 - 0) - (X2 - X2)*(0 - 0)\}/2.0 \tag{31}$$

Similarly, area formed by $P0(X2,0)$, $P2(X2,Y2)$ and $P3(X3,Y3)$ is

$$\text{Area 2} = \{(X2 - X2)*(Y3 - 0) - (X3 - X2)*(Y2 - 0)\}/2.0 \tag{32}$$

Similarly, area formed by $P0(X2,0)$, $P4(X4,Y4)$ and $P5(X5,0)$ is

$$\text{Area 3} = \{(X4 - X2)*(0 - 0) - (X5 - X2)*(Y4 - 0)\}/2.0 \tag{33}$$

Hence,

$$\text{Total area} = \text{area } 1 + \text{area } 2 + \text{area } 3 \tag{34}$$

area covered by $P0(X2,0)$, $P1(X1,0)$, and $P2(X2, Y2)$ is

$$\text{Fraction } 1 = \text{area } 1/\text{total area} \tag{35}$$

area covered by $P0(X2,0)$, $P2(X2,Y2)$ and $P3(X3, Y3)$ is

$$\text{Fraction } 2 = \text{area } 2/\text{total area} \tag{36}$$

and area covered by $P0(X2,0)$, $P4(X4, Y4)$ and $P5(X5, 0)$ is

$$\text{Fraction } 3 = \text{area } 3/\text{total area} \tag{37}$$

Therefore, the co-ordinate for center of gravity is

$$X = (XC1 * \text{Fraction } 1 + XC2 * \text{Fraction } 2 + XC3 * \text{Fraction } 3) \tag{38}$$

and

$$Y = (YC1 * \text{Fraction } 1 + YC2 * \text{Fraction } 2 + YC3 * \text{Fraction } 3) \tag{39}$$

Considering **Figure 4(c)** for describing the mathematical procedure for calculating the center of gravity for Case 3, the point of intersections may be defined as $P1(X1,0)$, $P2(X2,Y2)$, $P3(X3,Y3)$, $P4(X4, Y4)$ and $P5(X4,0)$. Let the co-ordinate of the center of gravity of the area bounded by the above five co-ordinates represented by thick lines be $P(X,Y)$. There are two triangles one of which can be shown by the co-ordinates $P1(X1,0)$, $P2(X2,Y2)$, and $P5(X4, 0)$and the other triangle can be shown with the co-ordinates $P3(X3, Y3)$, $P4(X4, Y4)$ and $P5(X4,0)$.

Now, the average of X value in triangle formed by $P1(X1,0)$, $P2(X2,Y2)$, and $P5(X4,0)$ is

$$XC1 = (X1 + X2 + X4)/3.0 \tag{40}$$

and the Y value in the same triangle formed by $P1(X1, 0)$, $P2(X2,Y2)$, and $P5(X4,0)$ is

$$YC1 = (0 + Y2 + 0)/3.0 \tag{41}$$

Similarly, X value in triangle formed by $P3(X3,Y3)$, $P4(X4,Y4)$ and $P5(X4,0)$ is

$$XC2 = (X3 + X4 + X4)/3.0 \tag{42}$$

and the Y value in the same triangle formed by $P3(X3, Y3)$, $P4(X4,Y4)$ and $P5(X4,0)$ is

$$YC2 = (Y3 + Y4 + 0)/3.0 \tag{43}$$

Area formed by $P1(X1,0)$, $P2(X2,Y2)$ and $P5(X4, 0)$ is

$$\text{Area 1} = \{(X1 - X4)*(Y2 - 0) - (X2 - X4)*(0 - 0)\}/2.0 \tag{44}$$

Similarly, area formed by $P3(X3,Y3)$, $P4(X4, Y4)$ and $P5(X4,0)$ is

$$\text{Area 2} = \{(X3-X4)*(Y4-0)-(X4-X4)*(Y3-0)\}/2.0 \quad (45)$$

$$\text{Therefore, Total area} = \text{area 1} + \text{area 2} \quad (46)$$

and the area covered by $P1(X1,0)$, $P2(X2,Y2)$ and $P5(X4,0)$ is

$$\text{Fraction 1} = \text{area 1}/(\text{area 1}+\text{area 2}) = \text{area 1}/\text{total area} \quad (47)$$

and the area covered by $P3(X3,Y3)$, $P4(X4,Y4)$ and $P5(X4,0)$ is

$$\text{Fraction 2} = \text{area 2}/(\text{area 1}+\text{area 2}) = \text{area 2}/\text{total area} \quad (48)$$

Therefore, the co-ordinate for center of gravity is

$$X = (XC1 * \text{Fraction 1} + XC2 * \text{Fraction 2}) \quad (49)$$

$$\text{and } Y = (YC1 * \text{Fraction 1} + YC2 * \text{Fraction 2}) \quad (50)$$

3. Study Area

This manuscript presents a fuzzy inference model for predicting *RF* using meteorological scan data from the United States Department of Agriculture (USDA) Soil Climate Analysis Network Station at Alabama Agricultural and Mechanical University (AAMU) campus. Meteorological data for 2003 and 2004 were collected and analyzed to determine the variables that are involved in rainfall occurrences. The Alabama Mesonet (ALMNet) has been the apex representing fourteen combinations of meteorological/soil profile stations and twelve soil profile stations distributed in 11 counties in southern Tennessee and north and central Alabama. The combination stations are also part of the USDA and Natural Resources Conservation Service (NRCS) scan network. Alabama Mesonet (ALMNet) is controlled and run by the Center for Hydrology, Soil Climatology and Remote Sensing (HSCaRS) of Alabama Agricultural and Mechanical University (AAMU).

4. Model Development

Although meteorological scan data were collected for two years, 2003 and 2004, the model was developed based on year 2004 data. These data were very well organized including soil related parameters. Data for Bragg Farm and Winford A. Thomas Agricultural Research Station (WTARS) were also collected, monthly data spread sheets were prepared, and graphs plotted to assist with pre-assessment of analysis and to generate ideas on climatic behavior.

Figure 5 shows the characteristics of rainfall for the month of August 2004 using data from the AAMU campus. Based on the observations of the graphs prepared for

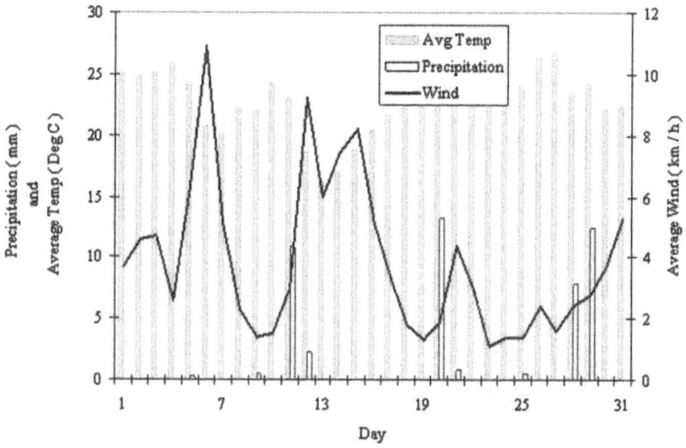

Figure 5. Rainfall pattern, wind and temperature (USDA scan data from AAMU campus for August 2004).

every month during the years 2003 and 2004 for AAMU, Bragg and WTARS farms, it was apparent that a value of *WS* and another value of *TP* when compared between the ith and $(i-1)$th day mostly resulted in a rainfall occurrence. Usually, the characteristic of rainfall occurrence usually takes place at the first or second day of the phenomena of increasing of wind speed and decreasing of temperature. Hence, the degree of association between *WS* and *TP* when compared between the ith and $(i-1)$th day causing *RF* occurrences was established. Based on analysis, it was observed that a *RF* occurrence has a positive relation with *TP* and *WS*. The observation further revealed that the relation of *RF* occurrence with *TP* and *WS* reflects expert knowledge. Hence, the values of *WS* and *TP* between ith and $(i-1)$th day and using them in the fuzzy inference model for the antecedent part of production rules was considered to be feasible. **Figure 6** shows the fuzzy inference model structure and the steps followed to determine the time and amount of *RF*.

This figure has been prepared by incorporating the consideration of threshold values as described in **Figure 7**. In the initial step of calculation, temperature, wind speed, and Relative Humidity were converted to yield the average daily values dividing by 24 (1 day = 24 h) to produce average temperature, average wind speed and average Relative Humidity.

A preliminary analysis showed that the variables described below had a significant influence over *RF* occurrences:

1) Relative Humidity (RH) of the ith day,

2) Humidity Increase (HI) is which is increase in Relative Humidity (RH) when compared between the ith and $(i-1)$th day, and

3) Product (*P*) of decreasing of *TP* and increasing of *WS*.

These three variables were taken into consideration and shown in **Figure 6** in the calculation process with seasonal variation as shown in tabular form in **Figure 7**. This variation was considered with two threshold values

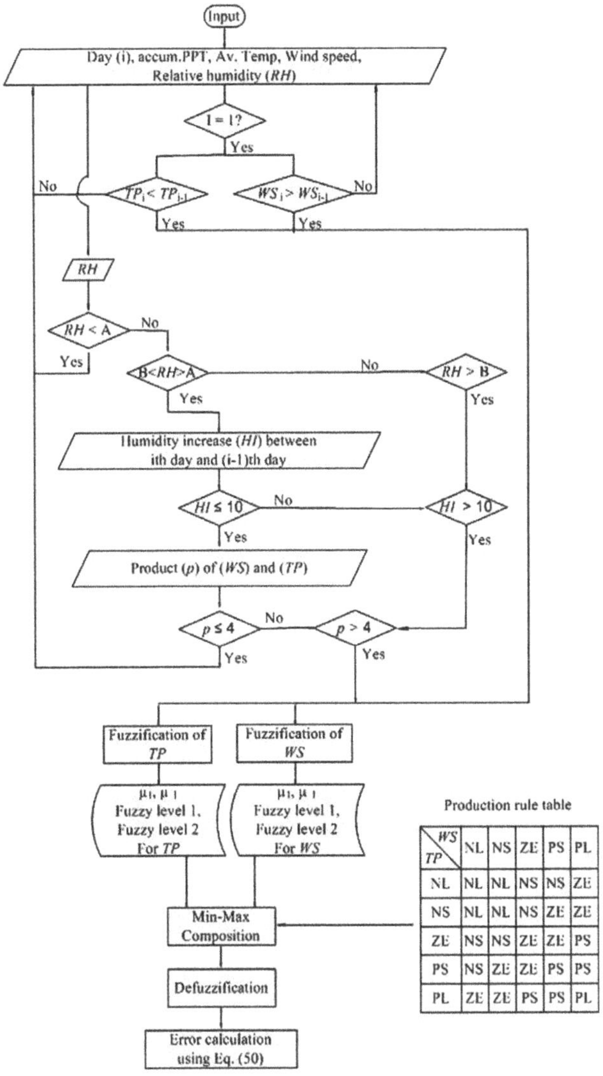

Figure 6. Model structure and steps in predicting timing and amount of Rainfall (*RF*).

for minimum and maximum limits as indicated by A and B in **Figure 7**:

1) Jan 1 to Apr 31
2) May 1 to Sep 30, and
3) Oct 1 to Dec 31

The threshold values were selected based on the calculation of results of the model.

In year 2004, there were 6 days out of 132 total rainy days when the actual amount of *RF* was more than 50 mm. Considering the uniformity of data range, and to avoid very unusual phenomena, the highest volume of *RF* was considered to be 50 mm for the maximum value of predicted *RF* for defuzzification process (refer **Figure 4**). The error was calculated using the following equation:

$$\frac{1}{n}\sum_{i=1}^{n} abs\left[\left(RF_{a_i} - RF_{c_i}\right)/RF_{a_i}\right]*100 \quad (51)$$

Here, n is the number of days of rainfall occurrences, RF_{a_i} is actual amount of rainfall, and RF_{c_i} is the calculated amount of rainfall.

5. Results and Discussions for Fuzzy Set Theory

5.1. Selection of Variables

Fuzzy variables of *WS* and *TP* between the ith and $(i - 1)$th day were good choices for the development of a model for predicting *RF*. In reality, fuzzy inference models involve with variables which are perceived by experts as responsible for the consequence part of the production rule. This means a fuzzy inference model reflects the scenario of thinking and decision-making process by expert knowledge. The fuzzy variables were chosen following the assessment on graphs prepared on the basis of monthly data from AAMU for 2003 and 2004. Selection of variables of *TP* and *WS* between the ith and $(i - 1)$th day was considered for this model as a better approach. The final results indicated that the selection of these two variables was suitable for the development of the model and that they showed a good agreement when used in the antecedent part of the production rule.

5.2. Selection of Fuzzy Levels for the Inference Part of the Production Rule Table

Selection of the fuzzy levels in the inference part of the production rules is a cumbersome process by trial and error method. Twenty five (5 × 5) fuzzy variables for the inference part were shown in the table in **Figure 3**. A method for iterating the fuzzy variables for *RF* was followed in the computer program that selected the one yielding the lowest percentage of error based on Equation (51). Depending on the scenario of the system, fuzzy levels in the inference part of the production rule must have either an ascending or a descending nature. Skill and logical approachability are required for determining fuzzy variables for the consequent part with respect to the fuzzy variables in the antecedent part of the production rule. The production rule table shown in **Figure 6** was the best set of fuzzy levels for *RF* that yielded the lowest error value of 12.35%.

5.3. Maximum Value of *RF*

Real *RF* data showed that the AAMU campus USDA Soil Climate Analysis Network weather station had only 6 occurrences of more than 50 mm *RF* in 2004. The maximum *RF* was 93 mm which is very unusual and rare for the same location. Moreover, if the actual amount of *RF* is considered to be more than 50 mm, the region of maximum *RF* [around PL of **Figure 4(c)**] will have unealistic and lesser density of number of data compared to

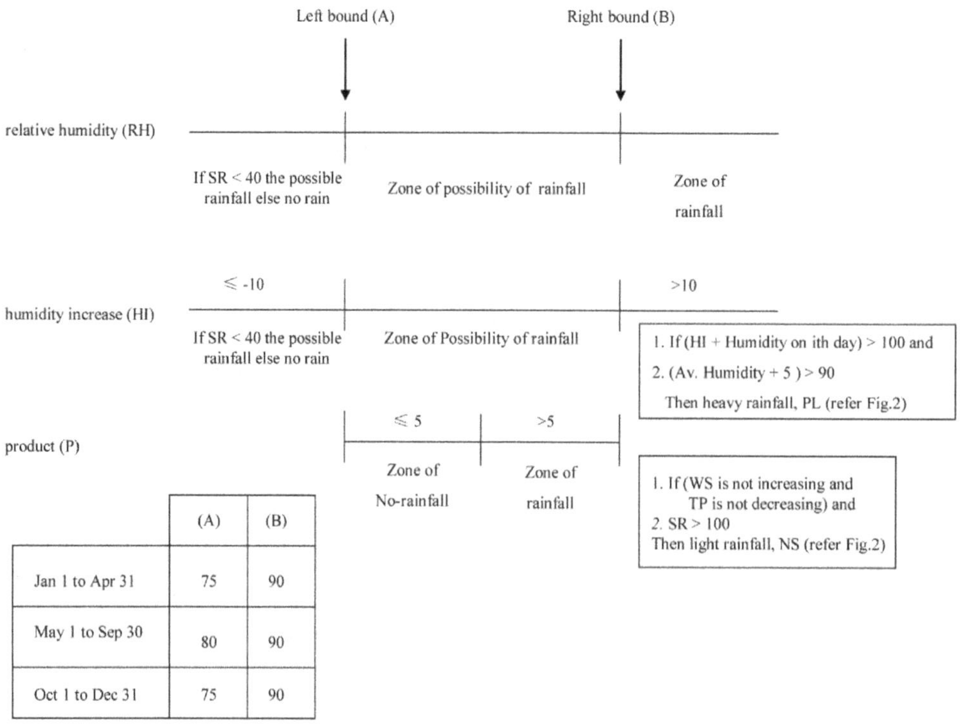

Figure 7. Threshold values and ranges of the factors for predicting RF for the improved model.

density of data in the region of NL, NS, ZE, and PS. Hence, considering the uniformity of data distribution among the ranges of NL, NS, ZE, PS, and PL the maximum value of predicted *RF* to be 50 mm was justifiable.

5.4. Selection of Threshold Values for Predicted Value of RF

Based on the fundamental logic of this research that a value of *WS* and another value of *TP* when compared between the ith and $(i-1)$th day may result in *RF*, their fuzzy levels, production rules, and ranges of variables, showed dependency on three other possible factors. These factors need to be considered with their threshold values for matching the actual and calculated amount of *RF*. These factors are 1) average daily Relative Humidity (RH); 2) Humidity Increase (HI) between the ith and $(i-1)$th day; and 3) Product (*P*) of *TP* and *WS* between the ith and $(i-1)$th day. **Figure 7** represents two boundary values (A) and (B) for RH. The zone between (A) and (B) is the range for a possible *RF* and the zone beyond (B) is the zone for *RF* regardless of any other consideration, whereas the RH of less than (A) is the zone for no *RF*. When the value of HI is more than 10 and it is within the boundary values of (A) and (B) then it becomes the zone for *RF*. The zone for the value of HI of less than 10 is again the zone for a possible *RF* occurrence. This possibility is further considered to occur when the value of product (*P*) of *TP* and *WS* is greater than 4. But if the value of *P* of *TP* and *WS* is less than 4, then it was considered that there would be no *RF* occurrence. Introducing these three threshold values with consideration of three different seasons as described in **Figure** 7, the model showed good agreement between the actual amount of *RF* and predicted value for *RF*. **Figures 8** to **10** show the actual and predicted values of *RF* using 2004 scan data from the USDA Soil Climate Analysis Network Station at the AAMU campus. These figures illustrate the actual amount of *RF* and predicted value of *RF* during three different seasons as considered in this model and explained in **Figure 7**. The figures further show that the timeliness of the actual amount of *RF* and predicted value of *RF* almost perfectly match, but the amount of *RF* needs further research to yield better agreement between actual and predicted values of *RF*. Therefore, further research planned to develop an approach for auto-generation of the production rules by iteration method and selecting the particular production rule table that yields the lowest percentage of error.

6. Methodology for Reliability Analysis

It consists of following steps:

Step 1- Calculate mean value $(\bar{x})$ and standard deviation (σ_x) for each of the parameters affecting rainfall, namely, relative humidity (RH) and humidity increase (HI) from the following equations:

$$\bar{x} = \frac{\sum x_i}{n} \tag{52}$$

Where, n = number of samples

$$\sigma_x = \sqrt{\frac{\sum(x_i - \bar{x})^2}{n-1}} \quad (53)$$

This step is done for all the sets of data from various sources at various locations-AAMU 2004, BRAGG 2004, WATARS 2004, AAMU 2005, BRAGG 2005 and WATARS 2005.

Step 2- Calculate the following theoretical probabilities for RH (assuming normal distribution) from information in step 1 as follows:

$P_{th1} = P(70 < X < 85)$ for January 1 to April 30

$P_{th2} = P(80 < X < 90)$ for May 1 to September30

$P_{th3} = P(75 < X < 90)$ for October 1 to December 31

X represents the random variable (Relative Humidity).

Where,

P_{th1} = theoretical probability for Event 1 for RH

P_{th2} = theoretical probability for Event 2 for RH

P_{th3} = theoretical probability for Event 3 for RH

Step 3- Count number of samples falling in the range for each event (1-3) and calculate the corresponding experimental probabilities as follows:

$$P_{e1} = n_1 / n_{teL}$$

$$P_{e2} = n_2 / n_{te2}$$

$$P_{e3} = n_3 / n_{te3}$$

Where,

n_1 = number of samples within the range of 70 - 85 (between January 1-April 30)

n_{te1} = total number of samples for event 1 (January 1-April 30)

n_2 = number of samples within the range of 80 - 90 (between May 1-Sept. 30)

n_{te2} = total number of samples for event 2 (May 1-Sept. 30)

n_3 = number of samples within the range of 75 - 90 (between October 1-December 31)

n_{te3} = total number of samples for event 3 = 115 (between October 1-December 31)

Step 4- Calculate the following theoretical probabilities for HI (assuming normal distribution) from information in step 1 as follows:

X represents the random variable (HI - Humidity increase)

$P_{th4} = P(X <= 10)$ for January 1 to April 30

$P_{th5} = P(X <= 10)$ for May 1 to September30

$P_{th6} = P(X <= 10)$ for October 1 to December 31

Where,

P_{th4} = theoretical probability for Event 1 - for HI

P_{th5} = theoretical probability for Event 2 for HI

P_{th6} = theoretical probability for Event 3 for HI

Step 5- Count number of samples falling in the range for each event (1-3) and calculate the corresponding experimental probabilities as follows:

$$P_{e4} = n_4 / n_{te4}$$

$$P_{e5} = n_5 / n_{te5}$$

$$P_{e6} = n_6 / n_{te6}$$

Where,

n_4 = number of samples of HI <= 10 (between January 1-April 30)

n_{te4} = total number of samples for HI for event 1 (January 1-April 30)

n_5 = number of samples of HI <= 10 (between May 1-Sept. 30)

n_{te5} = total number of samples for HI event 2 (May 1-Sept. 30)

n_6 = number of samples <= 10 (between October 1-December 31)

7. Results for Reliability Analysis

Results based on the reliability analysis are given in **Tables 1** and **2**. These tables lists all the statistical parameters for the two random variables connected with rainfall data ("RH" and "HI").

8. Discussion and Results

Tables 1 and **2** give statistical parameters (sample mean value, sample standard deviation and sample coefficient of variation (CV)) for relative humidity (RH) and humidity increase (HI). It is seen from these tables that the value of CV for the HI data is quite high (more than 1). CV is supposed to be less than 1 for good quality control.

Tables 3 and **4** give probabilities for RH and HI for various periods. The reason for considering these two parameters is because these are found to effect the rainfall to the maximum as discussed in this paper. **Tables 3** and **4** also show the comparison between theoretical and experimental probabilities for these two variables. It can be seen from these tables that they compare reasonably well. Another point to be noted is that all the experimenttal probabilities fall within 1 standard deviation (σ) from the mean value *i.e.* ($\mu - \sigma$ and $\mu + \sigma$) which represent about 63% of the uncertainty which reflects well on the data that is collected and theoretical analysis performed.

To calculate the probability of rainfall one can multiply the probabilities of the events for a particular period for RH and the same period for HI. For example for the period of January 1-April 30 (for RH values in the range of 70 - 80), the probability of rainfall is about 20%. This This number is calculated by multiplying the two probabilities considering they are independent events. Similarly, probabilities can be calculated for other ranges of RH and HI.

Table 1. Statistical Parameters for random variable (RH) connected with rainfall.

Source of Data	Name of Variable	Events	Statistical parameters		
			Mean $(\bar{x})$	Standard deviation (σ_x)	Coefficient of Variation (CV)
AAMU 2004	RH	P (70 < RH < 85)	67.89	15.55	0.23
		P (80 < RH < 90)	80.60	6.73	0.08
		P (75 < RH < 90)	78.76	10.92	0.14
BRAGG 2004	RH	P (70 < RH < 85)	66.53	14.81	0.22
		P (80 < RH < 90)	77.92	7.17	0.09
		P (75 < RH < 90)	76.13	11.57	0.15
WATARS 2004	RH	P (70 < RH < 85)	70.70	13.76	0.19
		P (80 < RH < 90)	81.25	6.52	0.08
		P (75 < RH < 90)	79.86	11.07	0.14
AAMU 2005	RH	P (70 < RH < 85)	70.77	13.96	0.20
		P (80 < RH < 90)	78.41	7.81	0.10
		P (75 < RH < 90)	73.53	11.23	0.15
BRAGG 2005	RH	P (70 < RH < 85)	69.42	13.17	0.19
		P (80 < RH < 90)	74.60	8.28	0.11
		P (75 < RH < 90)	71.07	10.82	0.15
WATARS 2005	RH	P (70 < RH < 85)	73.06	12.50	0.17
		P (80 < RH < 90)	78.36	7.73	0.10
		P (75 < RH < 90)	75.35	10.59	0.14

Table 2. Statistical Parameters for random variable (HI) connected with rainfall.

Source of Data	Name of Variable	Periods/Events	Statistical parameters		
			Mean $(\bar{x})$	Standard deviation (σ_x)	Coefficient of Variation (CV)
AAMU 2004	HI	January 1-April 30	6.15	9.83	1.60
		May 1-September 30	2.62	4.04	1.54
		Oct.1-December 31	3.95	6.78	1.72
BRAGG 2004	HI	January 1-April 30	5.86	9.24	1.58
		May 1-September 30	2.55	4.27	1.67
		Oct.1-December 31	4.12	6.95	1.69
WATARS 2004	HI	January 1-April 30	5.66	8.87	1.57
		May 1-September 30	2.44	4.09	1.68
		Oct.1-December 31	4.01	6.53	1.63
AAMU 2005	HI	January 1-April 30	5.82	9.79	1.68
		May 1-September 30	2.42	3.80	1.57
		Oct.1-December 31	4.15	7.02	1.69
BRAGG 2005	HI	January 1-April 30	5.48	9.34	1.70
		May 1-September 30	2.34	3.63	1.55
		Oct.1-December 31	4.03	6.66	1.65
WATARS 2005	HI	January 1-April 30	5.33	8.95	1.68
		May 1-September 30	2.24	3.63	1.62
		Oct.1-December 31	3.90	6.87	1.76

Table 3. Theoretical and Experimental Probabilities for Relative Humidity (RH) Connected with Rainfall.

Source of Data	Events	Theoretical Probabilities	Experimental Probabilities	Limit 1 Probability for $\mu-\sigma$	Limit 1 Probability for $\mu+\sigma$	Is experimental and theoretical probabilities fall between the limits?
AAMU 2004	P(70 < X < 85)	0.310	0.243	0.158	0.841	Yes
	P (80 < X < 90)	0.454	0.450	0.158	0.841	Yes
	P (75 < X < 90)	0.483	0.560	0.158	0.841	Yes
BRAGG 2004	P (70 < X < 85)	0.301	0.233	0.158	0.841	Yes
	P (80 < X < 90)	0.340	0.393	0.158	0.841	Yes
	P (75 < X < 90)	0.424	0.461	0.158	0.841	Yes
WATARS 2004	P (70 < X < 85)	0.371	0.248	0.158	0.841	Yes
	P (80 < X < 90)	0.486	0.520	0.158	0.841	Yes
	P (75 < X < 90)	0.490	0.562	0.158	0.841	Yes
AAMU 2005	P (70 < X < 85)	0.368	0.376	0.158	0.841	Yes
	P (80 < X < 90)	0.351	0.428	0.158	0.841	Yes
	P (75 < X < 90)	0.377	0.345	0.158	0.841	Yes
BRAGG 2005	P (70 < X < 85)	0.364	0.361	0.158	0.841	Yes
	P (80 < X < 90)	0.226	0.294	0.158	0.841	Yes
	P (75 < X < 90)	0.318	0.299	0.158	0.841	Yes
WATARS 2005	P (70 < X < 85)	0.427	0.404	0.158	0.841	Yes
	P (80 < X < 90)	0.350	0.432	0.158	0.841	Yes
	P (75 < X < 90)	0.430	0.398	0.158	0.841	Yes

Table 4. Theoretical and Experimental Probabilities for Humidity Increase (HI).

Source of Data	Events	Theoret-ical Probabi-lities	Experim-ental Probabili-ties	Limit 1 Probability for $\mu - \sigma$	Limit 1 Probability for $\mu + \sigma$	Is experimental and theoretical probabilities fall between the limits?
	January1-April 30	0.652	0.754	0.158	0.841	Yes
AAMU 2004	May 1-Sept. 30	0.966	0.941	0.158	0.841	No
	Oct.1-December 31	0.814	0.835	0.158	0.841	Yes
	January1-April 30	0.673	0.765	0.158	0.841	Yes
BRAGG 2004	May 1-Sept. 30	0.959	0.924	0.158	0.841	No
	Oct.1-December 31	0.801	0.875	0.158	0.841	No
	January1-April 30	0.688	0.759	0.158	0.841	Yes
WATARS 2004	May 1-Sept. 30	0.968	0.931	0.158	0.841	No
	Oct. 1-December 31	0.820	0.852	0.158	0.841	No
	January1-April 30	0.665	0.778	0.158	0.841	Yes
AAMU 2005	May 1-Sept. 30	0.977	0.944	0.158	0.841	No
	Oct. 1-December 31	0.798	0.848	0.158	0.841	No
	January1-April 30	0.686	0.776	0.158	0.841	Yes
BRAGG 2005	May 1-Sept. 30	0.982	0.957	0.158	0.841	No
	Oct. 1-December 31	0.815	0.837	0.158	0.841	Yes
	January1-April 30	0.699	0.778	0.158	0.841	Yes
WATARS 2005	May 1-Sept. 30	0.983	0.951	0.158	0.841	No
	Oct. 1-December 31	0.813	0.850	0.158	0.841	No

9. Conclusion

Selection of variables and the fundamental logic of the values *TP* and *WS* was an attempt to identify amount of *RF* and its time of occurrence as the consequent part of the fuzzy inference model. Introducing the idea of threshold values of a) RH of the *i*th day, b) HI when compared between the *i*th and ($i-1$)th day, and c) *P*, product of *WS* and *TP* appeared to be an appropriate attempt for the model to match the actual *RF* occurrences. Iteration of the fuzzy levels with logic both for antecedent and consequent parts was found to be efficient. Further research has been planned to attain the maximum possible matches of time and amount of *RF* between actual occurrences and the one predicted by the model. A methodology has been developed for reliability analysis to predict rainfall.

REFERENCES

[1] M. Hasan, T. Tsegaye, X. Shi, G. Schaefer and G. Taylor, "Model for Predicting Rainfall by Fuzzy Set Theory Using USDA-SCAN Data," *Agricultural Water Management*, Vol. 95, No. 12, 2008, pp. 1350-1360.

[2] M. Hasan, M. Mizutani, A. Goto and H. Matsui, "A Model for Determination of Intake Flow Size-Development of Optimum Operational Method for Irrigation Using Fuzzy Set Theory (1). System Nogaku," *Journal of Japan Agricultural System Society*, Vol. 11, No. 1, 1995, pp. 1-13.

[3] D. S. Wilks, "Multisite Generalization of a Daily Stochastic Precipitation Generating Models," *Journal of Hydrology*, Vol. 210, No. 1-4, 1998, pp. 178-191.

[4] L. A. Carrano, B. J. Taylor, E. Y. Robert, L. L. Richard and E. S. Daniel, "Fuzzy Knowledge-Based Modeling and Regression in Abrasive Wood Machining," *Forest Products Journal*, Vol. 54, No. 5, 2004, pp. 66-72.

[5] T. M. Brown-Brandl, D. D. Jones and W. E. Woldt, "Evaluating Modeling Techniques for Livestock Heat Stress Prediction," Paper No. 034009, 2003 *ASAE Annual Meeting*, 2003. http://www.frymulti.com/abstract.asp?aid=14084&t=2

[6] K. W. Wong, P. M. Wong, T. D. Gedeon and C. C. Fung, "Rainfall Prediction Model Using Soft Computing Technique. Soft Computing-A Fusion of Foundations, Methodologies and Applications," Springer, Berlin/Heidelberg, 2003.

[7] S. Lee, S. Cho and P. M. Wong, "Rainfall Prediction Using Artificial Neural Networks," *Journal of Geographic Information and Decision Analysis*, Vol. 2, No. 2, 1998, pp. 233-242.

[8] L. M. Pant and G. Ashwagosh, "Fuzzy Rule-Based System for Prediction of Direct Action Avalanches," *Current Science*, Vol. 87, No. 1, 2004, pp. 99-104.

[9] S. Abe and L. Ming-Shong, "Fuzzy Rules Extraction Directly from Numerical Data for Function Approximation," *IEEE Transactions on System, Man and Cybernetics*, Vol. 25, No. 1, 1995, pp. 119-129.

[10] A. L. Zadeh, "Fuzzy Logic," *Information and Control*, Vol. 8, No. 3, 1965, pp. 338-353.

[11] T. Hasan and S. Zenkai, "A New Modeling Approach for Predicting the Maximum Daily Temperature from a Time Series," *Turkish Journal of Engineering and Environmental Science*, Vol. 23, No. 3, 1999, pp. 173-180.

[12] C. C. Lee, "Fuzzy Logic in Control Systems: Fuzzy Logic Controller-Part I," *IEEE Transaction of System, Man and Cybernetics*, Vol. 20, No. 2, 1990, pp. 404-418.

Some Remarks to Numerical Solutions of the Equations of Mathematical Physics

Ludmila Petrova
Department of Computational Mathematics and Cybernetics, Moscow State University, Moscow, Russia

ABSTRACT

The equations of mathematical physics, which describe some actual processes, are defined on manifolds (tangent, a companying or others) that are not integrable. The derivatives on such manifolds turn out to be inconsistent, *i.e.* they don't form a differential. Therefore, the solutions to equations obtained in numerical modelling the derivatives on such manifolds are not functions. They will depend on the commutator made up by noncommutative mixed derivatives, and this fact relates to inconsistence of derivatives. (As it will be shown, such solutions have a physical meaning). The exact solutions (functions) to the equations of mathematical physics are obtained only in the case when the integrable structures are realized. So called generalized solutions are solutions on integrable structures. They are functions (depend only on variables) but are defined only on integrable structure, and, hence, the derivatives of functions or the functions themselves have discontinuities in the direction normal to integrable structure. In numerical simulation of the derivatives of differential equations, one cannot obtain such generalized solutions by continuous way, since this is connected with going from initial nonintegrable manifold to integrable structures. In numerical solving the equations of mathematical physics, it is possible to obtain exact solutions to differential equations only with the help of additional methods. The analysis of the solutions to differential equations with the help of skew-symmetric forms [1,2] can give certain recommendations for numerical solving the differential equations.

Keywords: Two Systems of Reference; Nonintegrable Manifolds and Integrable Structures; Solutions of Two Types; Discrete Transitions; Observable Formations

1. Specific Features of Solutions to Equations Describing Actual Processes

Let us take the simplest case: the first-order partial differential equation

$$F\left(x^i, u, p_i\right) = 0, \quad p_i = \partial u / \partial x^i \tag{1}$$

The exact solution (the solution that depends only on variables, *i.e.* it is a function) can be obtained in the case when the derivatives obeying the equations made up the differential.

Let us construct the differential made up of derivatives that obey the differential equation

$$\mathrm{d}u = \theta \tag{2}$$

where $\theta = p_i \mathrm{d}x^i$ (the summation over repeated indices is implied). It should be noted that $\theta = p_i \mathrm{d}x^i$ is a skew-symmetric differential form of the first degree.

It appears that, in the general case when differential Equation (1) describes any physical processes, the form $\theta = p_i \mathrm{d}x^i$ made up of derivatives of differential equation is not a differential.

For the form $\theta = p_i \mathrm{d}x^i$ be a differential, its differential has to be equal to zero.

The differential $\mathrm{d}\theta$ of the form $\theta = p_i \mathrm{d}x^i$ can be written as $K_{ij} \mathrm{d}x^i \mathrm{d}x^j$, where

$$K_{ij} = \partial p_j / \partial x^i - \partial p_i / \partial x^j$$

are the components of the differential form commutator. From Equation (1) it does not follow (explicitly) that the derivatives $p_i = \partial u / \partial x^i$, which obey to the equation (and to given boundary or initial conditions), are consistent, that is, their mixed derivatives are commutative. The components of the commutator K_{ij} is nonzero. Therefore, the differential form commutator and the differential of the form θ are nonzero. This points to the fact that the differential expression made up of the derivatives of differential equation is not a differential. That is, the derivatives of differential equation do not made up a differential (without additional conditions).

This means that the solution u to Equation (1) obtained from such derivatives is not a function of variables x^i only. This solution will depend on the commutator K_{ij} with nonzero value related to inconsistence of derivatives.

To obtain the solution that is a function (*i.e.*, the derivatives of this solution made up a differential), it is necessary to add the closure condition (vanishing the form differential) for the form $\theta = p_i \mathrm{d}x^i$ and for the relevant dual form (in the present case the functional F plays a role of the form dual to θ):

$$\begin{cases} \mathrm{d}F\left(x^i, u, p_i\right) = 0, \\ \mathrm{d}\left(p_i \mathrm{d}x^i\right) = 0. \end{cases} \tag{3}$$

If to expand the differentials, one gets a set of homogeneous equations with respect to $\mathrm{d}x^i$ and $\mathrm{d}p_i$ (in the $2n$-dimensional space):

$$\begin{cases} \left(\partial F/\partial x^i + p_i \partial F/\partial u\right)\mathrm{d}x^i + \mathrm{d}p_i\, \partial F/\partial p_i = 0 \\ \mathrm{d}p_i \mathrm{d}x^i - \mathrm{d}x^i \mathrm{d}p_i = 0 \end{cases}$$

It is well known that *vanishing the determinant* composed of coefficients at $\mathrm{d}x^i$, $\mathrm{d}p_i$ is the solvability condition of the set of homogeneous differential equations. This leads to relations:

$$\frac{\mathrm{d}x^i}{\partial F/\partial p_i} = \frac{-\mathrm{d}p_i}{\partial F/\partial x^i + p_i \partial F/\partial u} \tag{4}$$

Relations (4) specify the integrating direction, which defines an integrable structure, that is, a pseudostructure (in its metric properties) on which the form $\theta = p_i \mathrm{d}x^i$ turns out to be closed one, *i.e.* this form becomes a differential.

On the pseudostructure, which is defined by relation (4), the derivatives of differential Equation (1) constitute the differential $\delta u = p_i \mathrm{d}x^i = \mathrm{d}u$ (on the pseudostructure), and this means that the solution to Equation (1) becomes a function. Solutions, namely, functions on the pseudostructures, are so-called generalized solutions. The characteristics, characteristic surfaces, singular points, potential surfaces, and others are examples of pseudostructures or their formations.

It should be underlined the following. The solutions, which are functions, are obtained only under additional condition that determines integrable structures. This additional condition, as one can see, is *vanishing the determinant*. Such an additional condition is a condition of degenerate transformation. The degenerate transformation executes the transition from tangent nonintegrable manifold of differential equations to integrable structures (pseudostructures). (The Legendre transformations are examples of such a transformation.) The realization of additional conditions (that can be caused by any degree of freedom) leads to the realization of integrable structure and the transition from the solutions, which are not functions, to generalized solutions (functions).

One can see that the generalized solutions cannot be obtained by numerical modelling the differential equation only on original tangent manifold.

The first-order partial differential equation has been analyzed. Similar functional properties have all differential equations describing actual processes. Below it will be shown that the sets of differential equations that describe actual processes possess such properties.

2. The Properties and Peculiarities of Solutions to the Equations of Mechanics and Physics of Continuous Medium

The equations of mechanics and the physics of continuous media (material systems such as gas-dinamical and cosmological systems, the systems of charged particles and others) are the equations that describe conservation laws for energy, linear momentum, angular momentum, and mass [3] (The set of Navier-Stokes equations is an example [3]). Such conservation laws can be referred to as the balance conservation laws since these laws establish a balance between variations of physical quantities and appropriate external action.

The equations of conservation laws are differential (or integral) equations that describe the variation of functions corresponding to physical quantities like the particle velocity (of elements), temperature or energy, pressure and density. Since these physical quantities relate to one material system, the connection between them has to exist. This connection is described by state functional that specifies the material system state. The action functional, entropy, the Pointing vector, Einstein's tensor, wave function and others can be regarded as examples of such functionals [4]. From the equations of conservation law, it follows the evolutionary relation for state functional, which enables one to disclose the properties and peculiarities of the solutions to the equations of mechanics and the physics of continuous media.

2.1. Evolutionary Relation

When studying the solutions to partial differential equations the conjugacy of derivatives with respect to various variables was analyzed. When describing physical processes in continuous media (in material systems) one obtains not one differential equation but a set of differential equations. And in this case it is necessary to investigate the conjugacy of not only derivatives but also the conjugacy (consistency) of the equations of this set.

The equations are consistent if they can be contracted into identical relations for the differentials, *i.e.* for closed forms.

Let us now analyze the consistency of the equations that describe the conservation laws for energy and linear momentum.

In the accompanying frame of reference, which is tied to the manifold built by the trajectories of particles (elements of material system), the equation for energy is written in the form (see example [5])

$$\frac{\partial \psi}{\partial \xi^1} = A_1 \tag{5}$$

Here ξ^1 are the coordinates along the trajectory, ψ is the functional of the state, A_1 is the quantity that depends on specific features of the material system and external (with respect to the local domain) energy actions onto the system.

Similarly, in the accompanying frame of reference, the equation for linear momentum appears to be reduced to the equation of the form

$$\frac{\partial \psi}{\partial \xi^\nu} = A_\nu, \quad \nu = 2, \cdots \tag{6}$$

where ξ^ν are the coordinates along the direction normal to the trajectory, A_ν are the quantities that depend on the specific features of material system and the external force actions.

Equations (5) and (6) can be convoluted into the relation

$$d\psi = A_\mu d\xi^\mu \quad (\mu = 1, \nu) \tag{7}$$

where $d\psi$ is the differential expression $d\psi = (\partial\psi/\partial\xi^\mu)d\xi^\mu$.

Relation (7) can be written as

$$d\psi = \omega \tag{8}$$

here $\omega = A_\mu d\xi^\mu$ is a skew-symmetric differential form of the first degree.

[In the case of the Euler and Navier-Stokes equations a concrete form of relation (8) and its properties were considered in papers [5]].

Since the equations of balance conservation laws are evolutionary ones, the relation obtained is also an evolutionary relation.

Relation (8) was obtained from the balance conservation law equations for energy and linear momentum. In this relation the form ω is that of the first degree. If the balance conservation law equation for angular momentum be added to the equations for energy and linear momentum, this form in the evolutionary relation will be a form of the second degree. And in combination with the equation of the balance conservation law for mass this form will be a form of degree 3.

Thus, in the general case, the evolutionary relation can be written as

$$d\psi = \omega^p \tag{9}$$

where the form degree p takes the values $p = 0, 1, 2, 3$.

Evolutionary relation obtained from the equations of the balance conservation laws possesses some peculiarity. This relation proves to be nonidentical since the differential form in the right-hand side of this relation is not a closed form, and, hence, this form cann't be a differential like the left-hand side.

Let us analyse the relation (8).

The evolutionary relation $d\psi = \omega$ is a nonidentical relation as it involves the *unclosed* skew-symmetric differential form $\omega = A_\mu d\xi^\mu$. The form ω isn't a close form since its differential $d\omega$ is nonzero. The differential $d\omega$ can be written as $K_{\alpha\beta} d\xi^\alpha d\xi^\beta$, where

$$K_{\alpha\beta} = \partial A_\beta / \partial \xi^\alpha - \partial A_\alpha / \partial \xi^\beta$$

are the components of the differential form commutator built of the mixed derivatives (here the term connected with the nonintegrability of the manifold has not yet been taken into account). The coefficients A_μ of the form ω can be obtained either from the equations of the balance conservation law for energy or from that for linear momentum. This means that in the first case the coefficients depend on the energetic action and in the second case they depend on the force action. In actual processes energetic and force actions have different nature and appear to be inconsistent. The commutator of the form ω constructed from the derivatives of such coefficients is nonzero. This means that the differential of the form ω is nonzero as well. Thus, the form ω proves to be unclosed and cannot be a differential.

[The skew-symmetric form in evolutionary relation is defined on the manifold made up by trajectories of the material system elements. Such a manifold is a deforming manifold. The commutator of the skew-symmetric form defined on such manifold includes an additional term connected with the differential of the basis. This term specifies the manifold deformation and hence is nonzero. Both terms in the commutator (obtained by differentiating the basis and the form coefficients) have a different nature and, therefore, cannot compensate one another. This fact once more emphasize that the evolutionary form commutator, and, hence, its differential, are nonzero. That is, the evolutionary form remains to be unclosed].

Hence, without the knowledge of a particular expression for the form ω, one can argue that for actual processes the evolutionary relation proves to be nonidentical.

The nonidentity of the evolutionary relation means that the initial equations of conservation laws turn out to be inconsistent, and hence, they are not integrable. The solutions to these equations will not be functions without additional conditions. They will depend on the commutator, which is nonzero due to inconsistence of the conservation law equations.

The solutions that are functions can be obtained only under additional conditions when the identical relation can be obtained from nonidentical evolutionary relation. This will point to the consistency (but only local, under additional condition) of the conservation law equations and the local integrability.

The identical relation can be obtained from nonidentical evolutionary relation only in the case when the closed exterior form, which is a differential, is obtained from unclosed evolutionary form. This is possible only under degenerate transformation, namely, under the transformation that does not conserve the differential, since, the evolutionary form differential is nonzero, whereas the differential of closed form is equal to zero. The additional conditions (which are the conditions of local integrability) are the conditions of degenerate transformation.

The additional conditions are caused by any degrees of freedom. The vanishing of functional expressions such as the determinant, Jacobian and so on corresponds to the additional conditions. These conditions can be realized under changing the evolutionary relation, which is self-variable.

If the conditions of degenerate transformation are realized, from the unclosed evolutionary form with nonvanishing differential $d\omega^p \neq 0$, one can obtain the differential form closed on pseudostructure. The differential of this form equals zero. That is, it is realized the transition

$$d\omega^p \neq 0 \rightarrow (\text{degenerate transformation}) \rightarrow \begin{cases} d_\pi \omega^p = 0 \\ d_\pi {}^*\omega^p = 0 \end{cases}$$

where ${}^*\omega^p$ is the dual form (which is a metric form). The condition $d_\pi {}^*\omega^p = 0$ is an equation for a certain pseudostructure π on which the differential of evolutionary form vanishes: $d_\pi \omega^p = 0$. This points to the fact that the pseudostructure is realized, and the closed (inexact) exterior form ω_π^p is obtained on pseudostructure.

On the pseudostructure, from evolutionary relation $d\psi = \omega^p$ it is obtained the identical relation $d_\pi \psi = \omega_\pi^p$, since the closed exterior form ω_π^p is a differential of some differential form. (This relation will be an identical one as the left and right sides of the relation contain differentials). The identity of the relation obtained from the evolutionary relation means that on pseudostructures the original equations for material system (the equations of conservation laws) become consistent and integrable.

Pseudostructures constitute the integrable surfaces (such as characteristics, singular points, potentials of simple and double layers, and others) on which the quantities of material system desired (such as the temperature, pressure, density) become functions of only independent variables and do not depend on the commutator (and on the path of integrating). This are generalized solutions. They may be found by means of integrating (on integradle structures) the equations of conservation laws for material systems.

Since generalized solutions are defined only on realized integrable structures (pseudostructures), they or their derivatives have discontinuities in the direction normal to integrable structure [6].

One can see that the integrable structures are obtained from the condition of degenerate transformation of the evolutionary relation. The conditions of degenerate transformation (a vanishing of such functional expressions as determinants, Jacobians, Poisson's brackets, residues) are connected with the symmetries, which can be due to the degrees of freedom of the material systems under consideration (for example, the translational, rotational and oscillatory degrees of freedom of material system).

The degenerate transformation is realized as the transition from the noninertial frame of reference to the locally inertial one, *i.e.* the transition from nonintegrable manifold (for example, tangent or accompanying) to integrable structures and surfaces.

Thus, one can see that the solutions to the set of equations, as well as in the case of a single equation, may be of two types: the solutions that are not functions, that is, they depend not only on independent variables, and generalized solutions, which are functions, and are obtained only under realization of additional conditions (which determine integrable structures or surfaces). The specific feature is the fact that they are definded on different spatial objects. Such solutions cannot be obtained by continuous modelling the differential equations only on a single spatial odject.

Before turning back to the problems of numerical solving the differential equations, it should call attention to the physical meaning of the solutions to these equations.

2.2. Physical Meaning of the Solutions to the Mathematical Physics Equations

The physical meaning of the solutions to the mathematical physics equations can be understood by the analysis of the evolutionary relation.

The evolutionary relation includes the functional that specifies the system state. Sinse this relation is nonidentical, it is impossible to obtain the state functional from this relation. This points to the absence of the state function and nonequilibrium state of the material system under consideration. The solutions of the first type just describe such nonequilibrium state. In this case, the commutator describes the internal force that induces the nonequilibrium state of material system. The solutions of the second type (genealized solutions, which are functions) are obtained under realization of additional conditions when the closed exterior forms is obtained from the unclosed evolutionary form, and the identical relation is realized. From such relation one can get the state functional and find the state function. This fact will point to the transition of material system into the locally equilibrium state.

The transition of the material system from nonequilibrium state into the locally-equilibrium one means that the unmeasurable quantity described by the nonzero commutator of the unclosed evolutionary differential form, which acts as an internal force, transforms into the measurable quantity. In material system, this reveals as the emergence of certain observable formations, which develop spontaneously. Such formations and their manifestations are fluctuations, turbulent pulsations, waves, vortices, and others [7].

It appears that the transition from the solutions of the first type to the generalized solution corresponds to the transition of material system from the nonequilibrium state to the locally equilibrium one that is accompanied by the emergence of a certain (observable) formation in material system. The discontinuous functions that correspond to generalized solutions just describe such observable formations.

Thus we obtain that the discrete realization of generalized solution points to the emergence of a certain (observable) formation in material system that is described by discontinuous functions corresponding to generalized solutions.

It may be also noted that the type of solutions to the equations describing material systems is of great significance for mechanics and physics of continuous media. In mechanics and physics of continuous media the same equations are considered (the equations of conservation laws for energy, linear momentum, angular momentum, and mass). The set of Navier-Stokes equations is an example [5]. However, the approaches to solving these equations in mechanics and physics are different. In physics the interest is expressed in only generalized solutions that are invariant ones and describe measurable physical quantities (but not the process itself), and noninvariant solutions are ignored (even if they have a physical meaning). The aim of mechanics of continuous media is to describe the process of the continuous media evolution. And in this case the numerical methods of solving differential equations are commonly used without studying the integrability conditions of these equations. The question of searching for invariant solutions that are realized only under additional conditions is commonly not posed. That is, one considers the solutions that are not functions.

Such restricted approaches, both in physics and mechanics, lead to nonclosure of relevant theories and this has some negative points. In mechanics without finding the generalized solutions it is impossible to describe such processes as the emergence of vorticity, turbulence and others. The physical approach enables one to find allowed invariant solutions, however, in this approach there is no way to say at what time instant of evolutionary process one or another exact solution was realized. This does not also discloses the causality of phenomenon described by these solutions. It is evident that in mechanics, as well in physics, it is necessary to seek for solutions of both types. In particular, in the case of gas-dynamic system such an approach had been studied in paper [5].

3. On the Problem of Numerical Solving the Differential Equations

As it was noted, the equations that describe actual processes are definded on manifolds (tangent, accompanying), which are nonintegrable. If to model the equations on such a manifold, one can obtain, without additional conditions, the solutions of only first type, *i.e.* the solutions that depend on the commutator with nonzero value caused by inconsistency of derivatives or equations in the set of equations. It should be emphasized once more that such solutions have physical meaning, namely, they describe the nonequilibrium state induced by the physical processes proceeded in the system. The generalized solutions, which are functions and describe discrete formations, cannot be obtained by modelling the equations only on original manifold, since they are obtained on integrable structures that do not belong to original nonintegrable manifold. Therefore, to obtain the generalized solutions by numerical simulation, one must use two systems of reference. One more problem of obtaining the generalized solution relates to the fact that the integrable structures with generalized solution are not initially given, and they are realized spontaneously in the process of integrating under the realization of additional conditions, namely, the integrability conditions. (As additional conditions it may serve, for example, the characteristic conditions, the dynamical conditions of the consistency of equations in the set of equations [8] and so on). To obtain the integrable structures, it is necessary to trace for the realization of additional conditions, which define the integrable structures, in the process of numerical integrating the equations on the original manifold. This gives a possibility to obtain the instant of realization the generalized solution.

In this case, the transitions from inexact solutions to generalized ones describe the process of emergence of any observable formations (in particular, such as waves, vorticity and others), which intensity is definded by generalized solution.

As it was noted, in mechanics and physics the interest is expressed in various types of the solutions to equations. The methods of numerical solving the equations relate to this fact.

In mechanics this is the method of direct numerical simulating the equations, which is fulfilled on tangent manifold (being nonintegrable) and enables one to obtain only inexact solutions.

In physics this is the method of solving equations when the equations are provided with the integrability conditions (the conditions of consistency) and this enables one to obtain integrable structures or surfaces, that is, to go out onto cotangent integrable manifold and obtain exact solutions (the methods of characteristics, symmetries, eigen-functions and others are examples of such methods). The analytical methods may be such methods.

It is possible that the integration of these both methods will allow solve some problems on numerical integrating the mathematical physics equations.

It should be emphasized that on integrable structures the variables of the function desired do not coincide with the variables of original manifold (since they belong to different spatial objects). Thus, the coordinates of the equations for characteristics are not identical to the independent coordinates of the initial manifold, on which the initial equation is defined.

It should be emphasized once more that the existence of two types of solutions has a deep physical sense. This peculiarity of the equations enables one to describe the process of origin of discrete formations such as waves, vorticities, turbulent pulsations [7] and so on.

It has been shown that the origin of discrete formations is described by the transition from the solutions of the first type, which depend on a certain commutator, to the generalized solution that is a function. To describe the process of origin of discrete formations, it is necessary, firstly, have the solutions of the first type, which can be obtained only by numerical modelling of the equation on the original nonintegrable manifold (it is impossible to find such a solution by analytical method), and, secondly, have the solution of the second type (generalized solution). This solution can be obtained only on integrable structure that is definded by the integrability conditions being realized. Here there is a delicate point. The allowable generalized solutions can be obtained by analytical methods if the integrability conditions are imposed on the equations. However, in this case it is impossible to define the instant of realization of generalized solution and thereby to describe the process of the discrete formation emergence. The description of evolutionary processes is possible only either by numerical methods, but with two frames of reference, or by using simultaneously numerical and analytical methods.

4. Conclusions

It has been shown that the equations, which describe actual processes, have the solutions of two types, which are defined on different spatial objects, and, therefore, cannot be obtained by continuous numerical simulations of the equations. The existence of two spatial objects on which the solutions are defined gives rise to the problems on which the attention must be focused while numerical solving the mathematical physics equations.

It should be also emphasized that the methods of numerical solving the equations with account for the existence of two types of solutions can allow describing the evolutionary processes such as the emergence of any discrete formations.

REFERENCES

[1] L. I. Petrova, "Exterior and Evolutionary Differential Forms in Mathematical Physics: Theory and Applications," Lulu.com, 2008, 157 p.

[2] L. I. Petrova, "Role of Skew-Symmetric Differential Forms in Mathematics," 2010. http://arxiv.org/abs/1007.4757

[3] J. F. Clarke and M. Machesney, "The Dynamics of Real Gases," Butterworths, London, 1964.

[4] L. I. Petrova, "Physical Meaning and a Duality of Concepts of Wave Function, Action Functional, Entropy, the Pointing Vector, the Einstein Tensor," *Journal of Mathematics Research*, Vol. 4, No. 3, 2012, pp. 78-88.

[5] L. I. Petrova, "Integrability and the Properties of Solutions to Euler and Navier-Stokes Equations," *Journal of Mathematics Research*, Vol. 4, No. 3, 2012, pp. 19-22.

[6] L. I. Petrova, "Relationships between Discontinuities of Derivatives on Characteristics and Trajectories," *Journal of Computational Mathematics and Modeling*, Vol. 20, No. 4, 2009, pp. 367-372.

[7] L. I. Petrova, "The Noncommutativity of the Conservation Laws: Mechanism of Origination of Vorticity and Turbulence," *International Journal of Theoretical and Mathematical Physics*, Vol. 2, No. 4, 2012, pp. 84-90.

[8] V. I. Smirnov, "A Course of Higher Mathematics, V. 4," Technology and Theory in the Literature, Moscow, 1957. (in Russian)

23

Hierarchical Linear Model of Monthly Rainfall with Regional and Seasonal Interaction Effects

Yonghua Zhu[1], Hongtao Lu[1], Zilin Zhu[2]

[1]Mathematics and physics department of North China Electric Power University, Beijing, China

[2]Information Engineering Department of Huazhong University of Science and Technology, Wuhan, China

ABSTRACT

According to the hierarchical characteristics of monthly rainfall in different regions, the paper takes the geographical factors and seasonal factors into the hierarchical linear model as the level effect. Through clustering methods we select two more representative regional meteorological data. We establish three-layer model by transforming the interactive structure date into nested structure data. According the model theory we perform the corresponding model calculations, optimization and analysis, accordingly to interpret the level effects, and residual test. The results show that most of the difference in Monthly Rainfall was respectively explained by Variables (Meteorological factors, seasonal effects, geographic effects) in different levels.

Keywords: Monthly Rainfall; Hierarchical Linear Model; Regional Effects; Interaction Effect Component

1. Questions and Data Description

For the defects of the past rainfall's regression, the literature [1] propose the regression model which take the factors and other effects into consideration. From the characteristics of monthly rainfall, we establish a two-layer model with the seasonal effect, in the model longitudinal data is grouped by month. Then through a series of operations, such as correlation analysis, data preprocessing, classification of seasonal effects, establishment of virtual indicators, gradually building models, the interpretation of fixed effects and random effects to complete the HLM2 model on monthly rainfall[1].

To dig the effect of monthly rainfall and various factors in different seasons and regions, we consider the data set: The meteorological data of Beijing, Tianjin and other 34 major cities in 1996-2009（monthly rainfall/(mm), average temperature/(℃), sunshine hours/(h), average relative humidity/(%), average air pressure/(100pa), hereinafter referred to as rainfall, temperature, sunshine, humidity, pressure. To take Beijing and Nanjing for example, get the following figure.

Figure 1 shows that, the monthly precipitation curve showed two features: the two regions' data show a certain cycle as a unit of year; overall, the Beijing's rainfall is always greater than Nanjing's. Therefore, in the study of the differences of rainfall, we not only should consider the impact factors and seasonal effects, but also need to consider the regional differences. Based on the previous two-layer model, we attempt to establish a three-layer model on the effects of a regional group and seasons.

2. Model and Analysis

2.1. Model

Level 1 model: the regression between rainfall and temperature, sunshine, humidity, air pressure. Outcome variables Yij represents the rainfall of the month j of the year I (i=1,2,…14, j=1,2,…12), x1ij, x2ij, x3ij, respectively, for the temperature, sunshine, humidity of the month j of the year i.

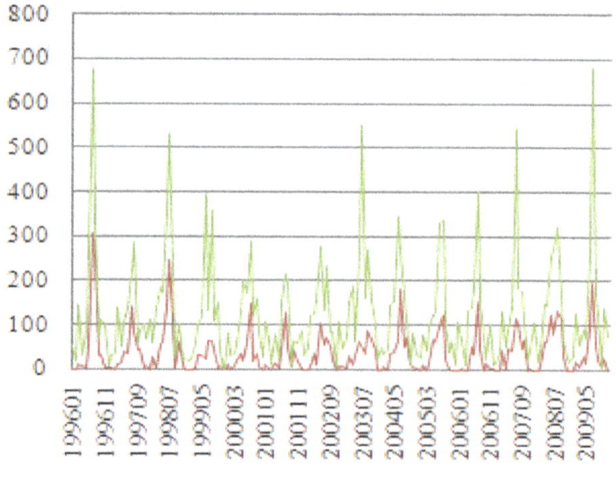

Figure 1. Beijing and Nanjing's monthly rainfall map(red: Beijing; blue: Nanjing).

Level 2 model: Create two new virtual season index CQ, X, to distinguish three kinds of seasonal effects (Winter, Spring, Summer). The combination of their values and other factors indicate the slope of the temperature, sunshine and humidity with rainfall in different seasons.

Level 3 model: Establish geographical index to explain the intercept and slop Level 2 model (model about the relationship between season with the coefficient of Level 1). Comprehensive three-tier model, it express there is regression (intercept and slope) about different degrees of effect of various factors and seasonal precipitation in different regions.

Basic data includes 31 cities, its geographical spread and the seasonal variations are large. If we want to establish the index about geographical differences, which measure the monthly rainfall of 31 cities, and make three-tier regression. It would be difficult. If we want to establish virtual index, at least we should build five. But in the HLM2's level 2, the maximum number of considered geographical effects is five. Even have built a three-tier model, the fixed coefficients and random coefficient which need to test will be large (in Level 3, there will be 35 items including the intercept, slope, and random items), so effects analysis is not easy to make. If we take some quantitative methods (AHP, quantitative weighting, expert scoring method, etc) to establish a index which can unified measure geographical differences of 31 cities. It's more difficult and difficult to estimate accurately the extent, because qualitative indicators are always randomness and fuzziness. Making scientific analysis and rigorous validation to its quantitative is another major issue [3].

The two effects which affect the rainfall are seasonal effect and geographical effects. But we find there are three seasonal effects in geography, and in a season there are also two geographical effects, two effects are interaction effect, rather than simply "students- class- school" nested structure. In this case Raudenbush (1993) developed a method; Level 2 is the definition of "unit" effect which is classified by two interacted factors. Level 1 represents the link between variables under the influence of the "unit". This model has only two layers, known as Hierarchical Cross-classified Linear Model, HCM2 [2]. Taking rainfall for example, HCM model mainly research which independent variable the seasonal level and regional level have, and the characteristics of two- factor interactions between seasonal and regional levels.

Here we attempt to establish a three-layer model to decompose the interaction structure of two-factor. We do hierarchical processing to the interactive structure as follows.

In the level 1 the data set formed by this method reach 48 groups (12 months * 4 cities) by the effect of six seasons; In level 2 model 6 units are influenced by two geographical effects. The three-tier model is the same as the interaction effect model; each city's monthly data is corresponding to seasonal effects and regional effects.

List 1. The virtual index of seasonal effects in Level 2.

Month	CQ	X
12,1,2 (winter)	0	0
3,4,5,9,10,11 (spring)	1	0
6,7,8 (summer)	0	1

List 2. Data' two- factor interactions between seasonal and regional levels.

Region	Cities	Winter	Spring and Autumn	Summer
North China	Beijing	1 2 12	3 4 5 9 10 11	6 7 8
	Tianjin	1 2 12	3 4 5 9 10 11	6 7 8
East China	Nanjing	1 2 12	3 4 5 9 10 11	6 7 8
	Hefei	1 2 12	3 4 5 9 10 11	6 7 8

List 3. Stratification of seasonal-geographical two-factor interacted structure.

Region	Season	City monthly weather data
	North-Winter	Beijing, Tianjin(12,1,2)
North China	North Spring	Beijing, Tianjin (3,4,5,9,10,11)
	North-Summer	Beijing, Tianjin (6,7,8)
	East-Winter	Nanjing, Hefei(12,1,2)
East China	East-Spring	Nanjing, Hefei (3,4,5,9,10,11)
	East-Summer	Nanjing, Hefei (6,7,8)

2.2. Zero Model

Level 1 model:

$$Y_{ijk} = P_{0jk} + e_{ijk}$$

Level 2 model:

$$P_{0jk} = B_{00k} + r_{0jk}$$

Level 3 model:

$$B_{00k} = G_{000} + u_{00k}$$

P_{0jk} is the average monthly rainfall in the region k and season j, e_{ijk} is the individual differences of monthly rainfall at the same region and season. B_{00k} is the average of all average seasonal rainfalls at region k, r_{0jk} is the variational degree between different seasons at the same region. G_{000} is the average rainfalls in all seasons at all regions, u_{00k} represent the variational relative to he mean at different regions. Zero model parameter estimation results are listed below.

Based on the principle of variance decomposition described above we can obtain follows, the group differences of monthly rainfall group differences account for 47.8%, The differences of monthly rainfall affected by seasonal effect account for 42.2%, regional impact account for 10.0% of the total differences. That shows the 52.2% differences of monthly rainfall are related to the geographical and seasonal effects. This suggest us we should add more explanatory variables to the level 1 and level 2 to explain more variance of levels.

2.3. Random Effects Model

Level 1 model:

$$Y_{ijk} = P_{0jk} + P_{1jk}x_{1ijk} + P_{2jk}x_{2ijk} + P_{3jk}x_{3ijk} + e_{ijk}$$

Level 2 model:

$$P_{0jk} = B_{00k} + r_{0jk}$$
$$P_{1jk} = B_{10k} + r_{1jk}$$
$$P_{2jk} = B_{20k} + r_{2jk}$$
$$P_{3jk} = B_{30k}$$

Level 3 model:

$$B_{00k} = G_{000} + u_{00k}$$
$$B_{10k} = G_{100}$$
$$B_{20k} = G_{200}$$
$$B_{30k} = G_{300} + u_{30k}$$

Compared to the zero model, the variance components of three-intercept of the random effects model were reduced by 18%, 11%, 17%. Clearly, this variance is explained by the various factors added to the level 1.

2.4. Optimalizing Full Model

Model Overview: the total number of level 1 units is 672= 4 cities*12 months*14 years (in addition to missing values, the total is 660); the total number of level 2 units is 48=4 cities*12 months, belonging to 48 different "regions- season"; the total number of level 3 units 2=2 regions.

Level 1 model:

$$Y_{ijk} = P_{0jk} + P_{1jk}x_{1ijk} + P_{2jk}x_{2ijk} + P_{3jk}x_{3ijk} + e_{ijk}$$

Level 2 model:

$$P_{0jk} = B_{00k} + B_{01k}CQ_{0jk} + B_{02k}X_{0jk}$$
$$P_{1jk} = B_{12k}X_{1jk} + r_{1jk}$$
$$P_{2jk} = B_{21k}CQ_{2jk} + r_{2jk}$$
$$P_{3jk} = B_{31k}CQ_{3jk} + B_{32j}X_{3jk}$$

List 4. Variance components' estimation of levels.

Random effects	Standard error	Variance components	df
Random item of Level1's intercept	44.79**	2006.39	46
Random item of Level 1	47.65	2270.68	--
Random item of Level 3	21.84**	477.12	1

List 5. The results of random effects models.

	Random effects	Standard deviation	Variance	df
Level 1	Individual random effects	40.64	1651.81	--
	Level 1 intercept	44.91**	2017.75	46
Level 2	Temperature corresponds to the slop	7.24**	52.51	47
	Sunshine corresponds to the slop	0.32**	0.10	47
Level 3	Level 2 intercept	19.91**	396.43	1
	The intercept of humidity corresponding to the slope	1.19**	1.42	1

Level 3 model:

$$B_{00k} = G_{001}D_k$$
$$B_{01k} = G_{010}$$
$$B_{02k} = G_{020} + u_{02k}$$
$$B_{12k} = G_{120}$$
$$B_{21k} = G_{211}D_k$$
$$B_{31k} = G_{310} + G_{311}D_k$$
$$B_{32k} = G_{320}$$

There are three parts affected by region in total: level 1's intercept, sunshine slope in spring and autumn, Humidity slope in spring and autumn. In other words, the geographical differences are obvious in the overall mean. In the spring and autumn, the differences of the impact of humidity and sunshine to rainfall are significant [4].

In **Figures 4-2** A stand for north, B stand for south, red stand for spring and autumn, blue stand for other seasons.

Obviously, there is a positive correlation between rainfall and humidity. In southern spring and autumn (chart B), humidity causes greater impact on rainfall.

In above figure, A stand for north, B stand for south, red stand for spring and autumn, blue stand for other seasons. There is a weak negative correlation between sunshine and rainfall in spring and autumn in two regions. The other seasons are messier and the relationship is unknown. The relationship can also be observed from the model and coefficients.

There is a great negative correlation between sunshine and rainfall in autumn. Other seasons were not significant and the north and south regions' differences were not significantly.

Compared to the zero models, level 1 model's random item variance changes little. [5]The random items of level 2 and level 3 models are different, but we can obviously find that the vast majority of random effects are explained by different levels' seasonal and geographical variables, and random item's variance is very small.

List 6. The results of fixed effects models.

	Fixed effects	Coefficient	Standard error	df
	Intercept G001	45.55**	3.19	652
Level 1 intercept	Winter and autumn G020	24.42**	2.84	652
	Summer G030	112.08*	6.42	652
Temperature slope	Summer G130	-10.42*	4.38	47
Sunshine slope	Winter and autumn G221	-0.35*	0.15	47
	Winter and autumn G320	1.08*	0.51	652
Humidity slope	Winter and autumn G321	2.21*	1.01	652
	Summer G330	4.81**	0.93	652

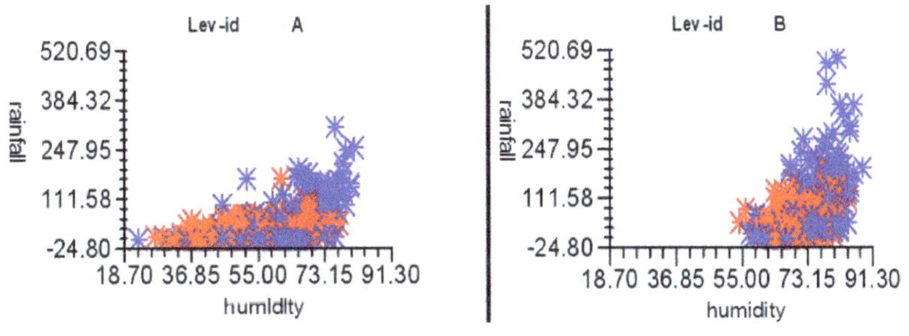

Figure 2. Scatter diagram about the relationship between humidity and rainfall in different regions.

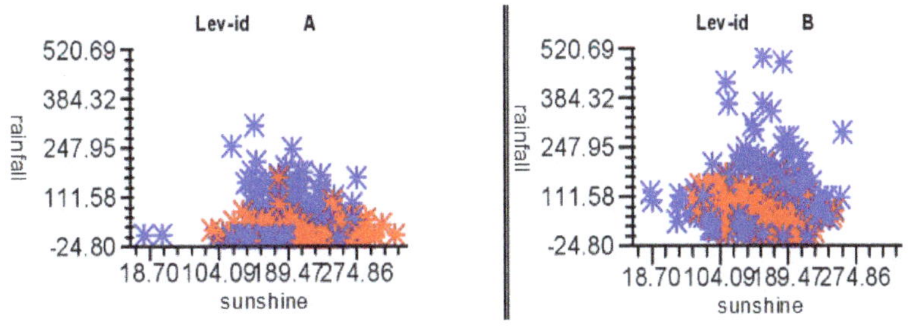

Figure 3. Scatter diagram about the relationship between sunshine and rainfall in different regions.

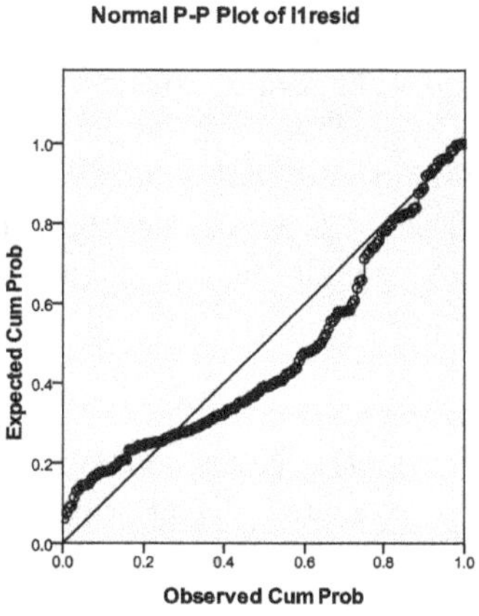

(a) Northern spring and autumn residuals

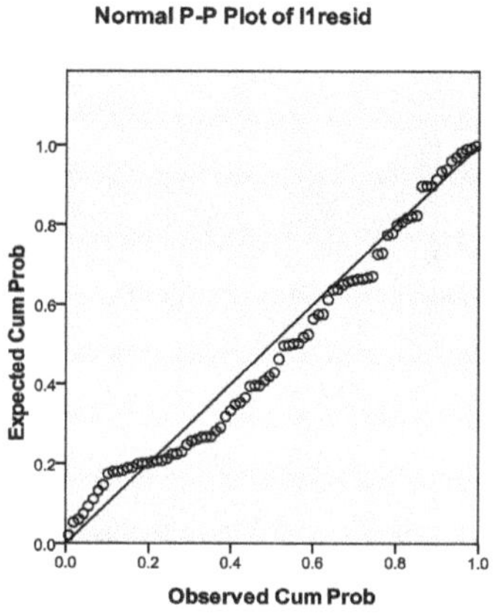

(b) Northern summer residuals

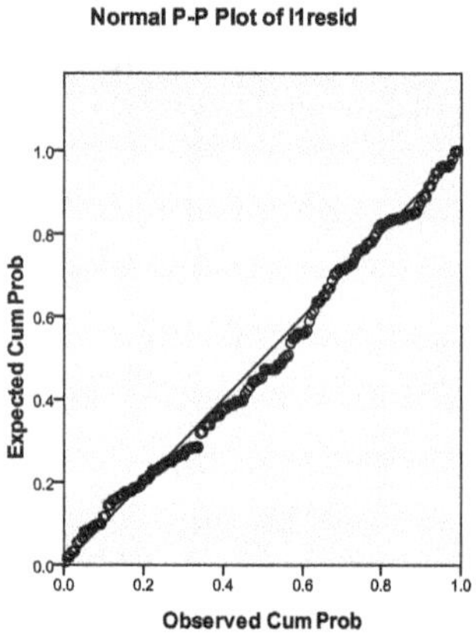

(c) Southern spring and autumn residuals

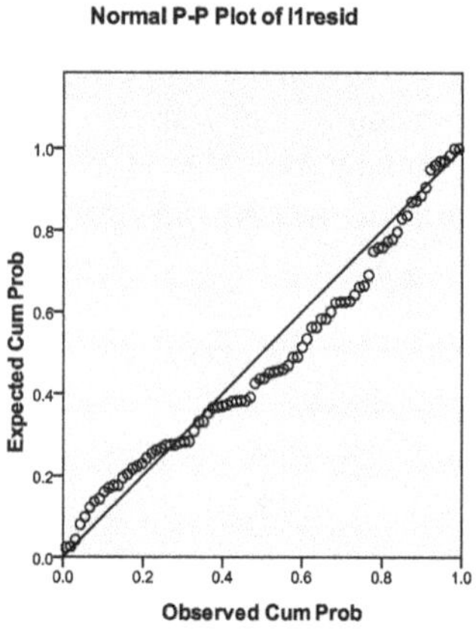

(d) Southern summer residuals

Figure 4. Level 1 model's residual comparison.

List 7. Variance components' estimation of levels.

	Random effects	Standard error	Variance	df
Level 1	Individual random effects	42.63	1817.86	--
Level 2	Temperature corresponds to the slop	6.23**	38.93	47
	Sunshine corresponds to the slop	0.36**	0.13	47
Level 3	Level 1's intercept corresponds to the slop in summer	7.47**	55.86	1

List 8. HLM3 Model summary and comparison.

	Model 1 (zero model)	Model 2 (random effects model)	Model 3 (final model)
σ2	2270.68	1651.81	1817.86
Level-2:	--	--	--
μ00p	2006.40	2017.75	--
μ11p	--	52.52	38.93
μ22p	--	0.10	0.13
Level-3:	--	--	--
μ00b	477.12	396.44	--
μ02b	--	--	55.86
μ03b	--	1.43	--
Total deviation	7100.66	6936.67	6865.15
Number of Parameters	4	10	12
Iterations	4	512	673
The total variance	4754.20	--	--

3. Model Summary

3.1. Model Comparison

Compared with the zero models, the random item's variance of level 1 changes little. The random items in level 2 and level 3 can't be comparable. But we can obviously find that the vast majority of random effects are explained by different levels' seasonal and geographical variables, and random item's variance is very small. Hierarchical interpretation of the effect of rainfall is significant, and seasonal and geographical explanatory variables play a good role in regression.

3.2. Residual Analysis

From the above residual plots we can get two features: Overall, the residuals of the south are closer to the normal distribution than those of the north, and spring season is closer to the normal distribution than summer season, because the southern seasonal effect and geographical effect are more significant than the northern on the whole, and the summer's effect is more significant than the winter and autumn's. This result the differences of rainfall in the South in summer has been more fully explained, assumption of the residuals closer to the normality.

Overall, HLM3 model's size of the model residuals is similar to the HLM2 models, the summer's residuals are bigger, but their relative offset is to a lesser extent.

4. HLM3 Model Conclusions

In this paper, we mainly research the rainfall HLM3 model under the seasonal and geographical effects. Here is a brief summary of geographical effects.

From the overall average level, the geographical differences of monthly rainfall are significant.

Most of monthly rainfall's differences between the groups can be explained by seasonal and geographical of variable levels.

In spring and autumn the degree of humidity and sunshine's influence on the rainfall has significant differences. Positive correlation with precipitation and humid ity, sunshine is a negative correlation;

In the summer there is a big negative correlation between temperature and precipitation, but in the other seasons it's not significant and the geographical differences are obvious.

From the data fit through the hierarchical model, although larger residuals in summer, its relative deviation is more minimum than the other seasons; the overall fit of the South is better than the North.

5. Summary

After taking different regions of precipitation into comparison, we take the effect of geographical factors as a higher level. Using cluster analysis, we select two regional more representative meteorological data. Translate structure of the interaction data into nested structure, and establish a corresponding three-level linear model (HLM3). In accordance with model theory we do the corresponding model calculation, optimization and analysis and reach some major conclusions. The explanatory variables of various levels (the meteorological factors, seasonal effects, geographic effects) can well explained the differences in monthly rainfall.

Hierarchical linear models have been widely used in social science fields. About the natural science problems we can make use of the professional knowledge and draw on some ideas and methods of appropriate social science to establish an appropriate model. With the development of a variety of technologies, in most cases the size of data will no longer be limited, a lot of data can be repeatedly observed and recorded, which cause the formation of the corresponding longitudinal data. Therefore, the hierarchical linear model will be widely used.

REFERENCES

[1] Y. H. Zhu and G. X. Jiang, "Hierarchical Linear Model and Its Research on Hierarchical Characteristics of Rainfall," *2011 International Conference on Multimedia Technology (ICMT 2011)*, pp. 2146-2150.

[2] S. W. Raudenbush, A. S. Bryk and Z. G. Guo, "Hierarchical Linear Models: Application and Data Analysis Method," Beijing: Social Sciences Academic Press, 2007, pp. 83-90.

[3] J. C. Wang, H. Y. Xie and B. F. Jiang, "Application of Hierarchical Linear Model——Methods and Applications," Beijing: Higher Education Press, 2007, pp. 27 30.

[4] L. Zhang, L. Lei and B. L. Guo, "Application of Hierarchical Linear Model," Beijing: Science and Education Press, 2003, pp. 28-40.

[5] X. Zhang and J. Y. Wang, "The Study for the Sample Size Problem about Hierarchical Linear Models," Statistics and Decision, Vol. 15, 2010, pp. 4-8.

Reduction in Complexity of the Algorithm by Increasing the Used Memory - An Example

Leonid Kugel, Victor A. Gotlib
The Faculty of Sciences, Holon Institute of Technology (H. I.T), Holon, Israel

ABSTRACT

An algorithm complexity, or its efficiency, meaning its time of evaluation is the focus of primary care in algorithmic problems solving. Raising the used memory may reduce the complexity of algorithm drastically. We present an example of two algorithms on finite set, where change the approach to the same problem and introduction a memory array allows decrease the complexity of the algorithm from the order $O(n^2)$ up to the order $O(n)$.

Keywords: Algorithm; Complexity Reduction; Memory Usage

1. Introduction

An algorithm efficiency understood as the time of its execution is the focus of primary care in the design and analysis of algorithms ([1, 2]). The lower bond of the execution time of an algorithm directly correlated with the order of complexity of the algorithm. A different approach to the solution to the problem allows sometimes to change the algorithm and to reduce its complexity by introduction an additional memory, for example ([3]). Such a method requires some further analysis of the problem at hand. We illustrate this case with the following example.

2. The Problem

Given two n-digit natural numbers ($n > 0$). One needs to find the number of matching digits at the same positions in both numbers, alongside with overall count of matching digits over the numbers. If a digit has already participated in a matching pair, it is ignored in further encounter. Consider, for example, two numbers 172345 and 287376. The amount of matching identical digits in the equivalent positions is 1 (this is the digit 3). The count of matching identical digits in the various positions is 2 (digit 2 and digit 7). If the same digit in the numbers appears more than once, the count is defined as the minimum number of occurrences in one of the positions of the numbers. For example: 22275 and 86322 specifies the repetition of the number 2 twice.

2.1. The First Approximate

To address the first part of the task (counting the digits on the same positions) we use quite a simple approach: check the number of the units digit in both numbers (the remainder of these numbers divided by 10 gives the number of units in those numbers), and if they are equal, then the corresponding counting variable is increased by 1. Then we "delete" the units digit in both numbers (integer divide by 10). If a match was encountered, the digits are not returned to the original numbers. The performance of this algorithm requires $2O(n)O(1)$ operations. The detailed first approximate algorithm is as follow:

Equals Digits In Position (num1, num2)
count_pos=0, tmp_num1=0, tmp_num2=0
while (num1> 0)
if (num1%10 = num2%10) {if digits match, we counting them without storing into temporary variables}
count_pos ← count_pos+1
else {if digits differ, we store them in temporary variables}
*tmp_num1 ← tmp_num1*10+num1%10*
*tmp_num2 ← tmp_num2*10+num1%10*
{In any case, we delete the "right" units digits}
num1 ← num1 mod 10
num2 ← num2 mod 10
end while
while (tmp_num1 > 0) {restoring original numbers, without replicable digits }
*num1 ← num1*10 + tmp_num1%10, tmp_num1← tmp_num1/10*
*num2 ← num2*10 + tmp_num2%10, tmp_num2← tmp_num2/10*
end while
return count_pos

end algorithm

Note that in the above algorithm to the presented problem, it is not significant that the variables *tmp_num*1 and *tmp_num*2 include digits of the original numbers in reverse order. One can just assign as desired.

For the second part of the solution, an auxiliary algorithm-function that receives two parameters: the number and the digit which will be used. The second algorithm checks if the digit is presented in the number. If so, the algorithm deletes its first occurrence from the right (of unit's digit) and returns the number lower by one order, else it returns the number unchanged.

```
DeleteDigit ( num, dig )
if ( num<10 )
if ( num==dig ) return 0
else return num;
tmp_digit←num % 10; { remainder of division by 10}
if ( tmp_digit == dig )
return num div 10
{ return number without the given digit}
    tmp_num← DeleteDigit ( num div 10, dig )
{ check number without units digit (one order lower)}
if ( tmp_num < num div 10 )
{ digit was deleted, and digit tmp_dig is missing
   Return the missing digit}
tmp_num ← tmp_num * 10 + tmp_digit
return tmp_num
else
return num
{ return the number without change}
end if
end algorithm
```

The number of actions needed to run this algorithm is about $O(n)O(1)$ runtimes in the worst case scenario (see [1, 2]). Using this algorithm, one can count the total number of occurrences of digits in different positions (matched digits in same positions were "deleted" in the first part of the solution)

```
countEqualsRegular ( num1, num2 )
count_eq ← 0;
while ( num1 > 0 )
digit ← num1 % 10
tmp_num2 ← DeleteDigit ( num2, digit )
if ( tmp_num2 ≠ num2 )
   {If deletion occurred, count as digits match}
count_eq ← count_eq + 1
num1 ← num1 div 10   {deleting reviewed digit}
return count_eq
end if
end while
end algorithm
```

The number of actions required to perform this algorithm with the auxiliary algorithm will be of order $O(n)O(n)O(1)$. Finally counting, for both parts the runtime will approach $\left(2O(n)+O(n^2)\right)O(1)$. In other words, the final value of the asymptotic limit of the above functions is of $O(n^2)$.

A question is, if one can to reduce the order of operations amount in order to solve the problem under consideration (see [3])? To answer is yes, but the using memory should be extended.

2.2. The second Approximate

The idea behind the second solutions is based on the idea of the quick sort method like "counting sort", where we count equal elements in the array, based on the property of the studied values limits. For the first part of the task we construct two auxiliary arrays, which are equal to the length of the given numbers, hence of length *n*. Thus, equal digits on matching positions give us the matching digits on same positions. By presenting the numbers as digits in arrays, we are not obliged to use digits standing on same positions.

For the second part of the problem we use the same idea, but considering the length of the given numbers. We know that all decimal numbers consist of digits from 0 to 9. Thus, creating an auxiliary array of size 10 we can place in it the amount of counted corresponding digits. Checking the corresponding numbers of such arrays for each of these numbers will give us the result of matching numbers ([3, 4]).

```
CountEquals2 (num1, num2)
{two arrays of length n for digits of first part of the solution. Arr_num1 [n], arr_num2 [n], the lengths of our numbers are given – n.
count_pos←0 count of position matches
count_eq ←0 count of total matches}
for ( i ← 1, i ≤ n, i←i+1 )
{ operations of order O(n)*5.
   Counting the number of matches }
if ( num1%10 = num2%10 )
   count_pos ← count_pos + 1
else
{storing the number of units digit in the proper variable }
arr_num1 [i]←num1%10
arr_num2[i]←num2%10
{ "deleting" digit of units }
num1←num1 div 10
num2←num2 div 10
end if
end for
n ← n – count_pos { the remained amount of digits in the original numbers. All the digits of our numbers are stored in the arrays, with the exception of. Those, matched by position. Create two arrays of 10 elements each and initiate its values with 0 (digits are not counted yet.) Then recount the amount of remaining different dig-
```

its with}
{counting arrays - [1], [2] }
arr_count1={ 0 }, arr_count2={ 0 }
*for (i←1, i ≤ n, i ← i+1 { operations of order O(n)*2 }*
arr_count1[arr_num1[i]]←arr_count1[arr_num1[i]] +1
arr_count2 [arr_num2[i]]←arr_count2[arr_num2[i]]+1
end for
{counting matches ([1], [2], [4]) }
*for (i←1, i ≤ n, i ← i+1) { operations of order O(n)*3 }*
if (arr_count1[i]>0 AND arr_count2[i>0])
count_eq←count_eq +min(arr_count1[i], arr_count2[i])
end for
return count_pos, count_eq
end algorithm

The number of operations required to perform the algorithm is of an order

$$5O(n)+2O(n)+3O(n)=10\cdot O(n).$$

The final value of the asymptotic function that limits our result from above is $O(n)$ comparing with $O(n^2)$ in the previous versions.

3. Conclusions

For small values of n, the second solution may require even more operations than the first solution. There is a range of values of the segment closest to the origin of the function $y=a\cdot n$ while above will limit the function by $y=b\cdot n^2$, where a and b are appropriate constants. However, when n tends to infinity (or became sufficiently large), the square function grows faster than linear obviously ([3]).

With this simple example, we wanted to show how the runtimes of different algorithms solving the same problem can be different drastically, when the problem is correctly formulated and the right model and built solution are properly matched.

REFERENCES

[1] T. H. Cormen, C. E. Leiserson, R. L. Rivest and C. Stein, "Introduction to Algorithms", 3rd Edition, MIT Press, Cambridge, MA, 2009.

[2] M. A. Weiss, "Data Structures and Algorithm Analysis in C," Addison-Wesley, 1997.

[3] G. L. Abdulgalimov and L. A. Kugel and N. A. Masimova, "To the Question About Teaching Designing of Information Systems and Data Analysis", Science and Education," No. 9, 2012. pp 81-82 (in Russian). http://elibrary.ru/item.asp?id=18251112

[4] G. L. Abdulgalimov, S. M. Yevstigneev and L. A. Kugel, "Analysis of the Data for Teaching the Basics of Programming", Proceedings of the X National Russian Conference "IT Education in the Russian Federation", 16-18 May 2012, Moscow, Moscow State University, pp 273-275 (in Russian). http://2012.xn----8sbacgtleg3cfdxy.xn--p1ai/upload/IT-EDUCATION-2012-book.pdf

Unsteady Incompressible Couette Flow Problem for the Eyring-Powell Model with Porous Walls

Haider Zaman, Murad Ali Shah, Muhammad Ibrahim
Faculty of Numerical Sciences, Islamia College University, Peshawar, Pakistan

ABSTRACT

This work is concerned with the influence of uniform suction or injection on unsteady incompressible Couette flow for the Eyring-Powell model. The resulting unsteady problem for horizontal velocity field is solved by means of homotopy analysis method (HAM). The characteristics of the horizontal velocity field and wall shear stress are analyzed and discussed. Pade approximants and Taylor polynomials are also found for velocity profile and are used to make the maximum error as small as possible. The graphs of the error for the Pade approximation and Taylor approximation are drawn and discussed. Convergence of the series solution is also discussed with the help of $\hbar$-curve and interval of convergence is also found.

Keywords: Unsteady; Couette Flow; Eyring-Powell Model; Pade Approximants; Porous Plates

1. Introduction

The study of non-Newtonian fluids has generated much interest in recent years in view of their numerous industrial applications, especially in polymer and chemical industries. The examples of such fluids includes various suspensions such as coal-water or coal-oil slurries, molten plastics, polymer solutions, food products, glues, paints, printing inks, soaps, shampoos, toothpastes, clay coating, grease, cosmetic products, custard, blood, etc. Some interesting studies of non-Newtonian fluids are given by Hayat *et al.* [1-5], Asghar *et al.* [6], Khan *et al.* [7,8], Cortell [9,10], Ayub *et al.* [11-13], Ariel *et al.* [14], Rajagopal [15-17], Erdogan [18], Siddiqui and Kaloni [19] and Fetecau [20]. Couette flow is an important type of flow in the history of fluid mechanics. Researchers have deep interest in this flow and they study it in many ways. Some important studies about this flow are as follows:

Fang [21] studied Couette flow problem for unsteady incompressible viscous fluid bounded by porous walls. Khaled and Vafai [22] considered Stokes and Couette flows due to an oscillating wall. Asghar *et al.* [23] discussed unsteady Couette flow in a second grade fluid with variable material properties. Hayat *et al.* [24] examined the axial Couette flow problem of an electrically conducting fluid in an annulus. Hayat and Kara [25] studied Couette flow of a third-grade fluid with variable magnetic field. Seth *et al.* [26] presented Couette flow problem for a porous channel. Bhaskara and Bathaiah [27] have analyzed Couette flow problem for flow through a porous straight channel with MHD and Hall effects. Das *et al.* [28] considered unsteady Couette flow problem in a rotating system. Ganapathy [29] presented a note on the oscillatory Couette flow in a rotating system. Guria [30, 31] discussed Couette flow problem for rotating and oscillatory flow. Sigh [32] found a periodic solution for oscillatory Couette flow.

The Eyring-Powell model [33] although more mathematically complex, has certain advantages over the Second grade, Maxwell, Power-law and Micropolar fluid models. Eyring-Powell model is derived from the kinetic theory of liquids rather than the empirical relations. It correctly reduces to Newtonian behavior for low and high shear stress. Recently, Eldabe *et al.* [34] and Zueco and Beg [35] discussed the non-Newtonian fluid flow under the effect of couple stresses between two parallel plates using Eyring-Powell model. Prasad *et al.* [36] studied momentum and heat transfer of a non-Newtonian Eyring-Powell fluid over a non-isothermal stretching sheet. Patel and Timol [37] presented a numerical treatment of MHD Eyring-Powell fluid flow. Sirohi *et al.* [38] studied Eyring-Powell fluid flow past a 90° wedge. Javed *et al.* [39] discussed flow of an Eyring-Powell non-Newtonian fluid over a stretching sheet. Noreen and Qasim [40] analyzed peristaltic flow of MHD Eyring- Powell fluid in a channel.

Keeping this all in view, in the present paper, the authors envisage studying the time-dependent Couette flow of incompressible non-Newtonian Eyring-Powell model with porous walls. The resulting unsteady problem is solved by means of homotopy analysis method (HAM) [41-58], which is very powerful and efficient in finding the analytic solutions for a wide class of nonlinear differential equations. The method gives more realistic series solution that converges very rapidly in physical problems. The convergence region for the series solution is found with the help of $\hbar$-curve. For a given amount of computational effort, one can usually construct a rational approximation that has smaller overall error in given domain than a polynomial approximation [59]. Our goal is to make the maximum error as small as possible. For this purpose, Pade approximants and Taylor polynomials are found. The graphs of the error for Pade approximants and Taylor polynomials are plotted and it is observed that maximum absolute error occurs at the end point $Y=1$. The graphs for the horizontal velocity profile and shear stress at the wall for injection/suction are drawn and discussed in detail. The tables for the initial slope and wall shear stress are also constructed and discussed. More significantly, the series solution clearly demonstrates how various physical parameters play their part in determining properties of the flow.

2. Mathematical Description of the Problem

Consider an unsteady, incomprssible, non-Newtonian, Couette flow problem for the Eyring-Powell model, in which the bottom wall is fixed and subjected to a mass injection velocity v_w and there is mass suction velocity v_w at the top wall, $v_w > 0$ correspond to injection and $v_w < 0$ correspond to suction. The top plate is stationary when $t < 0$, there is only mass transfer in the transverse direction, say y- direction. At $t = 0$, the top wall is started impulsively to a constant velocity U_0. The Eyring-Powell model is derived from the theory of rate processes, which describes the shear of a non-Newtonian flow. The Eyring-Powell model can be used in some cases to describe the viscous behavior of polymer solutions and viscoelastic suspensions over a wide range of shear rates. The stress tensor in the Eyring-Powell model for non-Newtonian fluids is given by [33]

$$T = \mu \nabla V + \frac{1}{\beta}\sinh^{-1}\left(\frac{1}{c}\nabla V\right), \tag{1}$$

where μ is the dynamic viscosity, β and c are the characteristics of the Eyring-Powell model. Taking the second order approximation of the function $\sinh^{-1}(1/c\nabla V)$ as

$$\sinh^{-1}\left(\frac{1}{c}\nabla V\right) = \frac{1}{c}\nabla V - \frac{1}{6}\left(\frac{1}{c}\nabla V\right)^3, \quad \left|\frac{1}{c}\nabla V\right| \ll 1. \tag{2}$$

The governing equation for this problem can be obtained as

$$\begin{aligned}&\frac{\partial u(y,t)}{\partial t} + v_w \frac{\partial u(y,t)}{\partial y} - \left(\nu + \frac{1}{\rho\beta c}\right)\frac{\partial^2 u(y,t)}{\partial y^2} \\ &+ \frac{1}{2\beta c^3}\left(\frac{\partial u(y,t)}{\partial y}\right)^2 \frac{\partial^2 u(y,t)}{\partial y^2} = 0,\end{aligned} \tag{3}$$

$$u(0,t) = 0, \; u(h,t) = U_0, \; u(y,0) = 0, \tag{4}$$

where $\nu = \mu/\rho$ is the kinematic viscosity, ρ is the density of the fluid, bottom wall is located at $y = 0$, top wall is located at $y = h$ and U_0 is the velocity at the upper wall. Equations (3) and (4) can be non-dimensionalized by defining

$$U = \frac{u}{U_0}, \; Y = \frac{y}{h}, \text{ and } T = \frac{t}{\tau_c} = \frac{t\nu}{h^2}. \tag{5}$$

Then Equations (3) and (4) become

$$\begin{aligned}&\frac{\partial U(Y,T)}{\partial T} + R_e \frac{\partial U(Y,T)}{\partial Y} - (1+m)\frac{\partial^2 U(Y,T)}{\partial Y^2} \\ &+ m\lambda\left(\frac{\partial U(Y,T)}{\partial Y}\right)^2 \frac{\partial^2 U(Y,T)}{\partial Y^2} = 0,\end{aligned} \tag{6}$$

$$U(0,T) = 0, \; U(1,T) = 1, \; U(Y,0) = 0, \tag{7}$$

where $R_e = v_w h/\nu$ is the Reynolds number,
$m = 1/\rho\beta\nu c$ is the fluid parameter and
$\lambda = \rho U_0^2/2c^2h^2$ is the local non-Newtonian parameter based on velocity of plate U_0. Using stream function relations with velocity [60] Equations (6) and (7) become

$$\begin{aligned}&f'(Y,T) + T\frac{\partial f'(Y,T)}{\partial T} \\ &+ R_e\, T f''(Y,T) - (1+m) T f'''(Y,T) \\ &+ m\lambda\, T^3 \left(f''(Y,T)\right)^2 f'''(Y,T) = 0,\end{aligned} \tag{8}$$

$$\begin{aligned}&f'(0,T) = \left.\frac{\partial f(Y,T)}{\partial Y}\right|_{Y=0} = 0, \\ &f'(1,T) = \left.\frac{\partial f(Y,T)}{\partial Y}\right|_{Y=1} = \frac{1}{T}, \quad T > 0,\end{aligned} \tag{9}$$

where, $f(Y,T)$ is the reduced stream function and prime denotes ordinary derivative w. r. t Y. When $T = 0$, Equation (8) becomes

$$f'(Y,T)=0\Rightarrow\frac{\partial f(Y,T)}{\partial Y}=0\Rightarrow f(Y,T)=\varphi(T)$$
$$\Rightarrow U(Y,T)=0\Rightarrow U(Y,0)=0.$$

where $\varphi(T)$ is some arbitrary unknown function of T.

3. Analytic Solution

To start with the homotopy analysis method it is very much important to choose an initial guess approximation and a linear operator. Therefore, due to the boundary conditions (9) it is reasonable to choose the initial guess approximation

$$f_0(Y,T)=\frac{e^{1-Y}+e^{1+Y}}{T\left(e^2-1\right)},\tag{10}$$

and the linear operator

$$\mathrm{L}(f)\equiv\frac{\partial^2 f(Y,T)}{\partial Y^2}-f(Y,T),\tag{11}$$

which satisfies the following property:

$$\mathrm{L}\left[C_1e^{-Y}+C_2e^{Y}\right]=0,\tag{12}$$

where C_1 and C_2 are arbitrary constants. If $p\in[0,1]$ is an embedding parameter and $\hbar_1$ is auxiliary non zero parameter then the so-called zero-order deformation equation is

$$\begin{aligned}&(1-p)\mathrm{L}\left[\phi(Y,T;p)-f_0(Y,T)\right]\\&=p\hbar_1\mathrm{N}\left[\phi(Y,T;p)\right],\end{aligned}\tag{13}$$

subject to boundary conditions

$$\begin{aligned}\phi'(0,T;p)&=\left.\frac{\partial\phi(Y,T;p)}{\partial Y}\right|_{Y=0}=0,\\ \phi'(1,T;p)&=\left.\frac{\partial\phi(Y,T;p)}{\partial Y}\right|_{Y=1}=\frac{1}{T},\end{aligned}\tag{14}$$

where

$$\begin{aligned}\mathrm{N}\left[\phi(Y,T;p)\right]&=\frac{\partial\phi(Y,T;p)}{\partial Y}+T\frac{\partial^2\phi(Y,T;p)}{\partial T\partial Y}\\&+R_eT\frac{\partial^2\phi(Y,T;p)}{\partial Y^2}-(1+m)T\frac{\partial^3\phi(Y,T;p)}{\partial Y^3}\\&+m\lambda T^3\left(\frac{\partial^2\phi(Y,T;p)}{\partial Y^2}\right)^2\frac{\partial^3\phi(Y,T;p)}{\partial Y^3},\end{aligned}\tag{15}$$

and when $p=0$ and $p=1$, then

$$\phi(Y,T;0)=f_0(Y,T),\quad\phi(Y,T;1)=f(Y,T),\tag{16}$$

As the embedding parameter p increases from 0 to 1, $\phi(Y,T;p)$ varies (or deforms) from the initial approximation $f_0(Y,T)$ to the solution $f(Y,T)$. Using Taylor's theorem and Equation (16), one obtains

$$\phi(Y,T;p)=f_0(Y,T)+\sum_{m=1}^{\infty}f_m(Y,T)p^m,\tag{17}$$

in which

$$f_m(Y,T)=\left.\frac{1}{m!}\frac{\partial^m\phi(Y,T;p)}{\partial p^m}\right|_{p=0},\quad(m\geq1).\tag{18}$$

Clearly, the convergence of the series (17) depends upon $\hbar_1$. Assume that $\hbar_1$ is selected such that the series (17) is convergent at $p=1$, then due to equation (16) we have

$$f(Y,T)=f_0(Y,T)+\sum_{m=1}^{\infty}f_m(Y,T).\tag{19}$$

For the *mth* order deformation problem, we differentiate Equations (13) and (14) *m*-times w.r.t p and then setting $p=0$ and finally dividing it by $m!$ the *mth*-order deformation equation for $m\geq1$ is given by

$$\mathrm{L}\left[f_m(Y,T)-\chi_mf_{m-1}(Y,T)\right]=\hbar_1\mathrm{R}_m(Y,T),\tag{20}$$

$$\begin{aligned}f_m'(0,T)&=\left.\frac{\partial f_m(Y,T)}{\partial Y}\right|_{Y=0}=0,\\ f_m'(1,T)&=\left.\frac{\partial f_m(Y,T)}{\partial Y}\right|_{Y=1}=0,\end{aligned}\tag{21}$$

where

$$\begin{aligned}\mathrm{R}_m(Y,T)&=\frac{\partial f_{m-1}(Y,T)}{\partial Y}+T\frac{\partial^2 f_{m-1}(Y,T)}{\partial T\partial Y}\\&+R_eT\frac{\partial^2 f_{m-1}(Y,T)}{\partial Y^2}-(1+m)T\frac{\partial^3 f_{m-1}(Y,T)}{\partial Y^3}\\&+m\lambda T^3\sum_{k=0}^{m-1}\frac{\partial^2 f_{m-1-k}(Y,T)}{\partial Y^2}\\&\sum_{l=0}^{k}\frac{\partial^2 f_k(Y,T)}{\partial Y^2}\frac{\partial^3 f_l(Y,T)}{\partial Y^3},\end{aligned}\tag{22}$$

$$\chi_m=\begin{cases}0, & m\leq1\\ 1, & m\geq2.\end{cases}\tag{23}$$

Following the HAM and trying higher iterations with the unique and proper assignment of the results converge to the exact solution:

$$\begin{aligned}f(Y,T)&\approx f_0(Y,T)+f_1(Y,T)\\&+f_2(Y,T)+\cdots+f_m(Y,T),\end{aligned}\tag{24}$$

using the symbolic computation software such as MATHEMATICA, MATLAB or MAPLE to solve the system of linear equations, (20), with the boundary conditions (21), and successively obtain

4. Convergence of the Analytic Solution

The auxiliary parameter $\hbar_1$ gives the convergence region and rate of approximation for the homotopy analysis method for above problem. For this purpose, the $\hbar$-curve is plotted for above problem. It is obvious from **Figure 1** that the range for the admissible values for $\hbar_1$ is $-0.5 < \hbar_1 < 0.5$. The solution series converges in the whole region of Y and T for $\hbar_1 = -0.1$ or $\hbar_1 = 0.1$.

5. Pade Approximation

Pade approximants make up the best approximation of a function in the form of a rational function of a given order. Pade approximation helps us in improving the ac curacy of approximate solution available in the form of a polynomial. Pade approximants are better approximation of a function than its Taylor series, they work even in those cases where Taylor series does not converge. Pade

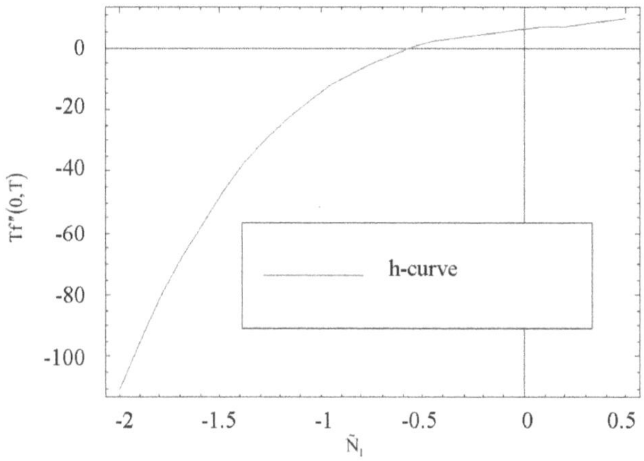

Figure 1. $\hbar_1$**-curve for the stream function** $Tf(Y,T)$ **at** $m = 0.001$, $\lambda = 0.001$, $T = 3$ **and** $R_e = 0.0001$.

$$
\begin{aligned}
f_1(Y,T) = {} & \frac{e^{1+Y}\hbar_1}{(e^2-1)^3} - \frac{2e^{3+Y}\hbar_1}{(e^2-1)^3} + \frac{e^{5+Y}\hbar_1}{(e^2-1)^3} + \frac{e^{1+Y}\hbar_1 m}{(e^2-1)^3} - \frac{2e^{3+Y}\hbar_1 m}{(e^2-1)^3} + \frac{e^{5+Y}\hbar_1 m}{(e^2-1)^3} \\
& + \frac{e^{1-Y}}{(e^2-1)^3 T} - \frac{2e^{3-Y}}{(e^2-1)^3 T} + \frac{e^{5-Y}}{(e^2-1)^3 T} + \frac{e^{1+Y}}{(e^2-1)^3 T} - \frac{2e^{3+Y}}{(e^2-1)^3 T} + \frac{e^{5+Y}}{(e^2-1)^3 T} \\
& - \frac{e^{1+Y}\hbar_1}{(e^2-1)^3 T} + \frac{2e^{3+Y}\hbar_1}{(e^2-1)^3 T} - \frac{e^{5+Y}\hbar_1}{(e^2-1)^3 T} + \frac{e^{1-Y}+e^{1+Y}}{(e^2-1)T} - \frac{e^{1-Y}\hbar_1 Y}{2(e^2-1)^3} + \frac{e^{3-Y}\hbar_1 Y}{(e^2-1)^3} \\
& - \frac{e^{5-Y}\hbar_1 Y}{2(e^2-1)^3} - \frac{e^{1+Y}\hbar_1 Y}{(e^2-1)^3} + \frac{e^{3+Y}\hbar_1 Y}{(e^2-1)^3} - \frac{e^{5+Y}\hbar_1 Y}{2(e^2-1)^3} - \frac{e^{1-Y}\hbar_1 mY}{2(e^2-1)^3} + \frac{e^{3-Y}\hbar_1 mY}{(e^2-1)^3} \\
& - \frac{e^{5-Y}\hbar_1 mY}{2(e^2-1)^3} - \frac{e^{1+Y}\hbar_1 mY}{2(e^2-1)^3} + \frac{e^{3+Y}\hbar_1 mY}{(e^2-1)^3} - \frac{e^{5+Y}\hbar_1 mY}{2(e^2-1)^3} + \frac{e^{1-Y}\hbar_1 Y}{2(e^2-1)^3 T} - \frac{e^{3-Y}\hbar_1 Y}{(e^2-1)^3 T} \\
& + \frac{e^{5-Y}\hbar_1 Y}{2(e^2-1)^3 T} + \frac{e^{1+Y}\hbar_1 Y}{2(e^2-1)^3 T} - \frac{e^{3+Y}\hbar_1 Y}{(e^2-1)^3 T} + \frac{e^{5+Y}\hbar_1 Y}{2(e^2-1)^3 T} - \frac{e^{3-3Y}\hbar_1 m\lambda}{8(e^2-1)^3} + \frac{3e^{1-Y}\hbar_1 m\lambda}{8(e^2-1)^3} \\
& + \frac{3e^{3-Y}\hbar_1 m\lambda}{8(e^2-1)^3} - \frac{3e^{5-Y}\hbar_1 m\lambda}{8(e^2-1)^3} + \frac{3e^{1+Y}\hbar_1 m\lambda}{8(e^2-1)^3} - \frac{11e^{3+Y}\hbar_1 m\lambda}{8(e^2-1)^3} - \frac{3e^{5+Y}\hbar_1 m\lambda}{8(e^2-1)^3} + \frac{e^{3+3Y}\hbar_1 m\lambda}{8(e^2-1)^3} \\
& + \frac{e^{3-Y}\hbar_1 mY\lambda}{2(e^2-1)^3} + \frac{e^{3+Y}\hbar_1 mY\lambda}{2(e^2-1)^3} + \frac{e^{3-Y}\hbar_1 R_e}{(e^2-1)^3} - \frac{e^{5-Y}\hbar_1 R_e}{(e^2-1)^3} + \frac{e^{3+Y}\hbar_1 R_e}{(e^2-1)^3} - \frac{e^{5+Y}\hbar_1 R_e}{(e^2-1)^3} \\
& - \frac{e^{1-Y}\hbar_1 YR_e}{2(e^2-1)^3} + \frac{e^{3-Y}\hbar_1 YR_e}{(e^2-1)^3} - \frac{e^{5-Y}\hbar_1 YR_e}{2(e^2-1)^3} + \frac{e^{1+Y}\hbar_1 YR_e}{2(e^2-1)^3} - \frac{e^{3+Y}\hbar_1 YR_e}{(e^2-1)^3} + \frac{e^{5+Y}\hbar_1 YR_e}{2(e^2-1)^3}
\end{aligned}
\tag{25}
$$

approximations are also used to enlarge the interval of convergence of approximate series solution [61]. A standard MATHEMATICA routine can be used to find Pade approximant for the function $f(Y)$. A $[2,2]$ Pade approximant for the solution in Equation (24) at $\lambda = 1.5$, $R_e = 0.01$, $m = 0.0005$, $\hbar_1 = -0.1$, $T = 0.5$ can be written as

$$R_{[2,2]}(Y) = \frac{12.2395 + 0.201978Y + 5.09203Y^2}{1 + 0.0165022Y - 0.0830199Y^2}, \quad (26)$$

Figure 2 depicts the graph of $f(Y)$ and its Pade approximant $R_{[2,2]}(Y)$. From **Figure 2** we observe that the difference between the HAM solution $f(Y)$ and Pade approximate solution $R_{[2,2]}(Y)$ is so small as to be invisible on this scale. The graph of the error $E_{R_{[2,2]}}(Y) = f(Y) - R_{[2,2]}(Y)$ over $[0,1]$ for the Pade approximant $R_{[2,2]}(Y)$ is shown in **Figure 3**. We note that the maximum absolute error occur at the end point, $E_{R_{[2,2]}}(1) \le 0.0205034$. The Taylor polynomials for $f(Y)$ of degree $N = 4$ and $N = 5$ at $\lambda = 1.5$, $R_e = 0.01$, $m = 0.0005$, $\hbar_1 = -0.1$, $T = 0.5$ obtained as

$$P_4(Y) = 12.2395 + 8.66081 \times 10^{-16} Y + 6.10815Y^2 - 0.100798Y^3 + 0.508761Y^4, \quad (27)$$

$$P_5(Y) = 12.2395 + 8.66081 \times 10^{-16} Y + 6.10815Y^2 - 0.100798Y^3 + 0.508761Y^4 - 0.0101178Y^5. \quad (28)$$

Figure 4 illustrates that the difference between $f(Y)$ and $P_4(Y)$ is invisible on this scale. **Figure 5** indicates the graph of the error $E_{P_4}(Y) = f(Y) - P_4(Y)$ over $[0,1]$ for the Taylor approximation $P_4(Y)$. It is observed that the largest absolute error occur at the end point, $E_{P_4}(1) \le 0.0067851$. **Figure 6** describes that the

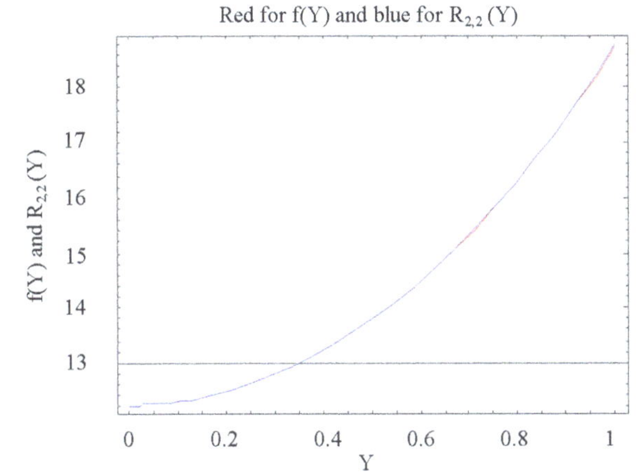

Figure 2. The graph of $f(Y)$ and its Pade approximation $R_{[2,2]}(Y)$.

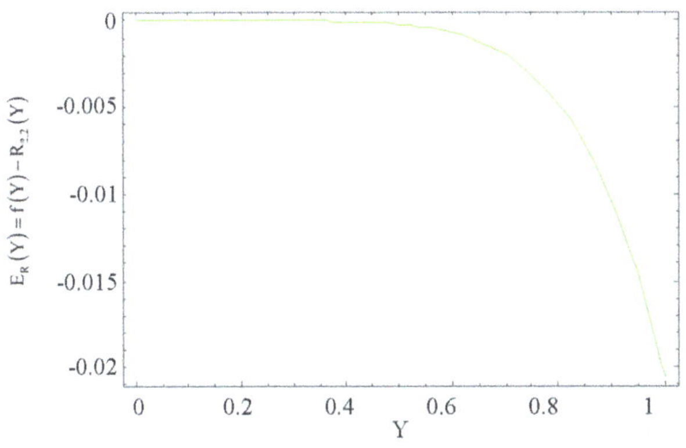

Figure 3. The graph of the error $E_{R_{[2,2]}}(Y) = f(Y) - R_{[2,2]}(Y)$ for the Pade approximation $R_{[2,2]}(Y)$.

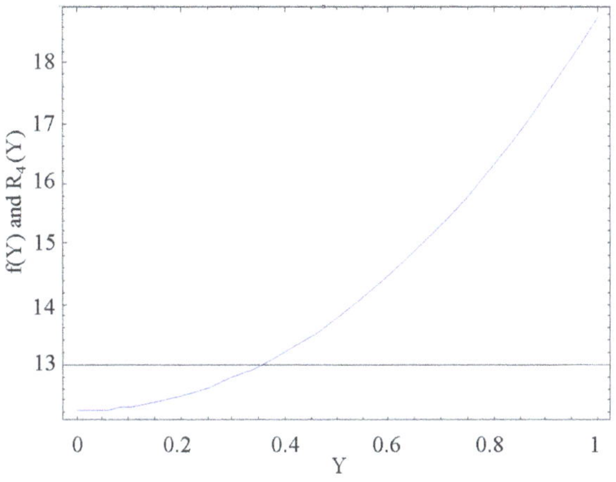

Figure 4. The graph of $f(Y)$ and its Taylor approximation $P_4(Y)$.

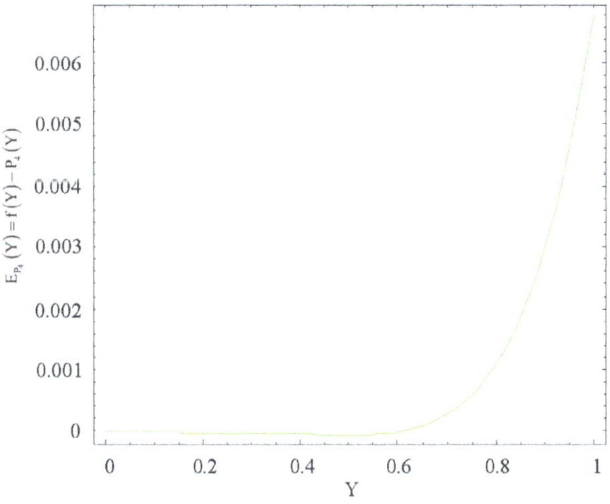

Figure 5. The graph of the error $E_{P_4}(Y) = f(Y) - P_4(Y)$ for the Taylor approximation $P_4(Y)$.

difference between $f(Y)$ and $P_5(Y)$ is also invisible on this scale. **Figure 7** explains the graph of the error $E_{P_5}(Y) = f(Y) - P_5(Y)$ over $[0,1]$ for the Taylor approximation $P_5(Y)$. The maximum absolute error occur at the end point, $E_{P_5}(1) \leq 0.0169029$. It is observed that the increase in the degree of Taylor polynomial increases the maximum absolute error.

6. Graphs and Discussion

In this part we discuss the graphs for the variation of the horizontal velocity profiles $Tf'(Y,T)$ and shear stress at the wall $Tf''(0,T)$ with distance from the wall Y for different values of Reynolds number R_e, local non-Newtonian parameter λ, fluid parameter m, homotopy parameter $\hbar_1$ and time T.

Figures 8 and **9** describe the variation of the horizontal velocity profiles $Tf'(Y,T)$ with Y for several values of m by keeping λ, R_e, $\hbar_1$ and T fixed. **Figure 8** shows that when there is mass injection $R_e > 0$ at

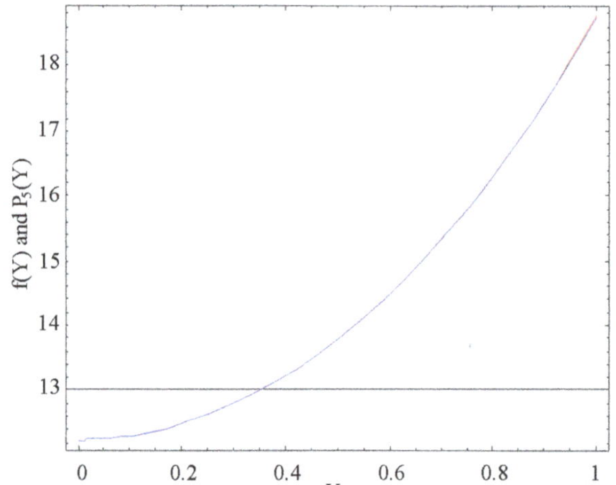

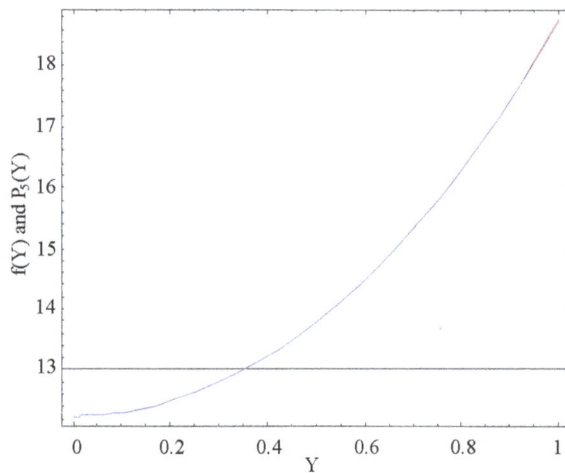

Figure 6. The graph of $f(Y)$ and its Taylor approximation $P_5(Y)$.

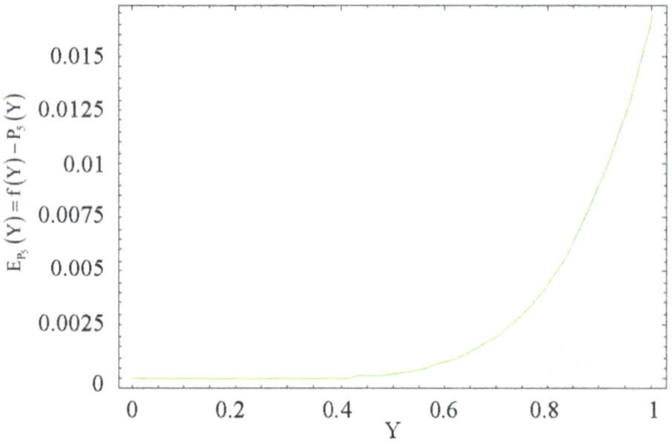

Figure 7. The graph of the error $E_{P_5}(Y) = f(Y) - P_4(Y)$ for the Taylor approximation $P_5(Y)$.

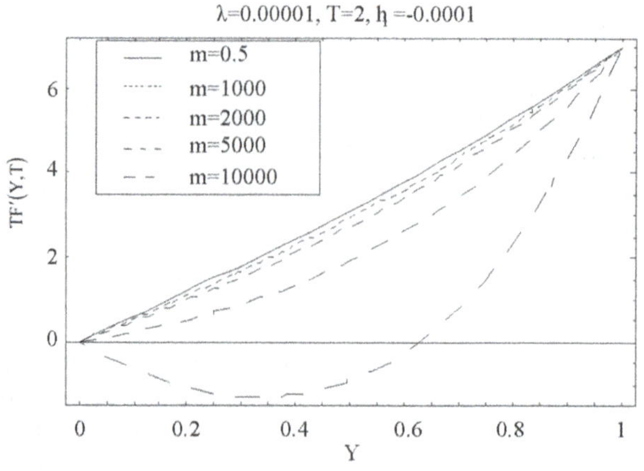

Figure 8. The graph of the horizontal velocity profiles $Tf'(Y,T)$ with Y for several values of m and $R_e = 0.0001$.

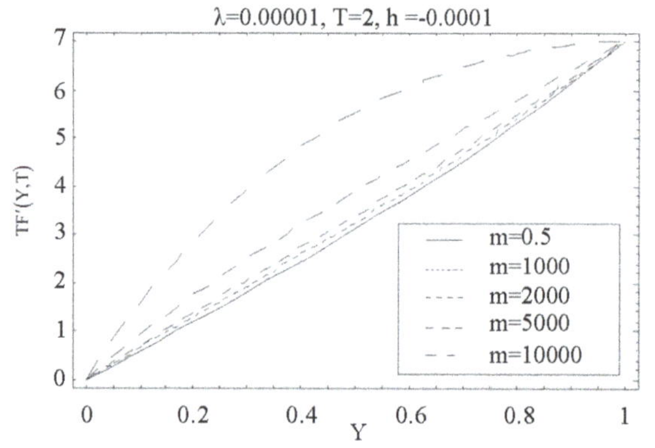

Figure 9. The graph of the horizontal velocity profiles $Tf'(Y,T)$ with Y for several values of m and $R_e = -0.0001$.

the bottom wall, with increase in fluid parameter m, horizontal velocity profiles $Tf'(Y,T)$ shows decreasing trend. **Figure 9** shows that when there is mass suction $R_e < 0$ at the top wall, with increase in m, $Tf'(Y,T)$ increases at all points. **Figures 10** and **11** indicate the variation of the horizontal velocity profiles $Tf'(Y,T)$ with Y for several values of λ by keeping m, R_e, $\hbar_1$ and T fixed. **Figure 10** shows that when there is mass injection $R_e > 0$ at the bottom wall, with increase in fluid parameter λ, horizontal velocity profiles $Tf'(Y,T)$ increases at all points. **Figure 11** shows that when there is mass suction $R_e < 0$ at the top wall, with increase in λ, $Tf'(Y,T)$ increases in magnitude but have negative values, an inverted behavior is observed, which is consistent with what we expected. **Figures 12** and **13** illustrate the variation of the horizontal velocity profiles $Tf'(Y,T)$ with Y for several values of time T, for fixed values of λ, m, R_e and $\hbar_1$. **Figures 12** and **13** are plotted for positive value of λ. **Figure 12** shows that for mass injection $R_e > 0$ at the bottom wall, with increase in T, horizontal velocity profiles $Tf'(Y,T)$ shows increasing trend in magnitude but have negative values. From **Figure 13** it is clear that for mass suction $R_e < 0$ at the top wall, with increase in T, $Tf'(Y,T)$ increases at all points and the reverse behavior is observed. **Figures 14** and **15** describe the variation of the horizontal velocity profiles $Tf'(Y,T)$ with Y for several values of time T, for fixed values of λ, m, R_e and h_1. **Figures 14** and **15** are plotted for negative value of λ. From **Figure 14** it is observed that for mass injection $R_e > 0$ at the bottom wall, with increase in T, horizontal velocity profiles $Tf'(Y,T)$ shows increasing trend in magnitude but have negative values. From Fig. 15 it is seen that for mass suction $R_e < 0$ at the top wall, with increase in T, $Tf'(Y,T)$ increases at all points and have positive values, that is, a reverse trend is observed. From the comparison of the **Figure 12** to **15** we observe that for positive and negative values of λ the variation of horizontal velocity profiles $Tf'(Y,T)$ is same. The **Figures 8** to **15** shows that mass transfer has a dominant effect on the horizontal velocity profiles $Tf'(Y,T)$. We observe from the graphs 8 to 15 that the

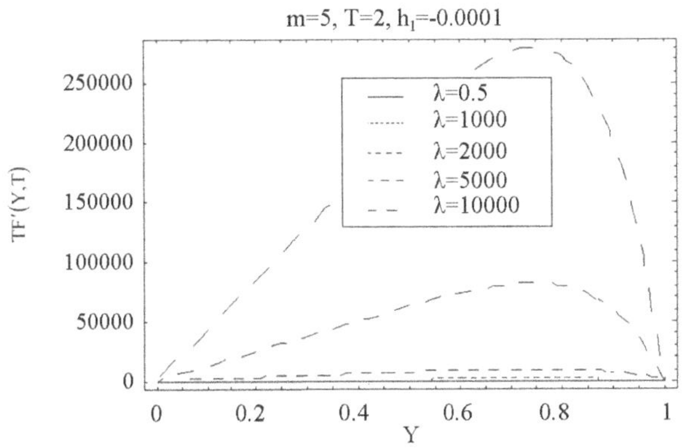

Figure 10. The graph of the horizontal velocity profiles $Tf'(Y,T)$ with Y for several values of λ and $R_e = 0.0001$.

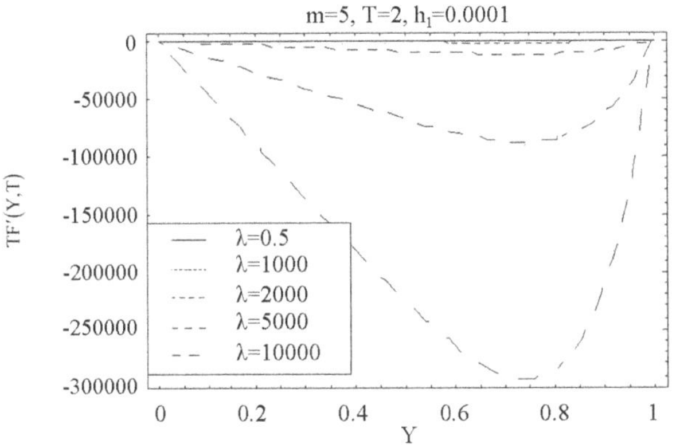

Figure 11. The graph of the horizontal velocity profiles $Tf'(Y,T)$ with Y for several values of λ and $R_e = -0.0001$.

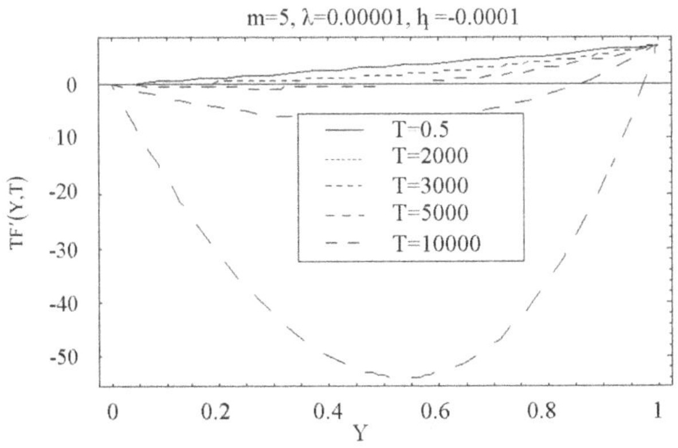

Figure 12. The graph of the horizontal velocity profiles $Tf'(Y,T)$ with Y for several values of T, for +ve value of λ and $R_e = 0.0001$.

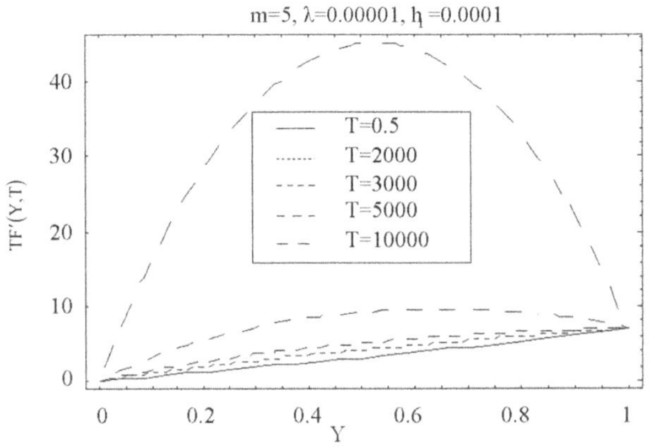

Figure 13. The graph of the horizontal velocity profiles $Tf'(Y,T)$ with Y for several values of T, for +ve value of λ and $R_e = -0.0001$.

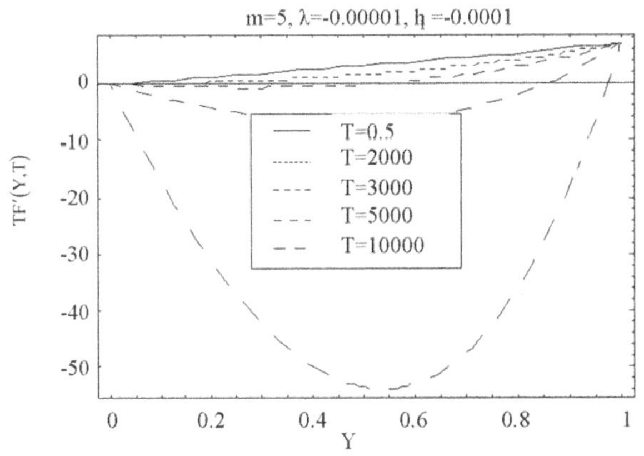

Figure 14. The graph of the horizontal velocity profiles $Tf'(Y,T)$ with Y for several values of T, for −ve value of λ and $R_e = 0.0001$.

fluid material parameters m and λ enhance the magnitude of the velocity profile. In **Figures 8** to **15** it is observed that the behavior of suction is the reverse of the injection in all the cases, which is a confirmation for the validity of our results. Graphs from 8 to 15 are plotted for large values of the parameter m, λ and T, because for small values it is observed that the curves of different profiles overlaps and behavior is not clear, whether it is increasing or decreasing.

Figure 16 and **17** elucidate the variation of the shear stress at the wall $Tf''(0,T)$ with the parameter λ for several values of m, for fixed values of R_e, T and $\hbar_1$. **Figure 16** is for mass injection $R_e > 0$ at the bottom wall and **Figure 17** is for mass suction $R_e < 0$ at the top wall. **Figure 16** shows that with increase in m, shear stress at the wall $Tf''(0,T)$ increases at all points for all values of λ and have positive values. **Figure 17** shows that with increase in m, shear stress at the wall $Tf''(0,T)$ increases in magnitude for all values of λ but have negative values, inverse behavior is observed. **Figures 18** and **19** illustrate the variation of the shear stress at the wall $Tf''(0,T)$ with the parameter m for several values of λ, for fixed values of R_e, T and $\hbar_1$. **Figure 18** shows that when there is mass injection $R_e > 0$ at the bottom wall with increase in λ, shear stress at the wall $Tf''(0,T)$ increases and positive for all values of m. From **Figure 19** it is observed that for suction at top wall, with increase in λ, $Tf''(0,T)$ increases in magnitude and have positive and negative values both for all values of m. **Figures 20** and **21** describe the variation of the shear stress at the wall $Tf''(0,T)$ with time T for several values of λ, for fixed values of R_e, m and $\hbar_1$. **Figure 20** shows that for mass injection $R_e > 0$ at the bottom wall, with increase in λ, shear stress at the wall $Tf''(0,T)$ increases and positive for all values of time T. **Figure 21**

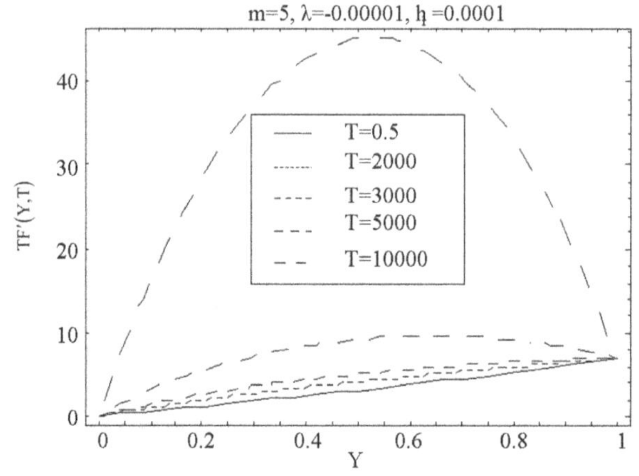

Figure 15. The graph of the horizontal velocity profiles $Tf'(Y,T)$ with Y for several values of T, for −ve value of λ and $R_e = -0.0001$.

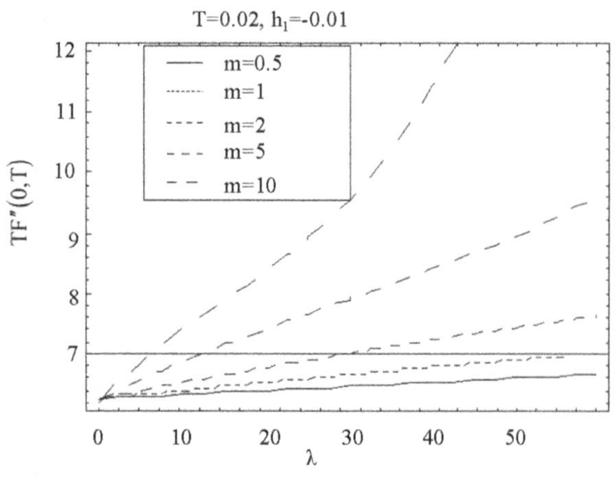

Figure 16. The graph of the shear stress at the wall $Tf''(0,T)$ with λ for several values of m and $R_e = 5$.

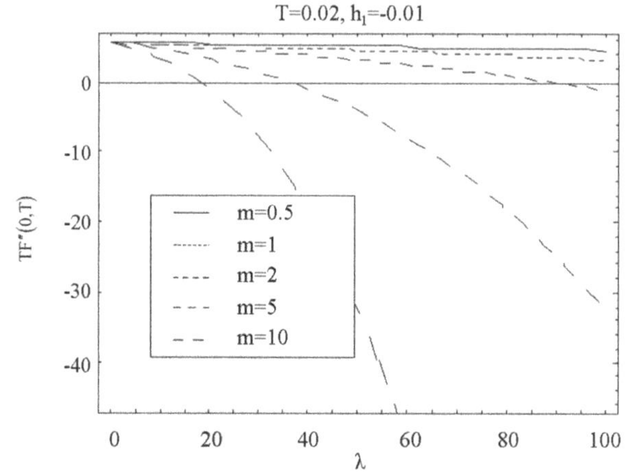

Figure 17. The graph of the shear stress at the wall $Tf''(0,T)$ with λ for several values of m and $R_e = -5$.

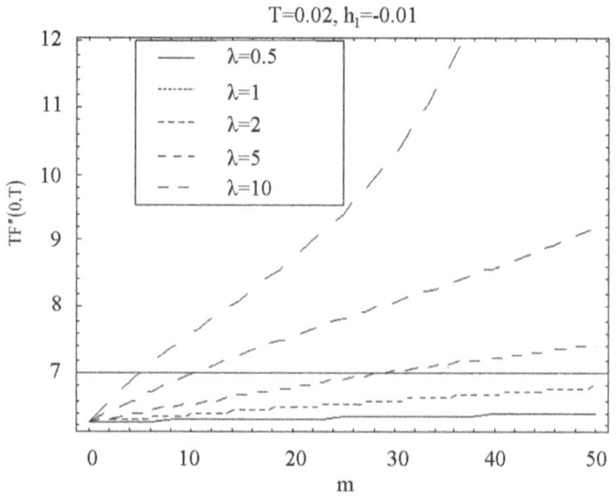

Figure 18. The graph of the shear stress at the wall $Tf''(0,T)$ with m for several values of λ and $R_e = 5$.

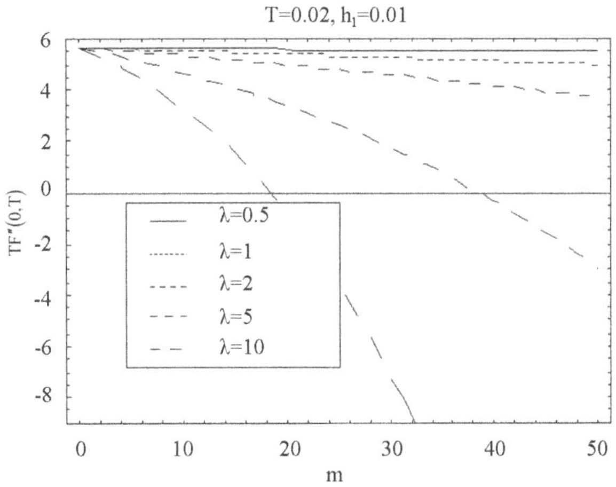

Figure 19. The graph of the shear stress at the wall $Tf''(0,T)$ with m for several values of λ and $R_e=-5$.

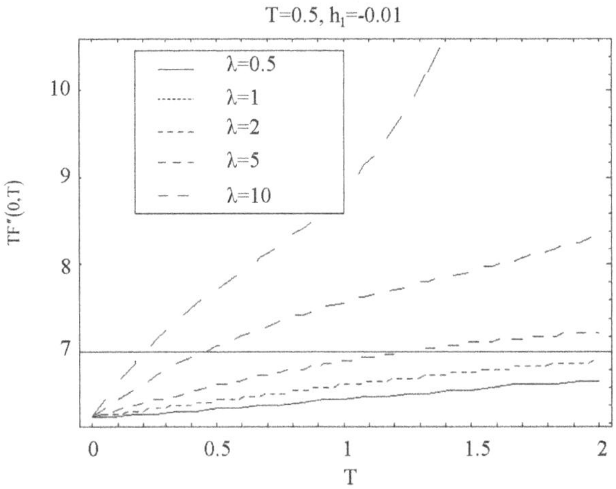

Figure 20. The graph of the shear stress at the wall $Tf''(0,T)$ with T for several values of λ and $R_e=5$.

shows that for mass suction $R_e<0$ at the top wall, with increase in λ, $Tf''(0,T)$ increases in magnitude and have positive and negative values both for all values of time T. In **Figures 16** to **21** it is observed that shear stress for suction has reverse behavior of injection.

7. Tables

Here **Tables 1-2** are prepared for the variation of the initial slopes $R=f'(0)$ and dimensionless shear stress at the wall $Tf''(0,T)$. These results are obtained for different values of $\hbar_1$ laying in the interval of convergence, for different order of approximations.

The diagonal Pade approximants can be used to investigate the mathematical behavior of the solution $f(Y)$ to determine the initial slope $f'(0)$. It can be seen from **Table 1** that for a fixed value of R_e, m, T and $\hbar_1$, with the increase in λ, the initial slope of $f(Y)$ for Pade approximants $R_{[1,1]}$, $R_{[2,2]}$ and $R_{[5,5]}$ increases.

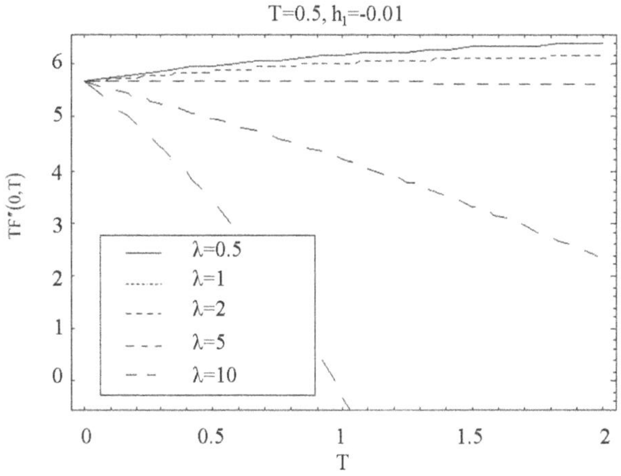

Figure 21. The graph of the shear stress at the wall $Tf''(0,T)$ with T for several values of λ and $R_e=-5$.

For Pade approximants $R_{[3,3]}$ and $R_{[4,4]}$ initial slope becomes negative and magnitude of the slope decreases with increase in λ.

It can be seen from **Table 2** that for mass injection $R_e>0$ at the bottom wall, for a fixed value of T and $\hbar_1$, with the increase in m the shear stress at the wall increases for all values of λ. It is observed that with the increase of mass injection shear stress at the wall also increases. It is also noted that for fixed value of T, $\hbar_1$, R_e and m, with the increase in λ shear stress at the wall increases. For all the parameters there is an increase is observed.

8. Conclusions

In this study, a series of solutions for the horizontal velocity field of unsteady incompressible Couette flow with Eyring-Powell model are constructed. The results are discussed under the effects of parameters m, λ, $\hbar_1$ and R_e through graphs and tables. We have following observations about the effects of pertinent parameters in the flow field on the horizontal velocity, shear stress at the wall and on initial slope of $f(Y)$.

- The solution series converges in the whole region of Y and T for $-0.5<\hbar_1<0.5$.
- We have considered the general Pade and Taylor approximations of $f(Y)$. The polynomials of the rational approximations are given in analytic form.
- We note that the difference between the HAM solution $f(Y)$ and Pade approximate solution $R_{[2,2]}(Y)$ is so small as to be invisible on this scale $[0,1]$.
- We observe that the maximum absolute error for Pade approximant and Taylor approximations occur at the end point $Y=1$.
- It is observed that increase in the degree of Taylor polynomial increases the maximum absolute error.

Table 1. Variation of the initial slopes $R = f'(0)$ at $R_e = 0.01$, $m = 0.0005$, $T = 0.5$ and $\hbar_1 = -0.1$ for various values of λ.

λ	$R_{[1,1]} = f'(0)$	$R_{[2,2]} = f'(0)$	$R_{[3,3]} = f'(0)$	$R_{[4,4]} = f'(0)$	$R_{[5,5]} = f'(0)$
0.1	5.18127	0.387940	−0.0330993	−0.000251122	0.0027086
0.2	5.18129	0.387971	−0.0330956	−0.000249615	0.00277907
0.3	5.18131	0.388002	−0.0330918	−0.000248108	0.00284854
0.4	5.18133	0.388033	−0.0330881	−0.000246601	0.00291704
0.5	5.18136	0.388065	−0.0330844	−0.000245096	0.00298457
0.6	5.18138	0.388096	−0.0330807	−0.000243591	0.00305117
0.7	5.18140	0.388127	−0.0330770	−0.000242086	0.00311684
0.8	5.18143	0.388158	−0.0330733	−0.000240582	0.00318161
0.9	5.18145	0.388189	−0.0330696	−0.000239079	0.00324548
1	5.18147	0.388221	−0.0330659	−0.000237577	0.00330848

Table 2. Variation of the dimensionless shear stress at the wall $Tf''(0,T)$ at $\hbar_1 = -0.01$ and $T = 0.01$.

R_e	m	$\lambda = 1$	$\lambda = 2$	$\lambda = 3$
		$Tf''(0,T)$	$Tf''(0,T)$	$Tf''(0,T)$
0.2	0.0	5.98596	5.98596	5.98596
	5.0	5.98892	5.99337	5.99781
	10.0	5.99188	6.00076	6.00961
	15.0	5.99484	6.00813	6.02137
	20.0	5.99780	6.01549	6.03309
0.4	0.0	5.98598	5.98598	5.98598
	5.0	5.98894	5.99339	5.99783
	10.0	5.99190	6.00078	6.00963
	15.0	5.99486	6.00815	6.02139
	20.0	5.99781	6.01551	6.03310
0.6	0.0	5.98600	5.98600	5.98600
	5.0	5.98896	5.99340	5.99784
	10.0	5.99192	6.0008	6.00965
	15.0	5.99488	6.00817	6.02141
	20.0	5.99783	6.01553	6.03312

- For positive and negative values of local non-Newtonian parameter λ, the variation of the horizontal velocity profiles is same.
- The fluid material parameters m and λ enhance the magnitude of the velocity profile.
- It is noted that the mass transfer has a dominant effect on the velocity profile and in all cases behavior of suction is the reverse of the injection.
- The curves of the velocity profile for small values of m, λ and T overlaps and behavior is not explainable.
- For mass suction and injection at the bottom and top wall shear, stress at the wall increases in all cases but has opposite sign.

Initial slope of $f(Y)$ for Pade approximants $R_{[1,1]}$, $R_{[2,2]}$, $R_{[5,5]}$ increases and for $R_{[3,3]}$, $R_{[4,4]}$ it decreases with increase in λ.

REFERENCES

[1] T. Hayat, M. Khan, A. M. Siddiqui and S. Asghar, "Transient Flows of a Second Grade Fluid," *International Journal of Non-Linear Mechanics*, Vol. 39, No. 10, 2004, pp. 1621-1633.

[2] T. Hayat, Z. Abbas and M. Sajid, "Heat and Mass Trans-

fer Analysis on the Flow of a Second Grade Fluid in the Presence of Chemical Reaction," *Physics Letters A*, Vol. 372, No. 14, 2008, pp. 2004-2408.

[3] T. Hayat, R. Ellahi and F. M. Mahomed, "Exact Solution of a Thin Film Flow of an Oldroyd 6-Constant Fluid over a Moving Belt," *Communications in Nonlinear Science and Numerical Simulation*, Vol. 14, No. 1, 2009, pp. 133-139.

[4] T. Hayat, R. Ellahi and F. M. Mahomed, "Exact Solutions for Thin Film Flow of a Third Grade Fluid Down an Inclined Plane," *Chaos, Solitons & Fractals*, Vol. 38, No. 5, 2008, pp. 1336-1341.

[5] T. Hayat, T. Javed and Z. Abbas, "MHD Flow of a Micropolar Fluid near a Stagnation Point towards a Non-Linear Stretching Surface," *Nonlinear Analysis: Real World Applications*, Vol. 10, No. 3, 2009, pp. 1514-1526.

[6] S. Asghar, M. Khan and T. Hayat, "Magnetohydrodynamic Transient Flows of a Non-Newtonian Fluid," *International Journal of Non-Linear Mechanics*, Vol. 40, No. 5, 2005, pp. 589-601.

[7] M. Khan, Z. Abbas and T. Hayat, "Analytic Solution for Flow of Sisko Fluid through a Porous Medium," *Transport in Porous Media*, Vol. 71, No. 1, 2008, pp. 23-37

[8] M. Khan, S. Hyder Ali, T. Hayat and C. Fetecau, "MHD Flows of a Second Grade Fluid between Two Side Walls Perpendicular to a Plate through a Porous Medium," *International Journal of Non-Linear Mechanics*, Vol. 43, No. 4, 2008, pp. 302-319.

[9] R. Cortell, "A Note on Flow and Heat Transfer of a Viscoelastic Fluid over a Stretching Sheet," *International Journal of Non-Linear Mechanics*, Vol. 41, No. 1, 2006, pp. 78-85.

[10] R. Cortell, "MHD Flow and Mass Transfer of an Electrically Conducting Fluid of Second Grade in a Porous Medium over a Stretching Sheet with Chemically Reactive Species," *Chemical Engineering and Processing: Process Intensification*, Vol. 46, No. 8, 2007, pp. 721-728.

[11] M. Ayub, H. Zaman, M. Sajid and T. Hayat, "Analytical Solution of Stagnation-Point Flow of a Viscoelastic Fluid towards a Stretching Surface," *Communications in Nonlinear Science and Numerical Simulation*, Vol. 13, No. 9, 2008, pp.1822-1835.

[12] M. Ayub, H. Zaman and M. Ahmad, "Series Solution of Hydromagnetic Flow and Heat Transfer with Hall Effect in a Second Grade Fluid over a Stretching Sheet," *Central European Journal of Physics*, Vol. 8, No. 1, 2010, pp. 135-149.

[13] M. Ayub and H. Zaman, "Analytical Solution of Orthogonal Flow Impinging on a Wall with Suction or Blowing," *Journal of Basic and Applied Sciences*, Vol. 6, No. 2, 2010, pp. 93-97.

[14] P. D. Ariel, T. Hayat and S. Asghar, "The Flow of an Elastico-Viscous Fluid Past a Stretching Sheet with Partial Slip," *Acta Mechanica*, Vol. 187, No. 1-4, 2006, pp. 29-35.

[15] K. R. Rajagopal, "A Note on Unsteady Unidirectional Flows of a Non-Newtonian Fluid," *International Journal of Non-Linear Mechanics*, Vol. 17, No. 5-6, 1982, pp. 369-373.

[16] K. R. Rajagopal, "On the Creeping Flow of the Second-Order Fluid," *International Journal of Non-Linear Mechanics*, Vol. 15, No. 2, 1984, pp. 239-246.

[17] K. R. Rajagopal and T. Y. Na, "On Stoke's Problem for a Non-Newtonian Fluid," *Acta Mechanica*, Vol. 48, No. 3-4, 1983, pp. 233-239.

[18] M. E. Erdogan, "Plane Surface Suddenly Set in Motion in a Non-Newtonian Fluid," *Acta Mechanica*, Vol. 108, No. 1-4, 1995, pp. 179-187.

[19] A. M. Siddiqui and P. N. Kaloni, "Certain Inverse Solutions of a Non-Newtonian Fluid," *International Journal of Non-Linear Mechanics*, Vol. 21, No. 6, 1986, pp. 459-473.

[20] C. Fetecau, "Cone and Plate Flow of a Second Grade," *Acta Mechanica*, Vol. 122, No. 1-4, 1977, pp. 225-230.

[21] T. Fang, "A Note on the Incompressible Couette Flow with Porous Walls," *International Communications in Heat and Mass Transfer*, Vol. 31, No. 1, 2004, pp. 31-41.

[22] A. R. A. Khaled and K. Vafai, "The Effect of the Slip Condition on Stokes and Couette Flows due to an Oscillating Wall: Exact Solutions," *International Journal of Non-Linear Mechanics*, Vol. 39, No. 5, 2004, pp. 795-809.

[23] S. Asghar, T. Hayat and P. D. Ariel, "Unsteady Couette Flows in a Second Grade Fluid with Variable Material Properties," *Communications in Nonlinear Science and Numerical Simulation*, Vol. 14, No. 1, 2009, pp. 154-159.

[24] T. Hayat, E. Momoniat and F. M. Mahomed, "Axial Couette Flow of an Electrically Conducting Fluid in an Annulus," *International Journal of Modern Physics B*, Vol. 22, No. 15, 2008, pp. 2489-2500.

[25] T. Hayat and A. H. Kara, "Couette Flow of a Third-Grade Fluid with Variable Magnetic Field," *Mathematical and Computer Modelling*, Vol. 43, No. 1-2, 2006, pp. 132-137

[26] G. S. Seth, Md. S. Ansari and R. Nandkeolyar, "Effects of Rotation and Magnetic Field on Unsteady Couette Flow in a Porous Channel," *Journal of Applied Fluid Mechanics*, Vol. 4, No. 2, 2011, pp. 95-103.

[27] R. Bhaskara and N. D. Bathaiah, "Halll Effects on MHD Couette Flow through a Porous Straight Channel," *Defense Science Journal*, Vol. 32, No. 4, 1982, pp. 313-326.

[28] S. Das, S. L. Maji, M. Guria and R. N. Jana, "Unsteady

MHD Couette Flow in a Rotating System," *Mathematical and Computer Modelling*, Vol. 50, No. 7-8, 2009, pp. 1211-1217.

[29] R. Ganapathy, "A Note on Oscillatory Couette Flow in a Rotating System," *ASME Journal of Applied Mechanics*, Vol. 61, No. 1, 1994, pp. 208-209.

[30] M. Guria, R. N. Jana and S. K. Ghosh, "Unsteady Couette Flow in A Rotating System," *International Journal of Non-Linear Mechanics*, Vol. 41, No. 6-7, 2006, pp. 838-843.

[31] M. Guria, S. Das, R. N. Jana and S. K. Ghosh, "Oscillatory Couette Flow in the Presence of an Inclined Magnetic Field," *Meccanica*, Vol. 44, No. 5, 2009, pp. 555-564.

[32] K. D. Singh, M. G. Gorla and H. Raj, "A Periodic Solution of Oscillatory Couette Flow through Porous Medium in Rotating System," *Indian Journal of Pure and Applied Mathematics*, Vol. 36, 2005, pp. 151-159.

[33] R. E. Powell and H. Eyring, "Mechanism for the Relaxation Theory of Viscosity," *Nature*, Vol. 154, No. 55, 1944, pp. 427-428.

[34] N. T. M. Eldabe, A. A. Hassan and M. A. A. Mohamed, "Effects of Couple Stresses on the MHD of a Non-Newtonian Unsteady Flow between Two Parallel Porous Plates," *Zeitschrift für Naturforschung*, Vol. 58, No. 2, 2003, pp. 204-210.

[35] J. Zueco and O. A. Beg, "Network Numerical Simulation Applied to Pulsatile Non-Newtonian Flow through a Channel with Couple Stress and Wall Mass Flux Effects," *International Journal of Applied Mathematics and Mechanics*, Vol. 5, 2009, pp. 1-16.

[36] K. V. Prasad, P. S. Datti and B. T. Raju, "Momentum and Heat Transfer of a Non-Newtonian Eyring-Powell Fluid over a Non-Isothermal Stretching Sheet," *International Journal of Mathematical Archive*, Vol. 4, No. 1, 2013, pp. 230-241.

[37] M. Patel and M. G. Timol, "Numerical Treatment of MHD Powell-Eyring Fluid Flow Using the Method of Satisfaction of Asymptotic Boundary Conditions," *International Journal of Computer Science*, Vol. 1, No. 2, 2011, pp. 71-78.

[38] V. Sirohi, M. G. Timol and N. I. Kalathia, "Numerical Treatment of Powell-Eyring Fluid Flow past a 90° Wedge," *Reg Journal of Heat and Mass Transfer*, Vol. 6, No. 3, 1984, pp. 219-223.

[39] T. Javed, N. Ali, Z. Abbas and M. Sajid, "Flow of an Eyring-Powell Non-Newtonian Fluid over a Stretching Sheet," *Chemical Engineering Communications*, Vol. 200, 2013, pp. 327-336.

[40] S. Noreen and M. Qasim, "Peristaltic Flow of MHD Eyring-Powell Fluid in a Channel," *The European Physical Journal Plus*, Vol. 128, No. 8, 2013, pp. 91-103.

[41] S. J. Liao, "Notes on the Homotopy Analysis Method: Some Definitions and Theorems," *Communications in Nonlinear Science and Numerical Simulation*, Vol. 14, No. 4, 2009, pp. 983-997

[42] S. J. Liao, J. Su and A. T. Chwang, "Series Solution for a Nonlinear Model of Combined Convective and Radiative Cooling of a Spherical Body," *International Journal of Heat and Mass Transfer*, Vol. 49, No. 15-16, 2006, pp. 2437-2445

[43] S. J. Liao, "An Analytic Solution of Unsteady Boundary-Layer Flows Caused by an Impulsively Stretching Plate," *Communications in Nonlinear Science and Numerical Simulation*, Vol. 11, No. 3, 2006, pp. 326-339.

[44] S. J. Liao, "On the Homotopy Analysis Method for Nonlinear Problems," *Applied Mathematics and Compu- tation*, Vol. 147, No. 2, 2004, pp. 499-513.

[45] S. J. Liao, "An Analytic Approximate Technique for Free Oscillations of Positively Damped Systems with Algebraically Decaying Amplitude," *International Journal of Non-Linear Mechanics*, Vol. 38, No. 8, 2003, pp. 1173-1183.

[46] S. J. Liao and A. Campo, "Analytic Solutions of the Temperature Distribution in Blasius Viscous Flow Problems," *Journal of Fluid Mechanics*, Vol. 453, 2002, pp. 411-425.

[47] S. J. Liao, "An Analytic Approximation of the Drag Coefficient for the Viscous Flow past a Sphere," *International Journal of Non-Linear Mechanics*, Vol. 37, No. 1, 2002, pp. 1-18.

[48] S. J. Liao, "An Explicit, Totally Analytic Approximate Solution for Blasius' Viscous Flow Problems," *International Journal of Non-Linear Mechanics*, Vol. 34, No. 4, 1999, pp. 759-778.

[49] S. J. Liao, "The Proposed Homotopy Analysis Technique for the Solution of Nonlinear Problem," Ph.D. Thesis, Shanghai Jiao Tong University, Shanghai, 1992.

[50] S. Abbasbandy, T. Hayat, R. Ellahi and S. Asghar, "Numerical Results of Flow in a Third Grade Fluid between Two Porous Walls," *Zeitschrift Fur Naturforschung A*, Vol. 64a, 2009, pp. 59-64.

[51] S. Abbasbandy and F. S. Zakaria, "Soliton Solution for the Fifth-Order Kdv Equation with the Homotopy Analysis Method," *Nonlinear Dynamics*, Vol. 51, No. 1-2, 2008, pp. 83-87.

[52] S. Abbasbandy, "Approximate Solution of the Nonlinear Model of Diffusion and Reaction Catalysts by Means of the Homotopy Analysis Method," *Chemical Engineering Journal*, Vol. 136, No. 2-3, 2008, pp. 144-150.

[53] S. Abbasbandy, "The Application of Homotopy Analysis Method to Solve a Generalized Hirota-Satsuma Coupled KdV Equation," *Physics Letters A*, Vol. 361, No. 6, 2007, pp. 478-483.

[54] S. Abbasbandy, "Homotopy Analysis Method for Heat Radiation Equations," *International Communications in Heat and Mass Transfer*, Vol. 34, No. 3, 2007, pp. 380-

387.

[55] S. Abbasbandy, "The Application of Homotopy Analysis Method to Nonlinear Equations Arising in Heat Transfer," *Physics Letters A*, Vol. 360, No. 1, 2006, pp. 109-113.

[56] H. Zaman and M. Ayub, "Series Solution of Unsteady Free Convection Flow with Mass Transfer along an Accelerated Vertical Porous Plate with Suction," *Central European Journal of Physics*, Vol. 8, No. 6, 2010, pp. 931-939.

[57] H. Zaman, T. Hayat, M. Ayub and R. S. R. Gorla, "Series Solution for Heat Transfer from a Continuous Surface in a Parallel Free Stream of Viscoelastic Fluid," *Numerical Methods for Partial Differential Equations*, Vol. 27, No. 6, 2011, pp. 1511-1524.

[58] T. Hayat, H. Zaman and M. Ayub, "Analytic Solution of Hydromagnetic Flow with Hall Effect over a Surface Stretching with a Power Law Velocity," *Numerical Methods for Partial Differential Equations*, Vol. 27, No. 4, 2010, pp. 937-959.

[59] J. H. Mathews and K. D. Fink, "Numerical Methods Using MATLAB," Printice-Hall Inc., Upper Saddle River, 2004.

[60] F. Labropulu and O. P. Chandna, "Oblique Flow Impinging on a Wall with Suction or Blowing," *Acta Mechanica*, Vol. 115, No. 1-4, 1996, pp. 15-25.

[61] F. Ahmad and W. H. Albarakati, "An Approximate Analytic Solution of the Blasius Problem," *Communications in Nonlinear Science and Numerical Simulation*, Vol. 14, 2008, pp. 1021-1024.

Permissions

The contributors of this book come from diverse backgrounds, making this book a truly international effort. This book will bring forth new frontiers with its revolutionizing research information and detailed analysis of the nascent developments around the world.

We would like to thank all the contributing authors for lending their expertise to make the book truly unique. They have played a crucial role in the development of this book. Without their invaluable contributions this book wouldn't have been possible. They have made vital efforts to compile up to date information on the varied aspects of this subject to make this book a valuable addition to the collection of many professionals and students.

This book was conceptualized with the vision of imparting up-to-date information and advanced data in this field. To ensure the same, a matchless editorial board was set up. Every individual on the board went through rigorous rounds of assessment to prove their worth. After which they invested a large part of their time researching and compiling the most relevant data for our readers. Conferences and sessions were held from time to time between the editorial board and the contributing authors to present the data in the most comprehensible form. The editorial team has worked tirelessly to provide valuable and valid information to help people across the globe.

Every chapter published in this book has been scrutinized by our experts. Their significance has been extensively debated. The topics covered herein carry significant findings which will fuel the growth of the discipline. They may even be implemented as practical applications or may be referred to as a beginning point for another development. Chapters in this book were first published by Scientific Research Publishing Inc.; hereby published with permission under the Creative Commons Attribution License or equivalent.

The editorial board has been involved in producing this book since its inception. They have spent rigorous hours researching and exploring the diverse topics which have resulted in the successful publishing of this book. They have passed on their knowledge of decades through this book. To expedite this challenging task, the publisher supported the team at every step. A small team of assistant editors was also appointed to further simplify the editing procedure and attain best results for the readers.

Our editorial team has been hand-picked from every corner of the world. Their multi-ethnicity adds dynamic inputs to the discussions which result in innovative outcomes. These outcomes are then further discussed with the researchers and contributors who give their valuable feedback and opinion regarding the same. The feedback is then collaborated with the researches and they are edited in a comprehensive manner to aid the understanding of the subject.

Apart from the editorial board, the designing team has also invested a significant amount of their time in understanding the subject and creating the most relevant covers. They scrutinized every image to scout for the most suitable representation of the subject and create an appropriate cover for the book.

The publishing team has been involved in this book since its early stages. They were actively engaged in every process, be it collecting the data, connecting with the contributors or procuring relevant information. The team has been an ardent support to the editorial, designing and production team. Their endless efforts to recruit the best for this project, has resulted in the accomplishment of this book. They are a veteran in the field of academics and their pool of knowledge is as vast as their experience in printing. Their expertise and guidance has proved useful at every step. Their uncompromising quality standards have made this book an exceptional effort. Their encouragement from time to time has been an inspiration for everyone.

The publisher and the editorial board hope that this book will prove to be a valuable piece of knowledge for researchers, students, practitioners and scholars across the globe.

List of Contributors

Shafiq Ur Rehman
Department of Mathematics, The University of Auckland, Auckland, New Zealand
Department of Mathematics, University of Engineering and Technology, Lahore, Pakistan

Mehwish Bari and Ghulam Mustafa
Department of Mathematics, The Islamia University of Bahawalpur, Bahawalpur, Pakistan

Luca Perotti, Daniel Vrinceanu and Daniel Bessis
Department of Physics, Texas Southern University, Houston, USA

A. A. James
Department of Mathematics, American University of Nigeria, Yola, Nigeria

A. O. Adesanya
Department of Mathematics, Modibbo University of Technology, Yola, Nigeria

J. Sunday
Department of Mathematical Sciences, Adamawa State University, Mubi, Nigeria

D. G. Yakubu
Department of Mathematics, Tafawa Balewa Federal University of Bauchi, Bauch State, Nigeria

Cristina I. Muresan
Department of Automation, Faculty of Automation and Computer Science, Technical, University of Cluj Napoca, Romania

Laiping Zhang and Wanhui Ji
Yinchan Energy College, Yinchuan, China

M. Sh. Mamatov, E. B. Tashmanov and H. N. Alimov
Department "Geometry", National University of Uzbekistan Named After M. Ulugbek, Tashkent, Uzbekistan

Peter C. L. Lin
Department of Mathematical Sciences & Financial Engineering Program, Stevens Institute of Technology, Hoboken, USA

Comlan de Souza
Department of Mathematics, California State University at Fresno, Fresno, USA

David W. Kammler
Department of Mathematics, Southern Illinois University at Carbondale, Carbondale, USA

Llambrini Sota
University "Pavaresia" in Vlora, Vlora, Albania

Fejzi Kolaneci
University of New York in Tirana, Tirana, Albania

Amnah S. Al-Johani
Department of Applied Mathematics, College of Science, Northern Borders University, Arar, Saudi Arabia
College of Home Economics, Northern Borders University, Arar, Saudi Arabia

Dongming Wei
Department of Mathematics, University of New Orleans, New Orleans, USA

Mohamed B. M. Elgindi
Texas A & M University-Qatar, Doha, Qatar

Rowsanara Akhter and M. Sharif Uddin
Department of Mathematics, Jahangirnagar University, Dhaka, Bangladesh

Mohammad Mokaddes Ali and Babul Hossain
Department of Mathematics, Mawlana Bhashani Science and Technology University, Tangail, Bangladesh

Wensheng Zhang and Jia Luo
Institute of Computational Mathematics and Scientific/Engineering Computing, LSEC Academy of Mathematics and Systems Science, Chinese Academy of Sciences, Beijing, P. R. China

Ning Ruan
School of Science, Information Technology and Engineering, University of Ballarat, Ballarat, Australia

Philippe R. Richard and Michèle Tes-sier-Baillargeon
Département de didactique, Université de Montréal, Canada

Michel Gagnon and Nicolas Leduc
Département de génie informatique et génie logiciel, École Polytechnique de Montréal, Canada

Josep Maria Fortuny
Departament de Didàctica de les Matemàtiques i de les Ciències Experimentals, Universitat Autònoma de Barcelona, Spain

Syed Yedulla Qadri
Department of Mathematics, St. Josephs PG College, Kurnool, India

M. Veera Krishna
Department of Mathematics, Rayalaseema University, Kurnool, India

Ivan Arango and Fabio Pineda
Mechatronics and Machine Design Group, Universidad EAFIT, Medellin, Colombia

Oscar Ruiz
Laboratorio de CAD/CAM/CAE, Universidad EAFIT, Medellin, Colombia

Wensheng Zhang and Yinyun Dai
Institute of Computational Mathematics and Scientific/Engineering Computing, LSEC Academy of Mathematics and Systems Science, Chinese Academy of Sciences, Beijing, P.R. China

Renu Chugh and Sanjay Kumar
Department of Mathematics, M.D.U, Rohtak, India

Mahbub Hasan
Department of Civil Engineering, Alabama Agricultural and Mechanical University, Normal, USA

Salam Md. Mahbubush Khan
Department of Mathematics, Alabama Agricultural and Mechanical University, Normal, USA

Chandrasekhar Putcha
Department of Civil and Environmental Engineering, California State University, Fullerton, USA

Ashraf Al-Hamdan
Department of Civil & Environmental Engineering, University of Alabama, Huntsville, USA

Chance M. Glenn
College of Engineering, Technology and Physical Sciences, Alabama Agricultural and Mechanical University, Normal, USA

Ludmila Petrova
Department of Computational Mathematics and Cybernetics, Moscow State University, Moscow, Russia

Yonghua Zhu and Hongtao Lu
Mathematics and physics department of North China Electric Power University, Beijing, China

Zilin Zhu
Information Engineering Department of Huazhong University of Science and Technology, Wuhan, China

Leonid Kugel and Victor A. Gotlib
The Faculty of Sciences, Holon Institute of Technology (H. I.T), Holon, Israel

Haider Zaman, Murad Ali Shah and Muhammad Ibrahim
Faculty of Numerical Sciences, Islamia College University, Peshawar, Pakistan

www.ingramcontent.com/pod-product-compliance
Lightning Source LLC
Chambersburg PA
CBHW080636200326
41458CB00013B/4653